T0256085

RATIONAL RABBIS

JEWISH LITERATURE AND CULTURE

Series Editor, Alvin H. Rosenfeld

Rational Rabbis

Science and Talmudic Culture

Menachem Fisch

INDIANA UNIVERSITY PRESS

Bloomington and Indianapolis

Manufactured in the United States of America

Library of Congress Cataloging-in-Publication Data

Fisch, Menachem.
Rational rabbis : science and talmudic culture / Menachem Fisch.
p.   cm. — (Jewish literature and culture)
Includes bibliographical references and index.
ISBN 0–253–33316–4 (cl : alk. paper)
1. Judaism and science. 2. Talmud Torah (Judaism). 3. Talmud—
Criticism, interpretation, etc. I. Title. II. Series.
BM538.S3F57   1997
296.3'75—dc21   97–7349

1 2 3 4 5 02 01 00 99 98 97

*To my father
Harold Fisch
who set this all in motion*

CONTENTS

PREFACE IX

INTRODUCTION XIII

Part 1

SCIENCE AS AN EXEMPLAR OF RATIONAL INQUIRY

The Problem   4

Heracliteans and Parmenideans   9

Science as Hypothesis   15

Problems and Progress   22

The Rationality of Goal-Directed Criticism   28

Part 2

THE JEWISH COVENANT OF LEARNING

1. The Great Tannaitic Dispute: The Jabne Legends and Their Context 51

   Traditionalism and Its Discontents   55

   The Jabne Reforms   63

   The Testimony of Eduyot   *71*

   "It Is Not in Heaven" (Deut. 30:12)   78

   Jabne's Antitraditionalist Manifesto   88

   A Traditionalist Response: Hillel and b'nei Beteira   *96*

2. The Changing of the Guard: Amoraic Texts and Tannaitic Legacies 111

   Discerning the Bavli's Point of View: A Schematic Overview   112

   Introducing the Bavli's Paradigm: Berakhot *19b   119*

   The Logic and Rhetoric of Transgenerational Negotiation   128

   Yerushalmi and Bavli Compared   133

   Antitraditionalism for the Advanced   142

   Giving away the Game, or The Gentle Art of Inaudible Instruction   150

3. Understanding the Bavli 163

Problem One: Explaining the Bavli's Double-Talk   *163*
Problem Two: Discerning a Role for the Mishna   *166*
The Mishna as a Formative Code   *171*
Rational Rabbis, or The Mishna as Textbook   *180*
"Turn Thee Around" (Bavli, Menaḥot 29b), or
Back to the Future   *188*

NOTES   197
GENERAL INDEX   251
INDEX OF TALMUDIC REFERENCES   258

PREFACE

In conscious defiance of the well-known prohibition against ever committing the Oral Torah to writing, the Judaic Sages of late antiquity compiled lengthy records of their learning. Commencing with the Mishna, the Tosefta and the tannaitic *midrashim* of the third century, the *Mekhilta* (to *Exodus*), the *Sifra* (to *Leviticus*), the *Sifré* (to *Numbers* and *Deuteronomy*), the Talmud of the Land of Israel, the Yerushalmi, the amoraic midrashic compilations of the fifth century, and culminating in the Talmud of Babylonia, the Bavli, the process of documentation inaugurated by R. Yehuda the Patriarch yielded the vast and unique talmudic canon of rabbinic Judaism. In no other collection of texts does one find as exacting an attempt by the authors themselves to reconstruct and record the plethora of debates and deliberations in which they participated. As such the corpus of talmudic texts is unparalleled in Western intellectual history. Given their enormous scope, there is no question that the rabbinic texts of late antiquity represent a unique record of a unique intellectual enterprise.

It is no wonder then that the literature of the talmudic era has been researched so intensively. But the major portion of this effort until now has focused primarily on framing and answering philological and historical questions. Excellent analyses have been written of the language of the rabbis; of the various strata, architecture, and literary structure of their texts; of their attitudes toward halakha and aggada; and of the social, political, economical, and cultural context of their lives, work, and writing. But as good and important as much of this work has been, it is incomplete. Conspicuously absent is an analysis of these texts from a philosophical perspective. This absence is all the more conspicuous when one contrasts the academic study of the rabbinic texts under consideration with that of Western science, the one sustained intellectual endeavor of Western culture that rivals that of the rabbis in scope and duration. Students of modern science have devoted enormous energy, and to great effect, to analyses of the ontological and epistemic assumptions of scientists, the logical structure of their explanations and descriptions, and their theoretical and experimental methods. These analyses are fundamentally philosophical; they have advanced so far that today the philosophy of science is considered a freestanding and important academic discipline.

This essay is a programmatic attempt systematically (if tentatively) to apply the perspectives unique to this discipline to the talmudic texts.

Originally subtitled: "A Preliminary Study of Talmudic Epistemology," *Rational Rabbis* is a first extensive, if hesitant attempt to take stock of an intellectual journey I embarked upon in earnest about eight years ago. It was initially motivated and given a particular direction by two essays I had read and was prompted by an invitation to participate in a conference organized by Bar Ilan University in honor of my father. The two essays were Jacob Neusner's "Why No Science in Judaism?" (which later formed the basis for his *The Making of the Mind of Judaism: The Formative Age*) and the last chapter of my father, Harold Fisch's *Poetry with a Purpose* entitled: "*Qohelet*: A Hebrew Ironist." Neusner's paper argued that the "mind" of talmudic Judaism was so different from that of Western science and philosophy as to render the former incapable of producing the latter. My father's essay offered a reading of *Qohelet*'s meditations on human wisdom as an ironic reductio ad absurdum of the very possibility of an anthropocentric, self-sufficient notion of rationality. I found the two essays equally intriguing and equally disturbing, and it occurred to me that they could be criticized jointly along the following lines. Contrary to the claim attributed by my father to *Qohelet*, science, I argued, provides a living example of a self-sufficient, humanly attainable, rational undertaking that is well aware of its own shortcomings. And, contrary to Neusner, the Talmud's manner of halakhic reasoning seemed to me to resemble quite closely the type of discourse I had learned to associate with the scientific method of trial and error. But if that is the case, one must assume that the talmudic sages allegedly responsible for the canonization of *Qohelet* could not have read it as ironically dismissing their very own position. I went on to propose a different reading of *Qohelet* better suited to their point of view. I summarized these initial thoughts first in my contribution to the volume commemorating the conference, and later in a short book published in Hebrew a year or so later.* The book was well received, although I knew even then that I had barely scratched the surface. The basic tripartite ground plan for the project—a claim about scientific rationality, a reconstruction of talmudic epistemology along those lines, and a reading of *Qohelet* that conforms the latter—remains unchanged, but it has become quite clear that it is impossible to realize within the boundaries of a single volume. *Rational Rabbis* attends exclusively to the first two parts of the project. With its completion, my earlier work on the talmudic

*Menachem Fisch, "The Perpetual Covenant of Jewish Learning," in E. Spolsky ed., *Summoning: Ideas of the Covenant and Interpretive Theory* (Albany: SUNY Press, 1993), pp. 91–114, and *'To Know Wisdom': Science, Rationality and Torah-study* (in Hebrew) (Tel Aviv: Hakibbutz Hameuchad, 1994).

literature is superseded. God willing, *Qohelet* will be published at a future date under separate cover.

In writing this book I have benefited greatly from several colleagues and friends. Special thanks go to Joseph Agassi, Hanina Ben-Menahem, Daniel Boyarin, Lorraine Daston, Noah Efron, Mayer G. Freed, Moshe Halbertal, William Kolbrenner, Aharon Oppenheimer, Hilary Putnam, David Ruderman, Avi Sagi, Ron Shapira, and Zvi Zohar, who all read earlier drafts of the entire manuscript and offered detailed and valuable criticism. Although I was not always able to accept each of their insightful suggestions, this book, which perhaps ended up reflecting their views less than they would have liked it to, still owes to them enormously. Many other individuals, teachers, colleagues, study partners, students, and friends have read or listened to parts of it, and their input too, over the years, has been invaluable. I refrain from naming the many I do remember for fear of offending the few I do not.

A study such as this requires a tolerant, challenging, and supportive intellectual environment. I have the good fortune and great privilege of belonging to two such remarkable institutions in which this sort of disciplinary border-crossing is not merely tolerated but genuinely encouraged. They are the Shalom Hartman Institute for Advanced Judaic Studies, Jerusalem, and the Cohn Institute for the History and Philosophy of Science and Ideas at Tel Aviv University. This work could not have been undertaken without them.

It is also a great pleasure to acknowledge my great debt to Janet Rabinowitch at Indiana University Press, whose help in seeing this work through from brief synopsis to its final form far exceeded the formal duties of an editor.

My father has always been my closest and keenest critic. To this book, however, his own lifelong transdisciplinary engagement with the Jewish bookshelf has been a model and an inspiration. *Rational Rabbis* is dedicated to him and to his work with love and admiration.

Finally, a very fond and special thanks to my wife Hanna, who insisted I try, and to her and to our children Yael and Eilon for enabling me to do so.

INTRODUCTION

This book concerns science and Judaism. Because this conjunction may mean different things to different people, it is perhaps important to explain at the outset what it is not intended to address. First, I shall not at all be concerned with the question of whether or not there are, or can be, contradictions between what scientists tell us about the world and what a committed Jew is supposedly obliged to believe about the world. Though often asked, I believe that this question is ill conceived and, predictably, so are the answers it most often inspires. It is unhelpfully apologetic to claim that there is and must always be perfect harmony between scientific 'truth' and revealed 'truth', or, conversely, that science and Torah-study examine complementary but mutually exclusive realms, and hence cannot conflict in principle. This is not to say that all apologetic reasoning is in principle wrong-headed, but to insist that in the case of Torah-study and science, apologetics make little sense and help clarify even less. It makes little sense to speak at any one moment of either conflict or confluence between scientific statements about the world and those allegedly implied in the talmudic literature because, as I propose to argue, Torah-study as it was largely conceived by the talmudic sages, is, in principle, like science, a continuing, open-ended inquiry. While the talmudic texts frequently consult science,[1] neither enterprise has in its power to assume in advance anything about the structure and laws of nature or to make final judgments about the physical world.[2] To subordinate Torah-study to a particular description of the physical world or, conversely, to deny scientists the right to pursue certain hypotheses would be dangerously detrimental to either undertaking.

Second, the present study does not address the historical question concerning the relationship between different schools of Judaism and the scientific knowledge and particular scientific endeavors of their age. In contrast to the previous question, this question is a legitimate, well conceived, and important one. It cannot be properly answered, however, until a prior, more fundamental question is answered.

My interest in this more basic question was prompted some years ago by the work of Jacob Neusner, especially his *The Making of the Mind of Judaism*. Neusner there raises and attempts to answer the following question: why "Judaism of the dual Torah" (that is to say, mainstream Judaism that viewed the Torah as comprising both oral and written components) in its institutions and formative intellects . . . made no contribution to the framing [of Western philosophy and science] which emerged elsewhere than from the mind and imagination of thinkers nurtured by the canonical writings of that Judaism?[3]

Neusner's question provokes thought, particularly in light of the enormous efforts that have been made in this century to trace the roots of modern science back to Christianity. It is a question that demands the attention not just of historians of ideas, but also of all committed Jews who also value science and philosophy. Still, this question has received little if any serious attention both within and without orthodox Judaism.

Neusner's fundamental assumption—that those nurtured by the canonical texts of Judaism made no contribution to science—clearly overgeneralizes. There are surely periods of Jewish history when Jews made important contributions to both the science and philosophy of their age— the High Middle Ages stands out in this respect and is by no means unique. But in those periods when modern science as we know it was nascent and developing, normative Judaism, in fact, contributed very little, just as Neusner argues. It is difficult to point to a single practicing Jew who actively contributed to the scientific or philosophical debates that accompanied the scientific revolution of the sixteenth and seventeenth centuries.[4] Thus, despite its overgeneralization, Neusner's question is still compelling.

His answer, however, is less compelling in my opinion. "My fundamental argument," he writes in the introduction to his book,

is that the paramount documents of Judaism not only preserved, but also inculcated, a particular way of forming propositions and also a particular mode of joining these . . . propositions together into sizable compositions of thought. The very means by which these modes of thought were transmitted and held together, the extraordinary power of analysis and argument characteristic of the normative documents—these explain also the incapacity of those same modes of thought to frame philosophy, including natural philosophy.

"The power," concludes Neusner, "is the pathos of the system."[5]

Neusner believes, then, that the apparent failure of Torah Judaism to render science attests to an elemental incongruity between what he calls the "logics" prevalent in its canonical writings and those that, in his view, inform Western scientific and philosophical discourse. The peculiar ways in which the Torah was traditionally studied, he argues, rendered those who studied it incapable of philosophical or scientific reasoning. Presumably, in his view, the reverse is equally true; for the same reason a scientist or philosopher would have great difficulties participating in talmudic discourse.

But incapacity is hard to demonstrate, let alone to prove. One might have expected Neusner at least to have bolstered his theory with examples of Jews originally trained in talmudic reasoning who upon turning to science or philosophy proved, in his opinion, to be systematically incompetent. Even then he would still have to demonstrate that their incompetence was owing to their talmudic training and sensibilities. Neusner provides no such examples, and it is doubtful that he could. It is hard to think of anyone who arguably failed at science or philosophy because of his or her talmudic education. In fact, there are many examples of Jews with traditional talmudic educations—past and present—who excelled in science and philosophy. This strongly suggests that training and talent for talmudic disputation do not exclude commitment and talent for scientific or philosophical debate. To wit, until recently few orthodox Jews regarded science and philosophy as viable pursuits, but that is an entirely different matter. Indeed, that is all that the historical record seems to indicate, namely, that the rabbis regarded Torah-study as a vocation much more meritorious than the study of nature, which in their opinion had its place, but only as a second or third choice. Regardless of our own preferences, this in itself will suffice to answer Neusner's question.[6]

But even if we reject Neusner's contention that incongruities between the logics underlying talmudic and scientific discourse actually incapacitate participants in the one adequately to participate in the other, the truth of its premise remains an open, compelling, and equally understudied question. Are the differences between the modes of analysis and inquiry exhibited in talmudic and scientific discourse as pronounced as Neusner maintains? Or, as Francis Bacon put it, are the study of God's Book of Words and His Book of Works as different as Neusner professes to demonstrate?

In contrast to Neusner, Rabbi Adin Steinsaltz and the late Amos Funkenstein argue in *The Sociology of Ignorance*[7] that there are essential similarities between the thought of the talmudic sages and Greek science, mathematics, and philosophy. Both cultures, they argue, developed independently what they call an "ideal of open knowledge." For those who embrace this ideal, knowledge is open in two senses. First, access to knowledge is not limited to members of any particular caste; in principle, anyone willing to make the effort can attain it. Second, the tools and methods for generating knowledge, and the criteria for judging knowledge claims, also all remain in the public domain. As the name of their book suggests, however, Funkenstein and Steinsaltz are primarily concerned with the social structures and strictures that serve to limit knowledge in different contexts. Because their real concern is ignorance, they barely develop their views on the relationship between the early Greek tradition of scientific learning and the talmudic tradition of Torah-study.

Ultimately, Funkenstein and Steinsaltz have written nothing that disproves Neusner's claims, although the way in which they historicize the two endeavors runs counter to the broader essentialism of Neusner's approach. Neusner never claims that the logics that inform the talmud are any more closed than those that inform science and philosophy. His only claim is that the modes of inquiry of each endeavor are so different that science was prevented from developing in communities that took Torah-study seriously. Funkenstein and Steinsaltz limit their comments to a comparison, equally essentialist, between Torah-study as envisaged in the talmudic texts and to science and philosophy in their Greek mode. Neusner, on the other hand, appears to be speaking ahistorically of talmudic and scientific reasoning in general. The present study, by contrast, compares a similarly transcultural, though decidedly nonessentialist, latterday theory of rational human endeavor, to a major vision of human intellectual accomplishment that arguably informed the talmudic texts at the time of their redaction. Regardless of the seemingly essentialist idiom of Neusner's analysis, he in fact also compares and contrasts the talmudic text with his own latter-day understanding of science and philosophy. Hence, to disprove the antecedent of his claim, one must unearth and compare the ontological and epistemological assumptions behind each of the two great intellectual endeavors: *talmud-Torah*—Torah-study—as it was originally conceived and understood by those responsible for the talmudic texts as we find them, and Western science as it is understood today. To a limited extent, such is the undertaking of the present book.

Such a comparison, I shall argue, reveals that Neusner's question, and to an greater extent the answer he proposes, reflects very commonplace, but mistaken, ideas about both science and Torah-study. It is a view that deems each a specialized intellectual endeavor with its own ways of thinking, its own logic, its own standards of success and failure, and its own epistemology. It hypostatizes 'scientific reasoning' and 'talmudic reasoning', and in doing so, conjures up a great divide between what goes on in a laboratory and what goes on in a *bet midrash*. Like C. P. Snow's famous formulation, the accepted view imagines each as the location of its own separate intellectual culture. All of this, I hope to demonstrate, misconstrues science and *talmud-Torah*. Different in their goals and methods as they undoubtedly are, both are human endeavors, and both are rational endeavors. Like all rational, human endeavors they share a common denominator, a common modus operandi. Just what scientists do, and just what the framers of the talmud believed they were doing, have both largely been mystified and mythologized. In each part of this book, I will view and compare, with a philosopher's eye, what framers of science do and what framers of the talmudic texts did and thought they were doing, in an attempt to demystify and demythologize each. When I have finished, it will be clear that Neusner's conclusion—that a rational scientist necessarily thinks and acts fundamentally differently than a rational rabbi—is simply mistaken.

It is also important to stress at the outset that the question that motivates this study is quite different from the question that motivates Neusner's. The comparative study presented in the following pages is *not* an attempt to explain the failure or success of Torah Judaism seriously to engage in science and philosophy. The ways in which Jewish communities and individuals related and responded to the science and philosophy of their day pertain to the social, political, and cultural contingencies of their place and time as much as they do to their theological and philosophical prejudices. And in so far as images of religious knowledge and knowing enter such explanations, they will inevitably be those held and employed by the communities and individuals under consideration. To paraphrase David Kraemer,[8] later generations related to the talmudic literature in their own ways and for their own reasons. They shaped and reinterpreted the rabbis' writings to fit their own circumstances much as the talmudic sages had shaped and reinterpreted earlier traditions to speak to the needs of *their* communities. I shall here be concerned only with the latter, seeking to retrieve as far as possible the images of knowledge, to use Yehuda Elkana's coinage, that arguably informed and guided the framers and redactors of the talmudic literature, rather than that of their later readerships. My objectives are, therefore, to expose and to analyze, and, at a later stage, to explain the frequently tacit, meditative discourse about the nature of humanly possible intellectual achievement that shaped the thinking of the framers of the talmudic canon. The present study shall therefore not address even the question of how the framers of these texts related and responded to the science of their own day.

To what end, then, one may ask, should the rabbis' writings at all be compared to science? In what sense can a latter-day understanding of science further our understanding of a corpus of essentially nonscientific texts written over fifteen centuries ago?

The answer is explained in detail in Part 1 and the beginning of Part 2 of this book, but is worth briefly summarizing at the outset. It consists of two unrelated hypotheses or theses. The first is an attempt, grounded in contemporary philosophy of science, to formulate a general, transcultural, or supracultural theory of progress and rationality applicable to all manner of human reflective endeavor. Science looms large in this context, but only as a paradigmatic *example* of a progressive and rational undertaking. The theory of progress and rationality I propose focuses on science and draws on philosophical discussions of science, but it aspires to be about much more. It is not a theory about what philosophers past and present assumed progressive and rational action to be but about what progressive and rational action *is*. Not surprisingly perhaps, it is also the view of rational endeavor to which I currently subscribe.

The second hypothesis, by contrast, is an attempt to expose notions of intellectual accomplishment and rationality that were self-consciously assumed by framers of the talmudic texts. It purports to lay bare an understanding of the rabbis' understanding of *their* project to which I currently subscribe. These, then, are the units of comparison: not simply scientific and talmudic discourse, but two theories or conceptions of rational endeavor—that (or those) which, in my opinion, is found, if for the most part subliminally, to inform science and other arguably rational undertakings, and that (or those) found, in my opinion, self-consciously to have informed much of the framing of the talmudic texts.

Clearly, in order to compare the views of rationality and progress that allegedly inform contemporary philosophy of science and the conceptions of Torah-study that motivate the talmudic texts, each of the two needs to be examined by itself. In doing so, I have incurred two main intellectual debts. My understanding of science, as will be evident from what follows, owes much to the works of R. G. Collingwood and Sir Karl Popper and his school, and to the important, and often heated philosophical debate they have provoked in recent years. My understanding of the rabbinic tradition of learning owes much to many, but most of all to my father Harold Fisch. It is to him and his inspiring work about what I shall later refer to as the Jewish covenant of learning that I dedicate this book.

In Part 1 a major strand shared by Collingwood's philosophy of history and Popper's philosophy of science is made the basis for a broader theory of rational and progressive human endeavor that is essential to the comparative task at hand. This in itself, however, is less straightforward an undertaking than I would have liked it to be. Although science has been the subject of keen philosophical reflection for almost as long as it has existed, just what scientists do when they do science is by no means universally or well understood—especially by those less versed in the philosophy of science. Indeed, one of the reasons why, in my opinion, the similarity between doing science and the way in which a prominent talmudic voice envisaged Torah-study had not been noticed before is precisely because both endeavors have been sorely misunderstood. A generally satisfactory, though, unfortunately, not yet generally accepted, explanation of the endeavor of science has only recently been provided, in my opinion, in the innovative work of Popper and his school. But the real value and importance of this explanation is only obvious when one views it in its historical and philosophical context. Only when seen in this context can one understand the problems that it solves and the advantages that it has over other explanations of the endeavor of science and similar undertakings. The best way to explain how science is done, in other words, is to review briefly and critically the recent problematic history of the ways in which doing science has been explained. Such is the methodological premise and strategy of Part 1.

Bearing the conclusions of Part 1 in mind, Part 2, the main bulk of this book, inquires into the epistemic assumptions that appear consciously to inform—at times implicitly and at times explicitly—the modes of inquiry and exegesis that dominate the talmudic texts. Needless to say, the idea is not to *apply* the theory of rationality and progress developed in Part 1 to the talmudic texts, but, by building on existing exegetical and historical studies of these writings, to compare the two endeavors by allowing the talmudic text, as it were, to speak for itself. It would be impertinent of me to claim to have been able fully to achieve such impartiality. There is no such thing as an innocent and truly impartial reading of any text, and I am well aware that my own philosophical convictions and religious commitment to talmudic Judaism inevitably predispose me in ways I cannot fully control. Hard as I have tried to resist finding in the rabbis' writings what I expect or, even worse, would like to find, I will no doubt have inevitably done so in certain respects and to some degree. But the same applies to any such study. And alternative accounts of the rabbis' mindset will inevitably be as potentially blinkered and apologetic as my own. Like truth, beauty, and ultimate justice, impartiality and hermeneutic neutrality should be treated, in principle, as regulative rather than as demonstrably attainable objectives, as ideals that no reader can ever justly profess to have fully achieved. I cannot, therefore, claim to do in Part 2 more than my honest best; to engage as best and as impartially as I can with existing alternative interpretations; and prudently to invite my critics similarly to engage with mine. This is another reason for putting my own philosophical cards on the table in considerable detail in Part 1 prior to embarking upon the main project of the book.[9]

So much for methodology. The main argument that this study purports to make and substantiate is for the presence in the Jewish texts of late antiquity of a major voice or school of talmudic thought whose views of human knowledge, learning, and intellectual accomplishment bear a striking resemblance to the latter-day theory of rationality argued for in Part 1. I term this approach to Torah-study 'antitraditionalist', by virtue

of its commitment to the critical appraisal rather than to the unquestioned reception of its inherited teachings. It is to the description, location, and detailed retrieval of this antitraditionalist voice that Part 2 is
devoted. It is a voice that shares with the modern constructive skeptic
the idea that no knowledge is set forever, that all knowledge needs to be
questioned for its deficiencies, that former rulings and sensibilities have
to be constantly reinterrogated. It is a voice that shares with this school
of thought a fundamental skepticism toward all first-order knowledge
claims, yet premises standards and criteria for deciding between rival
conjectures. Both are thoroughly skeptical yet not destructively so. On
the contrary, although they share a basic mistrust in the capacity of humans ever to attain the ultimate objectives of such grand undertakings as
science and Torah-study, they share a profound and philosophically main-
tainable belief in the capacity of humans to achieve real progress in such
endeavors, as opposed, say, to mere variation. And above all, the view of
rationality I argue for and the view of Torah-study I aspire to expose and
analyze share a basic openness, an intellectual modesty, and a genuine
pluralism that, in my humble opinion, render the retrieval of the latter
especially timely.

The systematic yet constructive skepticism of the approach developed
in Part 1 is a relative newcomer to the Western philosophical scene. This
is undoubtedly one reason why the antitraditionalist talmudic voice had
largely vanished from sight for so long. For reasons that will become apparent, it is also a voice that is not always easy to pick out even for a
trained philosopher. But the main reason for the antitraditionalist voice
of talmudic Judaism remaining to a large extent obscure is the way in
which the Babylonian Talmud, the most important and widely studied of
the rabbis' writings, systematically tones it down. And since this seems
to have been a policy that was adopted by some of the very same framers
to whom the Talmud's antitraditionalism owes its origin, it is a phenomenon that, once appropriately exposed, will require explanation. This,
then, is the plan, method, and thesis of the present study.

One further point needs to be made at the outset. There exists a basic
and important difference between what one would normally describe as
a scientific paper or treatise and the texts comprising the Jewish canon
under consideration. Whereas the former are usually written with a view
to systematizing and consolidating various scientific findings from a particular point of view, the latter endeavor in most cases not to develop one
particular perspective but to reconstruct and in part analyze the process
of debate between rival points of view. They are largely pluralistic and
usually only seek to decide issues related directly to ritual or law. Hence,
if we are to study the rabbis' writings in comparison to science, then the

comparison should be made not to individual scientific treatises but rather to studies that in similar fashion attempt historically or philosophically to reconstruct processes of scientific debate. In both cases we will then be in the realm not of particular research programs but of public discourse, its logics, the ends to which it is applied, the epistemological assumptions to which it pertains, and the ways in which, if at all, it succeeds in achieving closure.

The distinction, however, between the two levels of discourse with respect to the talmudic texts is far less evident than in science. The 'division of labor' between first-order scientific research and second-order historical and philosophical researches *into* science is sufficiently clear today to the extent that the two endeavors now constitute quite separate academic disciplines and are discussed in separate literatures. In the talmudic texts, however, the two levels of discourse are thoroughly and confusingly conflated. As a rule, these texts intensely collate, refashion, synthesize, and compress several hundred years of rabbinic discourse ranging over a variety of thinkers and schools of thought that operated in a variety of social, political, cultural, and intellectual environments, addressing an enormous range of legal, theological, and exegetical issues. Whatever second-order considerations informed these texts are to a large extent embedded in the editorial policies and narrative frameworks employed by their framers and redactors. Given the unusual level of complexity they present, the standards of philological, linguistic, and historical scholarship required for the complicated task of taking the talmudic texts apart at their appropriate stratified seams, and reconstructing and analyzing each subtext in relation to its proper place, time, and context are unusually, if understandably, high. Much excellent work has been and is being done along these lines by talmudic scholars past and present.

The concerns of the present study are different however. It is far less interested in the actual historical development of the rabbis' conceptions of Torah-study than in the theory or theories of knowledge and of knowledge acquisition that directed and motivated those responsible for the texts *in their finished form*. The subject matter and basic units of analysis this book aspires to address are the final versions of the texts that comprise the talmudic canon. The epistemological and methodological positions it attributes to the "rabbis" or the "sages" are, therefore, for the most part those of the latter-day framers, compilers, and editors of the texts as we find them. The reader is therefore cautioned to bear in mind that all talk in what follows of the controversy concerning Torah-study at Jabne, the halakhic and metahalakhic debates between the Houses of Hillel and Shammai, or the various statements and disagreements attributed to particular sages should always be taken, unless stated otherwise, as shorthand for the talmudic redactor's views of these matters. For the historian and philologist the texts in hand represent mythologizations to be appropriately distilled, sifted, and stripped of their latter-day attire; for the purposes of the present study they represent tacit manifestations of their framers' own political, social, religious, and epistemic concerns and agendas and are to be taken for what they are and for what they aspire to say. It will be necessary in what follows to return to this point on several occasions.

Finally, an almost technical matter of terminology. I use the terms *talmud-Torah* and Torah-study throughout this work to denote more than the learning process by which students acquaint themselves with and train themselves in their chosen fields of study. I take it in the wider contemplative, deliberative, and most of all investigative sense of term, as the meditative process by which the meaning of the word of God is determined. To put it differently, I use the term Torah-study analogously to study of nature rather than, say, to that of physics. Thus construed, Torah-study comprises two main related yet importantly different investigative efforts: the primarily *exegetical* task of understanding the word of the Holy Scriptures and the primarily *legislative* effort of developing and perfecting halakha—the law. Only for a community of learners who reduce the study of Torah to the careful reception of the exegetical and/or halakhic knowledge of old will the instructive and investigative modes of Torah-study be conflated.

While the main thesis of this book is relatively simple, it employs tools of analysis and addresses and examines texts, arguments, and strategies drawn from two separate and well-established disciplines: philosophy—in particular philosophy of science—and talmudic studies. I have done my best to address the nonspecialist reader as far as possible. Technical in-house terminology is either avoided altogether or, when necessary, explained. No specialized prior knowledge is assumed on behalf of my readers throughout. On the other hand, both parts of the book purport hesitantly to make novel and admittedly controversial claims. The theory of rationality and progress outlined in Part 1 and the assertion, presented and argued for in Part 2, that such a theory may be seen to have consciously informed the school of thought that came to dominate the notion of Torah-study exhibited in the talmudic texts are both humbly laid at the feet of the specialist critic who I sincerely hope will not confuse accessibility with simplemindedness.

PART 1

SCIENCE AS AN EXEMPLAR OF RATIONAL INQUIRY

Also He has set the mystery
of the world in their heart so
that no man can find out the
work which God has made
from the beginning to the
end.
—*Qohelet*, 3:11

The comparison I wish to explore between the philosophies that ground science and Torah-study is not a straightforward one. While science seeks to construct empirically adequate and logically consistent representations of external reality,[1] Torah-study, as practiced and preached by the Jewish sages of late antiquity, involved much more. Many of their midrashic writings—that is to say, texts and portions of texts devoted primarily to the interpretation and elaboration of Scripture—exhibit a keen concern for representational accuracy, analogous to science. In these texts the rabbis endeavor faithfully to represent and to interpret the given text of the Written Torah in much the same way as scientific theories endeavor faithfully to represent and to interpret the similarly given phenomena that they study. The well-known and widely exploited analogy, favored by many seventeenth- and eighteenth-century apologists, between the study of God's Book of Words and His Book of Works,[2] presupposes this type of affinity between the two enterprises. But not all talmudic discourse reflects so readily this sort of interpretive or representational concern. Most conspicuously divergent in this respect is the vast and multifarious body of essentially nonmidrashic, *halakhic* writings, in which their authors' main objectives are clearly to expose, construct, and develop a comprehensive and consistent system of law—an undertaking seldom conceived of or presented by the rabbis, at least not in its details, as an interpretative effort at all. Here, the analogy to science, if it exists at all, is far less obvious and cannot be easily established by direct comparison.

Rather than simply compare scientific and talmudic discourse, I wish first to view them, each in its own terms—science in the making, Torah-study in its talmudic reconstruction—as prized manifestations of deeper and more general theories of rational human endeavor, as premising, in other words, general convictions regarding the nature of humanly attainable achievement. The talmudic texts hardly speak in one voice about anything, and the epistemological grounding of *talmud-Torah* is no exception. What I intend to show is that central to, and to an extent formative of, this second-order polyphony is a major and explicit talmudic voice that proves not only comparable, but surprisingly similar to the view of human accomplishment on which, I believe, science and similar undertakings are grounded. In order to describe their similarities, however, we shall have to discuss each of them separately, and, in an important sense, differently.

The Problem

The second-order, philosophical appraisal of science is as old as science itself. The particular theory of the growth of scientific knowledge argued for in the following pages pertains to recent developments in this ongoing philosophical debate. It is not, therefore, an attempt to articulate how scientists of the past had in fact accounted for the nature of their undertakings. It is not an interpretation of the way science was conceived of by its practitioners. Rather, it is a latter-day philosopher's view of what science and similar undertakings can, in principle, be said to accomplish.

Science is regarded by many as a paradigm of rational and progressive endeavor. Indeed, it is further claimed, science prospers and progresses precisely because it is conducted so rationally. With the exception of purely accidental discoveries that allegedly crop up from time to time—Pasteur's discovery is frequently cited as a prime example—the success and prosperity of science is widely attributed to the no-nonsense rational manner in which scientists are believed to go about their business. But what is exactly meant by "the no-nonsense rational manner" in which scientists are said go about their business turns out to be notoriously unclear. Are we to attribute the rationality of science to the sort of goals scientists characteristically set themselves, or is it perhaps that scientific rationality has less to do with the goals so much as with the methods of attaining them that scientists typically seem to follow? Or is it their logic or their manner of reasoning, or perhaps their seemingly steadfast adherence to the bare facts, that should be deemed the crux of scientific rationality? Perhaps it is their willingness to theorize or, conversely, their willingness to refrain from theorizing too freely that makes the difference? Or, perhaps, in the last analysis, the very idea of seeking to articulate so general a standard in relation to an activity so generalized is misconceived? These questions, always at the core of attempts to understand science, became particularly pressing at the start of this century.[3]

The overthrow of Newtonian physics during the first two decades of the twentieth century generated a philosophical crisis of the first magnitude. Accounts of the growth and grounding of scientific knowledge that issued from enlightenment attitudes to science had to face up to the fact that their prime model of scientific rationality and progress had been discarded and replaced. The scientific worldview could no longer be simply described as the outcome of an accumulative process by which the great cathedral of knowledge is, as it were, gradually erected by joining one solid, well-confirmed brick to another. Yet almost every enlightenment depiction of science envisioned it in these or very similar terms. And all quite naturally took Newtonian physics as their prime example.

But the problem was more than that of losing one's pet example. The crisis was profoundly philosophical and went to the heart of these accounts. For one thing, philosophers at the turn of the century were required for the first time ever to account for a large body of knowledge which, on the one hand, was undoubtedly exemplary science yet at the same time was quite obviously untrue. In limited circumstances classical physics was, and still is, regarded approximately adequate. But classical physics can no longer be considered true in any but the most extremely pragmatist sense of the term.[4] In view of the relatively small velocities involved, the flightpaths of all of the National Aeronautics and Space Administration's (NASA's) space probes, for instance, are calculated on the basis of classical mechanics. This is done not because classical mechanics is *true* of such flightpaths, but because the margin of error it generates in these cases is considered negligible. In addition to having to come to terms with this new and seemingly oxymoronic category of disproved yet excellent scientific knowledge, philosophers were now also faced by major conceptual, epistemic, and ontological difficulties emanating from the new two theories that came in its stead. The physical sciences were undergoing a fundamental and all-embracing revolution that was shaking them to their very conceptual roots. Basic classical notions such as space, time, causality, and matter were being thoroughly modified to the point that the theories of relativity and quantum mechanics could not even be described as a developments of their Newtonian predecessor by any existing accumulative standard.

In contrast to the way earlier scientific upheavals were described in their time, this new revolution in physics could not be described as the simple replacement of a flawed theory by better grounded ones. Formerly, superseded theories, as prominent even as Ptolemaic astronomy,

Descartes' vortices, Newton's corpuscular theory of light, phlogiston theory, or Lamarckian evolution, had been Whiggishly dismissed as bad science—the unfortunate and allegedly avoidable consequences of methodological shortsightedness. But to describe Newtonian gravitation theory or classical electromagnetism in these terms was unthinkable. In one important respect classical physics was indispensable for the two new theories that replaced them. Although relativity theory and quantum mechanics could not be portrayed as developments of, or as logically entailing classical physics—since they clearly contradicted it—only in the context of a body of theory similar to classical physics could the problems be formulated to which the two new theories were presented as solutions. The motivation to innovate and rethink old beliefs depends on how pressing the problems that they give rise to are felt to be, which in turn depends on how compelling and fruitful they were thought to have been in the first place. In this respect classical physics was an outstanding scientific achievement even from the vantage point of the theories that eventually replaced it. However, to appreciate it as such, a totally new approach to the philosophy of science was called for, one that could account for the dynamics of such transformations.

Philosophers of science, who had so far concentrated on describing and extrapolating from what they took to be well-formed finished science, were caught quite unprepared for these developments. Einstein and his colleagues had rendered modern science an essentially historical entity whose progressive nature could no longer be accounted for in accumulative or derivative terms, or attested to simply by the example of illustrious past achievement. Science could no longer be explained and understood merely by analyzing its prized end products; it was swiftly being seen primarily as a process, as an essentially dynamic historical phenomenon.

More specifically, the rejection of classical physics in favor of relativity theory and quantum mechanics posed two new problems for the somewhat humbled philosophers of science at the start of this century: one concerning the rationality, and the other, the progressiveness of science. If scientists are unable to ever vouch for the truth of their conjectures, in what sense can their quest for truth be justified as a rational endeavor? And if they are unable ever to state whether one theory is more true than another, in what sense can scientific knowledge be said to grow or the quality of the picture science paints of the world improved in the course of their replacing old theories for new? In what sense can science be said to progress?

For a short while it was thought that the new developments in physics could be accommodated philosophically by the twentieth-century outgrowth of nineteenth-century instrumentalism known as 'logical positivism'.[5] The rise and inevitable fall of this short-lived philosophical program is an interesting story in itself, but of little consequence to our present concerns. For one thing the logical positivists had little if anything to offer regarding the two problems mentioned above. They concerned themselves exclusively with the scientific end product and utterly ignored the most pressing problem for philosophy of science: the problem of scientific change.

Logical positivists maintained that all knowledge claims must take one of two possible forms. They may either be claims to empirical knowledge that can be directly verified or falsified through experience or statements of purely formal character, like those of logic or mathematics. In their view, for a scientific knowledge claim to be considered meaningful it must be of the first sort. The second sort, comprising logic and mathematics, because it is entirely analytical[6] and, therefore, independent of the facts, renders no more than a series of tautologies (that is to say, analytical uninformative truths) that can never generate new scientific knowledge but at most assist in organizing the old. Science was thus identified with a subset of the class of meaningful cum verifiable propositions, namely, those that were actually verified by experience and logically structured by mathematical tautologies. Subsequently, all metaphysical musings, including statements about space, time, causality, and the like, were declared, to an extent self-refutingly, pure meaningless nonsense. Science progresses, on such a showing, either when new data are collected, or when old data are reorganized by means of new mathematical structures. Ideally speaking, the scientific lexicon, then, need not contain any terms other than those describing direct experience and those included in the vocabularies of formal logic and mathematics. Scientific theories, thus construed, are, in principle, purely descriptive devices incapable of explaining anything.

At first it had seemed that logical positivism had no trouble explaining the scientific revolution at the start of the century. When science, old and new, is appropriately stripped of all its superfluous 'metaphysical' accouterments, the shift from classical to modern physics, they argued, could be described satisfactorily as the expansion of the pool of empirical findings combined with its reorganization using mathematical structures quite different from those employed in classical physics. So described, the advent of relativity theory and quantum mechanics seemed not only not to challenge, but to actually illustrate their philosophy of science.

It was not long, however, before this form of positivism was severely criticized both for internal inadequacies and its incorrect interpretation of relativity and quantum theory. It will take us too far afield to take issue with the second criticism here; Michael Friedman, for one, force-fully debunks the positivist mythology about the supposedly formal va-cuity of relativity's space-time geometry in his *Foundations of Space-Time Theories*.[7] This is not merely a question of historical accuracy. If Einstein's revolution consisted of more than a formal redressing of verified empiri-cal findings, classical physics would have to be deemed empirically re-futed. And if that is the case, logical positivism can no longer rank it as bona fide science.

One basic condition of adequacy or viability for any account of science as a dynamic entity in flux, philosophical or other, is that it provide both retrospective criteria for the scientificality and quality of the superseded theories of yesterday and prospective criteria for scientific planning to-ward those of tomorrow. If science is exclusively identified with the theo-ries of today, it is deemed either accumulative or static—which, in the present context, adds up to much the same thing. And if science can no longer be considered accumulative, the criterion for being 'scientific' can-not be truth or truth-likeness. This is why, following the refutation of Newtonian physics, the problem of meaningfully demarcating the scien-tific and the non- or pseudo-scientific became so central. By equating the scientific and the verifiable, logical positivism declared its failure to even appreciate the problem of the day. And by focusing on demarcation in terms of static and supposedly objective meaning and logical form, it proved it!

The most important of the directly philosophical criticisms of logical positivism was a criticism of the notion of verification itself. This criti-cism, though not new, was contained most forcefully at the time in the philosophy of science proposed by Karl Popper. The contention that em-pirical data could in any way inductively render certified knowledge was based on a simple logical fallacy, claimed Popper in an argument fa-mously associated with his name. Contrary to the widely accepted, so-called hypothetico-deductive models of inductive support,[8] he argued, the fact that theory T logically entails the occurrence of an event E, cou-pled with the fact that E is in fact found to occur, does not, and can not teach us anything about the truth value of T. This is because although E may follow logically from T, E's actual occurrence could have been due to any number of causes other than those suggested by T! Thus, for ex-ample, although Kepler's laws of planetary motion are entailed with cer-tain reservations by Newton's laws of motion and theory of gravitation, the fact that they are found approximately to hold true cannot in the least be regarded as confirmation of Newton's theories, exactly because some-thing very different than the inverse square law might be responsible for their occurrence. Kepler himself, for instance, attributed them to a parallel push, rather than to a perpendicular pull, exerted on the planets by the sun, while general relativity denies the existence of a gravitational force altogether, attributing all "gravitational" phenomena, including the elliptic trajectories of the planets, to irregularities in the space-time manifold itself caused by the presence of the masses in question. In other words, argued Popper, much to the chagrin of the logical positivists of his day, the very notion of empirical confirmation, let alone verification, of any general hypothesis is in principle worryingly unfounded.

The disturbing implication of this important observation was that, despite what generations of scientists believed science could have, and at times actually had, achieved, science, in principle, would never be able to claim with confidence that any of its theories were true or even highly probable. This being the case, the two problems mentioned previously are seemingly rendered all the more glaring. If science is not the continued accumulation of true or at least well-confirmed knowledge, in what sense can it be said to progress? And if the discovery of truth cannot, in principle, be regarded as the business of science, what then is the business of science? And how can it go about its business if it cannot verify, or at least confirm, the hypotheses it generates? We shall have a chance to examine Popper's answers to these questions shortly. For our present purpose, however, the context in which this should be done is that of science studies today, rather than that of the logical positivism of the 1930s and 1940s in response to which they were originally formulated.

Heracliteans and Parmenideans

Owing to the enormous success of the Newtonian program during the eighteenth and nineteenth centuries, philosophers of science, as noted above, naturally took it as their task at the time to unearth the ontological and epistemological assumptions that inform classical physics and to elucidate the methods of research best suited to its further development. The subject matter of their studies was a fully developed body of what they took to be exemplar scientific knowledge. Their objective was philosophically to mine and analyze it for the sake of advancing lesser developed areas of scientific inquiry. It was an essentially descriptive enterprise directed nonetheless toward a normative goal, justified by the thought that it was supposedly founded upon a description of the best! The overthrow and eventual replacement of classical physics rendered this philosophical task to a great extent obsolete. All at once the pressing problem for philosophy became that of explaining scientific change. Philosophy of science could no longer regard science, even exemplary science, as a static entity identified merely by the immanent logical and conceptual structure of its finest theories and arguments. Philosophers were called upon to treat science as never before: as a historical, dynamic, and ever-changing process of inquiry, characterized not by the structure of its paradigms so much as by the nature of its paradigm shifts. Undoubtedly, the painful failure of logical positivism, that had all but ignored the question of scientific change, contributed much to this important refocusing of philosophic awareness.

Post-positivist studies of science that began to flourish in the late 1950s witnessed the emergence of two different approaches to the question of scientific change, approaches that may be usefully classified as Heraclitean and Parmenidean. In its fully developed, latter-day form, the Heraclitean or relativist approach literally denies philosophy the role of explaining science. Science, they declare, like any other form of organized human activity, is a culturally generated, socially and politically determined group undertaking. The social, institutional, and communal aspects of science are, therefore, best treated by trained social scientists, its epistemological and methodological aspects by evolutionists, decision theorists, and cognitive psychologists. Such is the inevitable fate of philosophy, insinuated the members of the self-proclaimed "Strong Programme in the Sociology of Science": to undertake initially speculative and thoroughly philosophical inquiries and to surrender them as soon as possible to the empirical methods of science proper. Science is philosophy come-of-age, and philosophy of science, it was argued, had now come of age. Had that been all the relativists had to offer, philosophers of science could well have ignored their attack. Few would deny that science is to an extent a culturally, if not politically, determined communal activity and that a comprehensive understanding of science requires a close study of its communal, institutional, social, and political life. Philosophy of science, so the argument would have gone, never claimed exclusivity in the study of all aspects of science. At the same time, however, major issues concerning the products and practices of scientific life remain primarily philosophical: theory structure, standards of facticity, the role of mathematics in science, the logic of scientific prediction and explanation, criteria for the acceptance and rejection of scientific theories, let alone metaphysical meditations about major scientific concepts such as space, time, causality, life, or truth. But latter-day relativists insistently denied philosophers even this. Taking their cue from the works of Michael Polanyi and Thomas Kuhn, they challenged the very legitimacy of any such undertaking. It is in principle pointless, they argued, to discuss any standard of human endeavor, scientific or other, in a manner unrelated to the specific cultural, social, and political context in which it is exercised. Criteria for rationality and progress, standards of reliable theorizing, competent test procedures, adequate patterns of explanation, guidelines for the acceptance and rejection of theories, even prerequisites for facticity are, according to this view, all social and political 'constructs', forever the mere contingencies of their particular place and time. It is, therefore, quite pointless to discuss them 'objectively', as philosophers supposedly do, independently of their particular contexts.

At the level of abstract doctrine philosophers responded with a measure of success. If all theoretical musing is so thoroughly context dependent, then, by the very same token, so is the doctrine of relativism itself. When stated in all generality, they urged, the Strong Programme is itself self-defeatingly reduced to a Heraclitean nonstarter by it own admission.[9] The real issue, however, has little to do with the second-order philosophical debate about the inner coherence of modern or postmodern relativism. The real challenge to traditional philosophy of science issues rather from the growing body of insightful first-order historical studies of science inspired by the Strong Programme. The philosophies of science proposed during the early seventeenth century by Francis Bacon and Rene Descartes were both radical, each calling upon science to begin de novo. But today, with regard to science, the days of philosophical radicalism are long past. For the past three hundred years or so, philosophers of science have directed their efforts, at least in the first instance, to the attainment of interpretative rather than constitutive goals, their task being primarily to make sense of science as they found it rather than, as Bacon and Descartes had suggested, to cast all existing science aside in order to establish it anew. In other words, as normative as it may view its undertaking, modern philosophy of science has as its object of research an existing, developing, essentially historical phenomenon. And even the staunchest absolutist can no longer ignore the convincing attempts by modern Heraclitean historians of science to contextualize almost every aspect of it. It is perhaps impossible to deter philosophers of science by arguing for relativism abstractly, but the type of history of science inspired by such arguments has certainly rendered their task exceedingly problematic.

While relativists see science adrift and unmoored, slapped this way or that by the ebb and flow of social, cultural, and political tides, Parmenideans or foundationalists seek to characterize science in ways that arguably transcend the contingencies of specific places and times. Scientists, they claim, typically adhere, if tacitly, to general and objective standards of progress and rational endeavor pertaining to some all-pervading, transcultural features of science that are manifestly immune to Heraclitean contextualizations. Until the turn of the century the truth or truth-likeness of scientific knowledge was considered the perfect candidate. The claims of scientific theories were either true or untrue regardless of the social, cultural, or political circumstances of their production or acceptance. If rationality could be equated with, or in some way reduced to, a person's capacity to attain certified knowledge, and progress with its accumulation, science could be convincingly portrayed as possessing a solid Parmenidean uncontextualizable, or supracontextual hard core. Latter-day Parmenideans, however, have been denied this option. Scientists simply have no such capacity, if only for the conclusive argument commonly attributed to Popper to the effect that the truth of scientific hypotheses is in principle humanly unrecognizable. A person may *believe* or *assume* that a certain theory is true, but, as we now realize, there exist no objective criteria for ever proving him or her right. And mere belief, argues the relativist, is forever conditioned by context. But if truth or truth-likeness can no longer be considered an achievable scientific goal, what other aspect of science is capable of sustaining a Parmenidean account?

As noted above, with the exception of logical positivism, twentieth-century philosophy of science was compelled to shift the focal point of its analyses from the product to the process of science, from what characterizes exemplar theories to what characterizes exemplar theory-choice. Philosophies of science were supported and appraised less for the type of scientific theories they commended than for the procedures they prescribed for the adoption, modification, or rejection of hypotheses. Appalled by what was described as Kuhn and his followers' attribution of scientific change to "mob psychology" and of scientific progress to a "bandwagon effect," philosophers of a Parmenidean leaning naturally attempted to locate the permanent transcontextual touchstone of their philosophical accounts in the scientific method. The most influential latter-day attempt of this kind was proposed during the late 1960s by Popper's former student, Imre Lakatos.

"Philosophy of science without history of science is empty," declared Lakatos, paraphrasing Kant, and "history of science without philosophy of science is blind."[10] Philosophy of science, he argued, provides the historian of science with a rationale; history of science, on the other hand, serves, in turn, to adjudicate between competing philosophies. He envisaged history of science as seeking to provide a "rational reconstruction" of the scientific past generated by the particular philosophy of science endorsed by the historian. At present, he submitted, aspiring historians of science have four such philosophical systems at their disposal: inductivism, instrumentalism, falsificationism (i.e., Popper's methodology, to be discussed below) and Lakatos's own theory entitled "The Methodology of Scientific Research Programmes", the details of which need not concern us here.[11] The four rival philosophies of science, he argued, entail rival "normative methodologies," each advocating different "sets of rules for the appraisal of ready, articulated theories." Historians, committed to different philosophies of science, will, therefore, subsequently carve up the historical record differently into two mutually exclusive groups of scientific episodes: those whose outcomes conform and those whose outcomes fail to conform to their preferred methodologies. Lakatos dubbed the two classes of historical episodes "internal" and "external" histories of science respectively. The idea is that while "internal history" represents scientific decisions that are well taken from the historian's point of view, and are, therefore, in no need of further explanation, the episodes comprising "external history," in deviating from the course that the historian believes science *should* have taken, require "external" explanation— namely, their arguable attribution to psychological, social, political, or religious causes. In Lakatos's opinion, internal history of science alone furnishes "a rational explanation of the growth of objective knowledge"—from the historian's particular philosophical point of view.[12]

Had Lakatos limited the lessons that he insisted could be drawn from this type of history to appraisals of scientific *progress*, his position may have been considered viable. Progress is, in principle, a category of retrospective evaluation and pertains primarily to the normative standards of its assessor. One may, of course, be interested in determining how past scientists assessed their own work, in which case their, rather than the historian's, preferred methodology should be made the standard of evaluation. On the other hand, it makes perfect sense to attempt to ponder the value of their decisions "objectively," as it were, from the point of view of one's own methodology. And there seems nothing fundamentally wrong in assuming that different "normative methodologies" may entail different accounts of the lasting value of particular scientific decisions. While Lagrange's analytical mechanics, for example, might be considered a substantial step forward from an instrumentalist's point of view,[13] it may, at the same time, seem quite inconsequential from an inductivist's perspective.

Lakatos's ambitious historicophilosophical project for science, however, aimed at more than providing tools for the assessment of scientific progress. "Internal history of science," he claimed, constitutes the *rational* part of science. Whenever a past scientist's decision to endorse or to reject a hypothesis is found to comply with the dictates of the historian's pet methodology when applied to the same data and background knowledge, it is considered by Lakatos to have been not only a step forward but to have been the rational thing to do. The four above-mentioned methodologies double for Lakatos as theories of scientific rationality, and the "internal histories" they generate, as "rational reconstructions of the history of science."[14]

In this, I submit, Lakatos went a gratuitous step too far. To equate the rationality of a scientist with the reconstructability by hindsight of his or her conclusions in the light of some latter-day methodology, unwarrantedly renders judgments of scientific rationality at once both wholly retrospective and dependent upon standards of procedure of which, in many cases, the scientist in question could not in principle have been aware. As a number of critics have pointed out,[15] in order to be ranked rational, a scientist who happened to have worked prior to the formulation of the historian's favored methodology is required by Lakatos to have been a canny *sleepwalker*—thoughtlessly arriving at what only hindsight will eventually judge to have been an appropriate decision—a view which seems to me to run counter to the very notion of rationality.[16] To act rationally, one would like to think, is first and foremost something a person is capable of *choosing* to do. Rationality should thus be construed as a *prospective*, rather than a merely retrospective, category of human endeavor, unrelated, in principle, to its successfulness or to the quality of its eventual results. It is also reasonable to insist that a necessary, though by no means sufficient, condition for any action to be considered rational is that it be performed consciously and intentionally. To rate thoughtless activity rational merely on the grounds of its meritorious outcomes would require us, absurdly in my opinion, to classify such praiseworthy feats as breathing in one's sleep, digesting one's food, or the beating of one's heart as bona fide rational activity.

But to insist that judgments of rationality are to turn on and pertain to the local and invariably context-dependent sensibilities, intentions, objectives, methodological guidelines, and background knowledge of the agent in question (rather than to those of the latter-day historian-philosopher) would seem to play straight back into the hands of total relativism. If neither the alleged truth or truth-likeness of scientific pronouncements nor the tacit method by which scientists allegedly operate can be viably considered supracontextual Parmenidean touchstones for theories of scientific rationality and progress, what can? How is it at all possible to elucidate the notions of rationality and progress with respect to science in a manner that avoids both the relativists' Strong Program and the type of blatant ahistorical hindsight preached by Lakatos and his followers? Put differently, to what extent is a general and consistent philosophical account of scientific accomplishment and scientific rationality possible which also does justice to the history of science as we now understand it? In what follows I shall argue that the writings of Karl Popper contain the foundation for a philosophical framework for answering these questions satisfactorily.

Science as Hypothesis

Although they acquire unprecedented centrality and are brought home with unprecedented force in Popper's writings, the arguments against induction, so widely associated with his name, were hardly original. Versions of these claims can be found long before 1934 in the writings of a broad variety of philosophers such as David Hume, William Whewell, Charles Sanders Peirce, and John Dewey, to mention but the most prominent. The novelty and crucial importance of Popper's philosophy of science lies far more in the solution he proposed than in the problem he raised. To appreciate this we need to look briefly at the philosophical discussion of modern skepticism beyond the pale of science proper.

Setting the tone for the discussion of skepticism to this day, Hume reasoned that wholesale, total doubt cannot be overcome to anyone's philosophical satisfaction. I call such skepticism wholesale because doubt is cast sweepingly and simultaneously on all we aspire to know at any given moment. If the skeptic's claims are admitted, and in Hume's opinion they are impossible to resist, the result is devastating, amounting to an epistemic meltdown[17] as it were, in which all epistemic holding on reality evaporates and to which any conceivable counterargument inevitably falls prey. However, submitted Hume, such wholesale, crippling reflexivity remains confined to the philosopher's study, and can have little effect beyond its door. Our psychological makeup is such that in everyday life wholesale skepticism is safely ignored. Outside their studies even the most skeptical of philosophers join the rest of humankind and generalize and project on the basis of past experience, assume that their memories are in general trustworthy, and rely without further ado on their senses, on their cognitive abilities, as well as on whole stocks of unquestioned belief. Michael Williams, in a recent book, appropriately dubs Hume's somewhat schizophrenic approach to wholesale skepticism 'biperspectivalism'.[18] He also shows quite convincingly that Hume's twofold contextualization is widely accepted by most parties to the latter-day debate.

Naturally, it has been the philosophical, study-bound part of Hume's twofold solution that has exercised most philosophers most keenly. Responses vary. There are those, such as Thomas Nagel, who agree with Hume that wholesale skepticism is in principle unanswerable.[19] Others attempt variably to show that the wholesale skeptic's challenge is provably incoherent. Stanley Cavell, for example, submits that the skeptic can be shown not to be able to mean what he seems to mean.[20] Williams himself argues interestingly that it is possible to expose and, at a tolerable cost, reject some of the theoretical assumptions the wholesale skeptic can be shown to premise. The problem is that while these options—the acceptance, refutation, proof of incoherence, or development of theoretical alternatives to Humean wholesale skepticism—are hotly debated, Hume's observations regarding human behavior outside the philosopher's study are largely accepted. Without underestimating the philosophical importance of, and urgency in, attending to wholesale skepticism and without disputing the claim that much of what we think and do reduces to unreflective, almost knee-jerk habit, it is quite wrong, in my opinion, to suppose that the only philosophical problem we face from skeptical quarters is that of the wholesale variety.

Well beyond the confines of the Humean study lies an enormous range of reflective, middle-range human activity that can hardly qualify as habitual and that urgently requires a retail rather than a wholesale response to the uncertainty that surrounds it. And it is here, rather than in Hume's philosopher's armchair, that science looms largest. The skeptical challenge to our construction of science issuing from the groundlessness of all inductive reasoning has nothing in principle to do with such wholesale, study-bound skepticism as doubting one's memory, pondering the reality of the world, questioning the soundness of one's senses or cognitive abilities, or, to take Hillary Putnam's well-known example, seriously contemplating the possibility that we may really be mere brains in a vat. On the contrary, the more confident we are that such wholesale doubt is in some way unfounded or safely cordoned off, the more our memory, senses, and reasoning capacities are deemed reliable, so the fact that science can be shown to be incapable of ever warranting the truth of its theories is rendered all the more perplexing! Even if, indeed especially if, wholesale skepticism is met satisfactorily, the problem of accounting for the rationality and advancement of science becomes all the more worrysome. For what we are dealing with are not merely *doubts* about our ability to empirically prove scientific theories, but alleged *proof* of our inability in principle to do so. It is not as though we had good arguments for believing that science could divine the truth and were worrying whether our eyes and minds were fooling us into believing them. On the contrary, it is *because of* our utmost confidence in our eyes and minds that we are able to *argue to our satisfaction* that we lack the ability to vouch for the truth of any scientific theory.

Heraclitean accounts of the rationality and advancement of science tend, one could say, to treat the problem as one of wholesale skepticism.

These accounts resemble elaborate applications of Humean biperspectivalism, in the course of which the skeptical side of the equation is met by accusations of incoherence. First they argue that it is meaningless to speak of the truth, rationality, or progress of science without reference to a specific contextual framework, and second, that within such frameworks the true, the rational, and the progressive are determined contextually and can therefore not be challenged from without. Scientists perform all kinds of discursive moves, some of them inductive, in accord with the culturally and socially determined rules of the particular language game they play. And these, they maintain, can only be challenged from within.

Parmenideans object to contextualizing the problem away. Hence, one can either accept or attempt to disprove the arguments against induction. I have not discussed the latter option in the present study and do not intend to do so. While there are still several analytical philosophers who are determined adequately to explicate the notion of inductive support— by grounding it on probability calculus for example—it is my clear contention that these attempts are futile and that the refutation of induction associated with Popper is sound and compelling.[21] Popper's own Parmenidean approach to science paves the way for a satisfactory account of rationality and progress while, at the same time, firmly denying all general claims of science to truth or truthlikeness. In recent years this type of approach has come to be known as critical rationalism. I myself prefer the label 'constructive skepticism' (as opposed to destructive skepticism) because the term 'rationalism' normally means something different.

Popper's theory of science develops from the following argument: although it is impossible to empirically determine the truth or truth-likeness of any general scientific hypothesis, by prudently criticizing, rather than seeking to verify our efforts, we are sometimes capable of ascertaining that we were wrong. If the occurrence of an event E is logically entailed by a conjectured theory T, we have already seen that the fact that E is found to obtain is logically inconsequential to our knowledge of the truth of T. However, if E turns out convincingly *not* to be the case, logic permits us to conclude that in certain respects T is false. (Owing to the argument schema known as *modus tollens*: if P entails Q, and if it is the case that not-Q, it follows that it is the case that not-P.) Scientific hypotheses cannot be verified empirically, but they can, it seems, be empirically refuted. Even this, of course, is not as simple as it seems. All a failed prediction can prove is that there is something seriously amiss with at least one of our premises. And the premises presupposed in scientific predictions commonly comprise much more than a conjectured theory and an immediate perception. The amount of auxiliary and background knowledge at stake can be enormous in these cases. Still, whenever an empirical prediction is unsuccessful, we do know for certain that something is amiss with our prior assumptions.

But how can such an *epistemologia negativa* inform a theory of epistemic *accomplishment*? In what sense can scientific knowledge be thought to grow, or the objectives of scientific inquiry be considered better attained, if all we can ever confidently claim to learn from experience is that we have been mistaken? To this the Popperian answers that although no amount of empirical evidence is capable of verifying or even confirming a general scientific hypothesis, the outcomes of prudent empirical tests enable us, in an important sense, to rate the worth of a hypothesis in comparison to its allegedly problematic predecessors and rivals. Even if we are unable ever meaningfully to assert that today's theories are better confirmed or steer closer to the truth than those they replaced, we are sometimes capable of ensuring that, to the best of our knowledge, they are in a meaningful sense *less problematic* than, and at least as equally comprehensive as, those of yesterday.[22] Although it is impossible to confirm Thomas Young and Augustine Fresnel's wave, or undulatory, theory of light for example (notwithstanding the fact that at the time it was widely regarded as fully verified), the fact that it had convincingly withstood empirical tests that earlier emission theories of light had flatly failed attests to its relative superiority. The decision, therefore, to replace particle optics, during the early nineteenth century, with Young and Fresnel's new theory represents for the Popperian both a substantial and rational step forward—an assessment that remains unimpaired by the fact that at the turn of the present century new problems were discovered that prompted physicists to reconsider their earlier dismissal of particle optics and to formulate a formerly unthinkable, though now equally accepted, combination of the two.

Viewed thus, the advancement of science can be meaningfully elucidated in terms of the dynamics of the problem seeking and problem solving of day-to-day scientific research. One is thus able to chart and evaluate scientific advance with sole reference to the "interrogative logic" of scientific trial and error, without having to relate it to an arguably objective metric of truth-likeness. It remains to be shown, however, if and how such an approach is indeed capable of viably meeting the relativist's challenge. To do so we need to take a closer look at Popper's proposals.

Popper himself is an outspoken realist.[23] He regards scientific theories as means to a specific, if in principle unattainable, end: the construction of true explanatory accounts of whatever is thought to require an explanation. Lacking the ability ever to prove or even confirm its conjectures, science, he argues, typically proceeds by deliberate and contrived *troubleshooting*. Summarizing his methodology repeatedly by means of the following schema:

$$P_1 \rightarrow TT \rightarrow EE \rightarrow P_2$$

Popper observed that scientific inquiries are sparked as a rule not by the accumulation of data—as inductivists would have it—nor by speculating upon our ideas—as would idealists, but rather by the coming to light of a Problem (P_1), by noticing that, or better, by contriving to discover that, the theory at hand is problematic, seemingly at variance with what are thought to be relevant facts. As a result, Tentative alternative Theories (*TT*) are put forward with a view (a) to adequately explaining the apparent failure of their predecessor, (b) to accounting equally well for all phenomena successfully accommodated by their predecessor, and (c) to succeeding where their predecessor had allegedly failed. By seeking both to explain *and* solve P_1, the candidate hypotheses are guaranteed to be potentially less problematic than their predecessor. Each of the *TT*s is then in turn subjected to as thorough tests as possible (Error *E*limination), until, one hopes, a better-suited successor is selected for the time being. In time, and as a result of further testing, the new hypothesis will presumably give rise to new problems (P_2), and the process will be repeated. Science thus progresses by repeated cycles, or rather spirals, of imaginative conjecture and keen refutation. From such a perspective, the distinctive mark of scientific rationality is not the power of science to secure and amass true or even well-confirmed knowledge. On the contrary, science is ideally rational exactly when it elects to treat each and every item of general knowledge as a tentative and potentially refutable conjecture, attempting persistently to lay bare its apparent deficiencies, and willing at all times to modify or replace even its most cherished convictions.

Science, in short, is perceived by the constructive skeptic as the ordered combination of a prudent and creative effort to expose the inadequacies that beset each of its theories, followed, when successful, by an equally creative effort to modify or replace them by better-suited alternatives. Scientific research is ideally conducted on the assumption that its desired goal—truth—is in principle unrecognizable and should therefore be regarded as no more than a regulative objective. It follows that science is most productive when consciously conducted in the light of a working metahypothesis that states that its current conclusions are in all probability mistaken and that all scientific knowledge is conjecture, including even the outcomes of successful refutations.

As noted previously, even when a given prediction is convincingly refuted by a laboratory finding, determining where to lay the blame will always remain an open question. The theory being tested could be mistaken but, by the same token, so could the refuting data themselves. All facts are theory laden, Popper frequently reminds us, and the conjecture at fault could very well be the theory or theories invested in constructing the measuring devices or interpreting its readings. It could even be the case that both theory and data are correct and that their apparent incongruity owes to an unknown external factor. The existence of the formerly unknown planet Neptune, for instance, was conjectured in this manner in order to compensate for the apparent discrepancies between the observed and calculated trajectories of Uranus. All that an apparent refutation can attest to is that something is seriously amiss, that a particular body of scientific knowledge is flawed as it stands. A contradiction found to obtain between a theory and the data it is intended to explain is undoubtedly a clear indication that a scientific problem is afloat, but it can contribute little if anything toward the correct diagnosis or prognosis of the nature of the malady. The analysis of a problem and of the appropriate ways to treat it is, for the constructive skeptic, as conjectural as any other assessment of the situation. In most cases, of course, certain of the hypotheses involved will have formerly survived more thorough tests than others, and will be naturally regarded as perhaps less likely to be mistaken. In such cases it would be reasonable in the first instance to mistrust the newcomers, particularly the very theory being tested. But that again is pure conjecture and is itself as liable to be refuted as any other. For the prudent Popperian no part of the scientific system of ideas under consideration is taken for granted, no part of it is guaranteed in advance, none of it is held immune to criticism.

Outside the scientific system of ideas in question, much, of course, that the Humean wholesale skeptic systematically calls into question *is* taken for granted by the Popperian. Constructive skeptics do not cast wholesale doubt on everything that can in principle be doubted. They are not motivated by doubting their existence, their sanity, or the existence of a world out there—all of which they take for granted. What they doubt systematically is the quality of the job they are doing. And because their skepticism is so limited as to exclude the possibility that they are hallucinating or being persistently misled by malicious demons, their wary and quizzical attitude to their own suggestions is all the more keen. Most importantly, whatever is spared systematic doubt does not encroach upon their specific scientific projects. It follows that according to this approach, all talk of Lakatosian "hard cores" and Kuhnian paradigms should be dispensed with.[24] Even in the case of a striking refutation, all science can claim to know remains, for the Popperian, blatantly conjectural.

In this manner, unlike all forms of traditional skepticism, the deep mistrust that is thus premised with regard to all general scientific knowl-edge-claims is not destructive or dismissive of the scientific enterprise it-self. On the contrary, by flatly rejecting the traditional equation of science with truth, the constructive skeptic is able to equate scientific rationality with systematic criticism, and to present a constructive and inspiring vi-sion of scientific advance and accomplishment grounded upon no more than the logic of scientific trial and error.

But science cannot lay an exclusive claim on rationality any more than Popper's theory of the way science *should* be understood and pursued can lay an exclusive claim on how all scientists past and present have in fact perceived and performed their work. For Popper, the only viable, if regu-lative, aim of science is truth, and subsequently, the only real problems encountered en route are empirical refutations and logical inconsisten-cies. This is too narrow a basis for a general account of rationality and progress for at least three reasons. First, it is incapable of accounting for the work of scientists who, either unknowing of or in knowing disagree-ment with Popper's philosophy, have self-consciously directed their ef-forts to the disclosure and solution of very different problems than those Popper is willing to recognize as scientific. There are any number of ex-amples of scientists who wittingly constructed, questioned, and modified their theories to the attainment of such 'non-Popperian' goals as simplic-ity, conciseness, mere effective predictive power, or sheer elegance. And in order to avoid the type of criticism previously leveled against Lakatos, any viable theory of rationality has to be sufficiently general, regardless of our own preferences, to enable an assessment of the efforts of such non- or anti-Popperian researchers in relation to the goals they set them-selves.

Second, if the discussion of rationality is rendered subordinate to one's metaphysical or metascientific convictions—as in Popper's case it is made subservient to his realism—how are we to discuss and decide *them* ra-tionally? The notion of rational activity in general, it seems to me, is basic and should be discussed as far as possible prior to and independently of metaphysics or ethics, unless we agree a priori to treat our metaphysical or ethical commitments as mere fancy or as unaccountable acts of faith. Popper himself comes uncomfortably close to this in his candid admis-sions of the irrefutable nature of his scientific realism.[25]

Third, following from the last point and most importantly to the con-cerns of the present study is the need to consider the wide variety of de-cidedly nonscientific realms of discourse and activity. Even if one insists on regarding all non-Popperian science as not only ill conceived but ir-rational as such, it would still be preposterous to demand that *all* poten-tially rational and advancing endeavors meet strict Popperian specifications. Only a fraction of our enterprising activity is undertaken with a view to devising feasible representations of external reality. If my car breaks down, or if my German turns out not to be good enough to ask for directions in Berlin, we are dealing with malfunctions that have little to do with representation, empirical refutation, or logical inconsistency. My car and my meager German are not intended to represent anything. Since they are not propositional by nature, they cannot be said to be true or false, and hence cannot ever be considered refuted or contradicted to begin with. On the other hand, they can very well be said to harbor problems. They may be found deficient or lacking respective of their intended functions, and as such could well be considered capable of improvement, upgrading, or rectification. In such cases the way Popperians conceive of scientific problems seems inadequately narrow. A general theory of rationality, capable of accommodating all manner of conscious human activity, requires us, therefore, to go a significant step beyond Popper. In order for his approach to apply beyond the confinements of his particular understanding of the empirical sciences, the very notions of problem and solution need to be generalized far more than he ever attempted.

Problems and Progress

A general appreciation of problems and problem solving, capable of accommodating the entire range of malfunctions, defects, inadequacies, and imperfections sought and negotiated by enterprising humans, may be obtained by turning from the limited realm of scientific theories to consider, in as general a manner as possible, any structure, theoretical, institutional, or material, consciously designed or adopted by human agents as means to specific ends. Let us term such devices *goal-directed systems*.[26] Scientific theories designed to accomplish any of the above-mentioned objectives, my car, and my faulty German are all goal-directed systems.[27]

In the broadest possible sense of the word, we may describe a problem, any problem, as the apparent failure of any such system to achieve its intended goals: a discrepancy, that is to say, between what a particular goal-directed system accomplishes, or is capable of accomplishing, and what it is meant, desired, or expected to accomplish. The teleological aspect of the matter is unavoidable simply because I fail to conceive of anything we would be prepared to view as a problem other than in relation to (the frustration of) some desired goal.

A problem is thus construed as a real feature of a goal-directed system, rather than as an epistemological category, a fact of the matter, rather than a cognitive state of perplexity or a part of speech.[28] A system may be problematic without anyone having been aware of the fact. Problems can therefore be discovered, guessed at, misconceived, or overlooked like any other property of a given system. Bertrand Russell, it seems reasonable to claim, discovered (rather than created) the inconsistency that had plagued set theory long before he formulated the famous paradox associated with his name. Defects and malfunctions do not spring into existence by being noticed, but rather need first to exist *in order* to be noticed! What I am arguing for, in other words, is for a realist position if only with regard to the goal-directed systems with which we elect to operate and their ability to perform their intended functions. Even the staunchest antirealist must allow his or her antirealist theory itself a measure of ontological autonomy if he or she wishes to present it as susceptible to critical appraisal and capable of improvement.

All this might sound excessively Platonic but in truth it is not. To maintain that the potential pros and cons of a given system—intellectual or other—exist independently of human awareness is by no means to grant the systems themselves or any of their features the status of timeless immutable Platonic forms. The game of chess, set theory, or common algebra need not be regarded as timeless entities any more than my car or my German. One may easily consider each of them as a human construct, created ex nihilo at a particular moment in the course of human history for one reason or another. However, *once created* their features, dispositions, and potential virtues and defects spring into existence independently of them being known at the time, to the effect that some of them may not be noticed until later, while others may very well never come to light at all.

What, we may now ask, gives rise to problems, and subsequently what is it to resolve a problem?

A goal-directed system may be born problematic, incapable of fully fulfilling its promise from the very start. Alternatively, an initially unproblematic system may be rendered problematic in time as a result of either a deterioration in its performance, a revision of the goals it was meant to achieve, or a change of external circumstances (or, of course, any combination of the three). But since we shall be concerned throughout this book primarily with systems of knowledge (representational and interpretational schemes of propositions, systems of law, etc.) that cannot be said to deteriorate of their own accord (like the aging of an appliance or an organism), I shall pursue that type of problem genesis no further.

Problems of the latter kind arise when the circumstances of a system change. And, again, one may differentiate between two types of situation. In either case the circumstantial or environmental changes requiring a revision of the system's goals will be due to external factors. However, the changes themselves may be introduced intentionally or unintentionally. A social institution, fully capable of doing its job, may be rendered problematic in the event of unforeseen demographic or political changes. Similarly, a formerly successful scientific theory may be rendered problematic by the coming to light of unpredicted new data within its initially intended domain of application. However, systems may be rendered problematic when applied deliberately outside their formerly envisioned domains, as when appliances are intentionally applied to tasks they were not initially designed to perform or when scientific theories are deliberately applied to phenomena of a kind different from those they were originally meant to accommodate. Such a distinction has no bearing on the analysis of problems or problem solving as such. However, since I shall be arguing for a theory of rational action as action in the face of problems, cases of the latter kind, where action seems to be deliberately taken with a view to raise rather than to resolve problems, will demand special attention.

If, as proposed, problems are taken as discrepancies between whatever a particular goal-directed system is capable of achieving and what it is meant or supposed to achieve, a problem will be said to have been (fully or partially) *resolved* exactly if the discrepancy is (fully or partially) obliterated. Not every resolution of a problem, however, may be regarded as a *solution* of it. Problems, in the sense proposed, may be resolved by either improving the system's performance or by appropriately modifying its goals. But a problem will be said to have been *solved* only when resolved in the former sense, when by suitable modification the system in question is rendered better capable of achieving its objectives. Subsequently, a problem may be said to have been partly solved if, by modifying the system, the discrepancy between its achievements and goals is to an extent diminished. When, on the other hand, the discrepancy is lessened or made to vanish by an ad hoc reformulation of the system's goals, the problem may be said to have been (fully or partly) *dissolved* or *mitigated*, but by no means solved. A scientific theory found incapable of accommodating the full range of phenomena to which it was originally meant to apply may be rendered less problematic or even unproblematic by an appropriate restriction of its domain of application. But we would not consider this a solution of the problem of accounting for the kinds of phenomena excluded by such a move. When, for example, the representation of the sun, planets, and moons by mass-points was replaced in Newtonian astrophysics by solid spheres that approximated their true dimensions, certain of the problems that beset the original theory were solved. But when, following the advent of relativity theory, classical mechanics was restricted to inertial systems moving relative to each other at low velocities, it was rendered unproblematic without solving any of the problems it was known to harbor. When standards are lowered, systems are rendered less problematic, but their initial standards are thereby not rendered more readily attainable. A medication may be rendered less problematic by restricting its prescription to those populations in which it was proved effective, but in doing so the goal of effectively combating the malady for which it was originally designed is on no count better achieved.

At times, problems are found to have been but apparent, as when in the face of danger, an organism (if one is willing to consider an organism a goal-directed system) discovers that it is capable of running faster or jumping higher than it was accustomed to. In the case of scientific theories Thomas Kuhn has termed this type of pseudo problem solving "normal science." Here again, one cannot say that a real problem was *solved*, but only that an apparent problem was found not to have been a problem in the first place. Discrepancies, as in Kuhnian "normal science," between what *are thought to be* a system's goals and capabilities, are at best indicative of problems relating to second-order systems of knowledge, designed not to attain the goals of the system in question, but to describe or to understand it. (This reinforces my insistence on the need to differentiate clearly between problems and their proposed representations, namely, questions.)

We are now in a position to define progress. A goal-directed system, I suggest, will be said to have *progressed* or to have been *ameliorated* from state A to state B exactly if problems it harbored in A are wholly or partly solved in B. Conversely, a system will be said to have *deteriorated* or *regressed* in moving from state A to state B exactly if it is more problematic in B than it was in A. Notice that in order to progress from A to B it is not enough for a system merely to be less problematic in B than in A since we require that progress involve the solution not merely the resolution of problems. The solutions to problems and by implication the measure of progress achieved as a result are thus regarded, like the problems themselves, as objective features of goal-directed systems. A problem is a property of a *static* system frozen in time as it were—the discrepancy, at a given moment, between what it can and what it is meant to achieve. A solution to a problem is a particular *change* in a system as a result of which its original goals are rendered more readily attainable. Problems are contextualized with reference to specific goal-directed systems fixed in time; solutions, and hence progress, are contextualized to specific systems that are modified in time with reference to fixed goals. And all three, of course, may go unnoticed or be mistakenly construed.

Matters are complicated, of course, when progress with respect to one set of objectives necessarily entails regress with respect to another. Modifications of a system designed to solve one set of problems may at the same time inevitably give rise to others. For example, certain mutations, while increasing an organism's gene distribution, could be found to render it more vulnerable to predators. In biology—if again one is at all willing to regard organisms as goal-directed systems—the problem seldom arises since evolutionary amelioration is normally evaluated with reference to a single goal, namely, gene distribution. But systems may be, and usually are, designed simultaneously to achieve a variety of goals with respect to which modifications may very well be at cross-purposes. Problems of this kind arise all the time. In some cases they are resolved by appeal to a hierarchy of goals determined in advance as part of a higher-level system. Scientific theories, for example, are supposed, among other things, to be both simple and comprehensive, but simplicity is usually sacrificed readily for the sake of comprehensiveness. However, more often than not such dilemmas do not resolve themselves so readily, as, for example, when meeting a deadline, lowering the cost, and guaranteeing the adequate performance of a product are at cross-purposes. What seems to be called for in such cases are quantitative rather than qualitative criteria of progress—that is, ways of somehow weighing the progress achieved with reference to one set of objectives against the necessary regress it involves with reference to others. But it is difficult to see a way of going about such a project in a general manner. Such a weighing of the pros and cons of a move is conducted as a matter of course with reference to considerations external to the system in question, usually with reference to the goals of another, perhaps more fundamental system.

Such an account of problems and problem solving with regard to goal-directed systems in general thus allows us to retain a Popperian rendition of progress without having to limit ourselves to Popper's particular view of the aim of science and the nature of scientific problems. It also enables us to speak equally generally of rationality, as we shall see shortly. But before moving on to rationality, I would like to stress the following two points. First, although such a construal enables one to speak of problems, of problem seeking and solving, and to appraise progress in an objective and general fashion, they have not been decontextualized. Problems, solutions, and the measure of progress thereby achieved remain, by definition, wholly related to specific, context-dependent, goal-directed systems. They may be analyzed and evaluated with reference to the way the system in question was conceived by its initial developers or users, free of the type of anachronism built into Lakatos's approach, for instance, by considering the goals the developers originally set for it, and by assessing *their* solutions to the problems *they* thought it harbored with reference to the prevailing attitudes, knowledge, and tools that *they* had had at their disposal. On the other hand, a person is free, if he or she so wishes, to assess another person's work on a given system from his or her own latter-day perspective, in which case the system in question, its apparent deficiencies, and its proposed modifications will be analyzed and evaluated with reference to the (equally context-dependent) attitudes, knowledge, and tools available to their latter-day assessor. And yet the standards and criteria that govern any such assessment remain univocally grounded in the objective, transcultural logic of problems and problem solving.

Second, such a theory of system amelioration renders all assessments of a system's progress value-neutral. Progress is said to have been achieved if and when a goal-directed system is found, as the result of modification, to be better suited to accomplish its intended objectives *whatever they may be*—and not, as commonly construed, when it is rendered better suited to accomplish only what are thought to be commendable, virtuous, or fitting objectives. This, of course, is what enables discussion of the measure of progress achieved independently of our own preferences. This feature of the proposed theory reinforces the former point, for had assessments of progress been made to turn chauvinistically on what *we* happen to value as meritorious objectives, the theory would be rendered justly vulnerable to relativistic objection. The grounding of all judgmental appraisals of progress upon the value-neutral, means-end logic of problem solving serves effectively to meet such criticism. But it is more than merely a tactical maneuver. The fact that a person working on a system apparently took notice of certain problems and deliberately undertook to solve them is a clear indication of the value he or she ascribed to the goals he or she intended it to achieve. In such cases of deliberate troubleshooting, our construal of progress fully retains (if tacitly) the value judgments of the designer or user of the system under consideration. What *have* been rendered value-neutral are the historian or philosopher's second-order erotetic assessments of the measure of progress thereby accomplished. These, we have seen, premise no more than the abstract logic of problem solving. Progress assessment does not, of course, *necessarily* premise prior intent. Problems are frequently solved accidentally and may well go unnoticed, as when a loose cog is jolted back into place by a bump on the road. Automatic cybernetic devices, to take another example, detect and right malfunctions as a matter of course wholly devoid of human intervention or awareness. Nonetheless, any retrospective depiction of such accomplishments will invariably involve *someone's* particular context-dependent perspective—be it that of the original designer-user, or that of a latter-day narrator. And in all such cases the logic of assessment will remain the same.

Such a construal of the types of problems that goal-directed systems are liable to harbor, the ways in which they can be solved or dissolved, and the notion of progress it entails is so general that it is hard to imagine what form an effective relativist counterargument could take. Can one seriously envisage circumstances in which a person would *not* consider problematic the apparent failure of a system to achieve its intended goals? And if not, in what way could a concept of progress based exclusively on the successful solution of such problems ever be deemed a mere context-dependent product of historical circumstance? If the answers to these two questions remains negative, as I believe they have to, we may claim with a measure of confidence to have articulated a general Parmenidean explication of progress which cannot be said to do violence to history, and we may now turn to a similar construal of rationality.

The Rationality of Goal-Directed Criticism

Any general discussion of rationality seems to be faced with the following problem, eloquently stated by John Kekes: "If rationality is given a rational defense it is question begging, and if it is not defended rationally, it is arbitrary."[29] However, in order to take issue with Kekes's dilemma, we must first make clear what rationality is a category *of.*

In the vast philosophical literature on rationality this crucial preliminary question is seldom raised. On the other hand, one finds endless talk of rational theories, concepts, beliefs, and standards, even discussions of the 'rational relationship' obtaining between theory and fact.[30] Now, if, as I believe, the notion of rationality is expected minimally to involve a conscious and deliberate use of one's brains, a theory, belief, or concept cannot in itself be deemed any more rational than a table or a chair. One may rationally or irrationally accept, deploy, invoke, or reject a concept or a theory, just as one may rationally or irrationally decide to employ an appliance or make use of an institution. But there seems to be no sense at all in rating theories, appliances, or institutions as immanently rational in themselves. Rationality, as I have been careful to use the term in previous sections, is *exclusively* an evaluative category of human conduct, not of its products, be they appliances, institutions, concepts, or theories. Hence, it becomes important to differentiate clearly between a statement or a group of statements, for instance, and the acceptance, deployment, application, or reliance upon statements in particular circumstances. The latter, and only the latter, it seems to me, in being actions, are in principle susceptible to assessments of rationality.

But if theories themselves are not expected to be rational, neither should theories of rationality! The *acceptance* of a theory of rationality should be rational by its own standards, but to require this is no more question-begging, arbitrary, or viciously circular than to require a treatise on good English to be written in (what it claims is) good English, or that a paper on logic should be argued through logically. What, in the last analysis, compels one to accept one theory rather than another is not the fact that it complies consistently with its own standards (although this should perhaps be required as a necessary condition for acceptance), but that it seems arguably to provide better means than its rival to whatever end it was proposed. This, I suggest, is not only a criteria of adequacy for the adoption of theories (theories of rationality included), but makes for the very essence of rationality itself.

As an evaluative category of human endeavor, assessments of rationality must necessarily relate to goals. There is no sense whatsoever in judging a person's conduct without reference to the ends he or she might have had in mind. Even the most outlandish behavior may prove, in certain circumstances, to have been the most effective way to achieve a goal, just as the most seemingly levelheaded behavior may at times be totally misplaced. Hence, the domain of rationality is the now familiar realm of goal-directed systems and is primarily concerned with the selection, evaluation, application, and improvement of particular means to particular ends. This, however, is still not to say very much. At most we have selected the types of conduct to which judgments of rationality should apply. But how, and on the basis of what, are such judgments to be made?

Without taking too much for granted, it seems reasonable to require at the outset that the type of conduct we would be prepared to rate as rational should be performed consciously and deliberately and require an appeal to one's reasoning faculties. In this way we eliminate the obviously irrational, such as the unconscious, unintended, or purely accidental eventuation of desired goals.[31] It would be wrong, however, to conclude thereby that any conscious deliberation invoked in the pursuit of goals is rational as such. For one thing, as Polanyi has convincingly argued, the performance of any task necessarily requires that a measure of tacit knowledge remain tacit.

> Subsidiary awareness and focal awareness are mutually exclusive. If a pianist shifts his attention from the piece he is playing to the observation of what he is doing with his fingers while playing it, he gets confused and may have to stop. This happens generally if we switch our focal attention

to particulars of which we had previously been aware only in their sub-
sidiary role.[32]

In other words, there are times, clearly, when such a conscious and deliberate shift of attention, however well intended, will necessarily result
in the frustration of one's objectives. I suggest therefore to further limit
the notion of rationality to situations in which the pursuit of a goal actually *requires* a conscious appeal to reason. Typically, these will be cases
in which either more than one seemingly satisfactory line of pursuit is
available or when the one selected proves insufficient. In the former case
the agent will be required to assess, compare, and subsequently choose
between alternative lines of approach, in the latter, to deal with an evidently problematic one. In both cases the pursuit of a goal requires one
to use one's brains. It goes, I think, without saying that in the former
case to act rationally is primarily to undertake to select the least problematic means available, and in the latter, in the absence of alternatives,
to undertake to determine, assess, and solve the problem. The domain of
rationality is hence further narrowed to that of *problematic* goal-directed
systems.

> *To act rationally, I propose, is to act upon problematic goal-directed systems by*
> *applying consciously and deliberately one's brains to the exposure and/or to the*
> *solution of the problems they harbor.*[33]

To reiterate, I suggested that we view problems as discrepancies between the intended and actual achievements of goal-directed systems and
view solutions to problems as changes introduced into systems whereby
they are rendered less problematic. Ideally, then, a rational agent aspiring
to accomplish an objective G by means of a system S will strive prudently
(a) to expose whatever problems exist in S with respect to G, (b) to understand and if possible to explain S's failure to fully eventuate G, and
(c) to introduce such changes in S as to render G more readily attainable.
To act rationally, in other words, is *to contrive deliberately to effect progress.*

As noted above, problems may be solved, and progress achieved unintentionally—as, for example, by automatic mechanical or organic feed-
back devices, so that not every successful step forward is a sufficient indication of rationality. Moreover, success cannot even be regarded as a
necessary indication of the workings of a rational agent. First of all, problems may turn out to be insoluble. The fact that attempts to rectify a system eventually prove futile is not to say that they were therefore irrational
or even ill conceived to begin with. Indeed, the mere fact that we now
know that they were futile is itself a step forward.

More importantly, a theory of rationality must allow for mistakes. To be rational cannot mean being infallible. The history of science is, in a sense, the history of the most splendid mistakes, yet it is also the history of perhaps the most rational of human undertakings. Rational agents should be allowed not only to fail but to err. I insist on this not merely in order that the theory of rationality here proposed take in as much of the history of science as possible[34] but on the grounds of the philosophical observation that we hardly ever know anything for sure. And if rational action is to be categorized as action taken on the assumption that the goal-directed systems we employ might be flawed, it would be absurd to insist that a rational reformer should even think of himself as flawless, let alone be required to be so. Second, and again contrary to Lakatos, we expect judgments of rationality to be essentially *prospective*. The actual solution of problems, and hence progress, can only be assessed by hindsight. Only time and further testing can assure us that a problem has, to the best of our knowledge, indeed been solved and progress in fact achieved. But we should be able to ascertain the rationality of a move before its completion and regardless of its outcome. To act rationally, we would like to think, is something an agent can decide to do rather than something we can be wise of only after the event. Hence, judgments of rationality should depend on what goes into them, rather than on what comes out. If not, if judgments of rationality are to depend on what we can only know tomorrow, there seems to be no sense in requiring rational agents to be consciously aware of what they are doing.[35] Hence, one should be allowed to fail or err and still be regarded as having acted rationally.

But how liberal are we prepared to be? To what extent can one's criticisms be off the mark, one's understanding of a system wrong headed, and one's suggestions for reform misplaced, yet still be considered rational? Are merely the good intentions of a reformer enough to rate as rational whatever he or she decides to do? And if not, what further criteria may we apply without invoking hindsight?

Let us take a closer look at the ideal case. Ideally, a rational reformer will first seek critically to expose the shortcomings of the system he or she is working on. This will require studying the system thoroughly, not merely as a competent user, but for the sake of testing it as rigorously as possible. Now, the difference between a competent driver and a test-driver, for example, is that unlike the former, the latter is required to be well acquainted with the *theory* thought to govern the system. If not, he or she will be testing randomly. To test a system rationally, I propose, is prudently to investigate its performance *in circumstances in which, to the best of our knowledge, it is most likely to fail.*[36] The difference between an arbitrary and rationally contrived test of a system, be it a theory, utensil, appliance, or institution, is that the latter is typically set up on the basis of an available second-order explanation of the past successes of the system and systems similar to it. If, as the result of such a test, the system is found problematic, a rational reformer will seek to put it right. And again, rather than aimlessly cast around for possible remedies by blind trial and error, he or she will contrive to *explain* the system's failure by seeking possible flaws in the initial theory that might have been overlooked by the designers of the system. This is true of any problematic goal-directed system. When an appliance, a social institution, or a scientific theory is found in certain circumstances not to live up to expectations, the rational reformer will first ask why. Perhaps the theory governing the system was wrong to start with and should be replaced or modified, as was, for example, Torricelli's suggestion that in view of the puzzling fact that standard vacuum pumps failed to "pull" liquids higher than certain levels, the accepted explanation of how they worked (by a pull exerted on the liquid by the vacuum) should be replaced by another (i.e., the push of atmospheric pressure). Or he or she might propose that the theory is basically correct but that some of the boundary conditions had been overlooked or miscalculated, giving rise to problems in the system. In this manner, as we have seen, the existence of Uranus and Neptune was conjectured and confirmed. In both cases, the rational reformer will then be in a position to suggest and argue for such modifications of the system that might put it right.

The sign of rational reform is, in short, a learned and deliberate appeal to some form of second-order knowledge of the system in question. But again, how liberal are we prepared to be with regard to a reformer's second-order reasoning? Are we to accept explanations of the failure of a system that make reference to the intervention of demons, ghosts, or witches? How are we to draw the line between learned and rational guesses and mere superstition? Can such a line be drawn other than by mere convention?

I believe that it can. What should be demanded of rational reformers, I propose, is not that they limit their deliberations to strategies regarded as solid and acceptable by the community, but that their attitude toward them be as rational as their attitude toward the problematic system they are working on. Namely, that they should, in all honesty, regard whatever second-order explanations they appeal to, be they even demons and witches for that matter,[37] as potentially open to criticism and reform as the system they are attempting to improve.

For the sake of simplicity, consider again a nontheoretical goal-directed system S such as a tool, scientific instrument, or social institution, designed to achieve a goal G. Seeking to test and subsequently to improve

S, a rational agent will need explicitly to make reference to two levels of theory. In the first instance he or she will devise tests of S on the grounds of what may be referred to as the designer's theory T_D, a theory that is intended to explain why, when, and how S is capable of eventuating G. T_D is itself a goal-directed system whose goal, though, is not G, but rather *the explanation of the achievement of G by S*. Notice that if S fails such a test and is found problematic (i.e., incapable of fully achieving the desired G), T_D is also necessarily rendered problematic.[38] The rational reformer will hence be faced with two different yet connected problems: one practical and one theoretical, corresponding respectively to the discrepancies between the achievements and goals of S and T_D, the solution of the latter being a prerequisite for the rational solution of the former. Having discovered a problem, the rational reformer will then propose ways of modifying S so as to render it better capable of eventuating G. But to do so rationally, I have argued, requires the employment of an even higher-level "reformer's theory" T_R, a third goal-directed system, introduced with the intention of both replacing T_D as an explanation of the eventuation of G and of explaining its failure.

Rational action thus comprises ideally the exposure of problems in S by explicit appeal to designer theory T_D and an attempt to solve them on the grounds of a higher-level reformer theory T_R. Three different goal-directed systems are hence involved in the reformatory process: S, the primary target for reform, designed initially to achieve G; T_D, the designer's theory, designed initially to explain S's achievement of G; and T_R, the reformer's theory, introduced in order to both replace T_D and to explain its apparent deficiencies.

When S is itself a theory rather than an appliance, the situation will be very much the same although T_D will normally be presented as a part of S and T_R as part of the modified version of S, S_M. In this case, teasing the two levels of explanation apart would have seemed less natural perhaps than in the case of an appliance or institution.

Rational reform thus mediates, as it were, between two successive states of affairs. The first comprises a system S thought effectively to achieve goal G by virtue of theory T_D; the second comprises a modified version of S, S_M, thought effectively to achieve G by virtue of T_R. The rational reformer's ultimate objective is, of course, to achieve progress on both levels, to be able to demonstrate that the problems found in S and T_D are at least partly solved in S_M and T_R, respectively. The rational grounds for this expectation is that T_R supposedly explains not only why S_M should achieve G, but why T_D failed to do so with regard to S in the first place.[39]

To return to the initial problem of safeguarding rational reform from the introduction of irrational strategies, the problem is now seen to narrow down to T_R, since both S and T_D are treated rationally in that they are both "troubleshot" and modified in the process. T_R, however, is not. T_R will be either an available theory or one introduced for the sake of solving the problem. In either case it seems sufficient to require that, in order to be considered rational, the reformer regard T_R as potentially refutable and modifiable as S and T_D, even though he or she is not engaged in the actual testing or improving of T_R.

We have, then, something like a tripartite 'recursive' theory of rational action as intentional reformatory action ideally comprising the following:

1. A deliberately critical study of a particular goal-directed system with a view prudently to expose and explain it problems,

2. an undertaking to solve whatever problems come to light, and

3. a willingness to do so on the grounds of second-order theories that are themselves, considered in principle, subjectable to (1) and (2).

Since T_R is itself not required to be subjected to tests in the process of reforming S, the above definition does not generate an infinite regress. It is not required of a rational agent that he or she first secure the second order theory before applying it but only that he or she hold it in principle open to criticism.[40]

So much for ideal reformers. Their efforts may prove futile—in which case T_R would normally be called into question—without detracting from the rationality of their move. Similarly, they may err considerably yet still be regarded rational. There seems to be no contradiction in terms between being mistaken, even stupidly mistaken, and being rational in the above sense. Rationality is thus construed, as required, as a wholly prospective category of exclusively human action—free of any appeal to hindsight and necessarily involving second-order reasoning; the latter is, as far as we can tell, peculiar to humankind.

The linkage the proposed theory establishes between rationality and progress—the former construed as action taken with the intention of achieving the latter—guarantees that assessments of rationality do not privilege particular goals. Rationality is certainly an evaluative category, but it is not, and in my opinion neither should it be, an ethical one. The rationality of an action, as I see it, should not be sanctioned by the *type* of goal desired but only by the manner in which it is pursued. Rationality, I have argued, turns not on the value, worthwhileness, or moral standard we ascribe to an agent's objectives, but on the ways in which their eventuation is critically assessed and improved. To be rational should not necessarily imply being good. One could correctly be thought of as rationally selecting the best ways to rob a bank or to perform a sexual assault. To act rationally wicked, I insist, is not a contradiction of terms. In such cases acting rationally is perhaps a measure of greater evil! Assessments of rationality are no more and no less than assessments of the extent to which an agent consciously strives to perfect the attainment of a given goal in the face of problems.[41]

Nonetheless, many might still insist that although assessments of rationality should not be subordinated to one's ethical judgments, there still exists a body of beliefs that rational agents should all be expected to hold, to the extent that any action allegedly taken in defiance of them be deemed irrational as such. Martin Hollis, for example, insists that it is necessary "to place an *a priori* constraint on what a rational man can believe about this world," that "there has to be [a] massive central core of human thinking *which has no history* [!] and [that] it has to be one which embodies the only kind of rational thinking there can be." To be able to talk of rationality, concludes Hollis, "[t]here has to be an epistemological unity of mankind."[42] Quentin Skinner has pointedly criticized this neo-Kantian "image of a rational bedrock," as he puts it, on the seemingly relativistic grounds that,

> even in the face of the clearest observational evidence, it will always be reckless to assert that there are any beliefs we are certain to form, any judgments we are bound to make, simply as a consequence of inspecting the allegedly brute facts. The beliefs we form, the judgments we make, will always be mediated by the concepts available to us for describing what we observed. And to employ a concept is always to appraise and classify our experience from a particular perspective and in a particular way.[43]

Contrary to Hollis, Skinner seeks to relativize "the idea of 'holding true' a given belief," deeming certain beliefs rational in certain cultures, even if they no longer strike us a "rationally acceptable."[44] Further down, however, he is forced to admit that "[w]e must be able to assume, in advance of our historical inquiries, that our ancestors shared at least some of our beliefs about the importance of consistency and coherence," among which he counts "the principle that, if one affirms the truth of a given proposition, then one cannot at the same time affirm the truth of the denial of the proposition," together with certain other assumptions about the inferential processes of using our existing beliefs to arrive at others.[45] As it turns out, Skinner's disagreement with Hollis is hence more a matter of degree than of principle. Reluctant to wholly relativize the notion of rationality, both writers seek to avert total relativism by salvaging a minimal "hard core" of arguably ahistorical, supracontextual beliefs allegedly shared by all human actors. Skinner insists that the hard core should only

include second-order beliefs—i.e., beliefs about appropriate belief formation. However, the theory of rational action developed in the present
study renders such a move superfluous. It is possible, we have seen, to
relativize *all* belief and still retain a normative, supracontextual modus
operandi for the assessment of rationality grounded upon an articulation
of the logic of troubleshooting of which the actors themselves need not
be aware. For human agents to be considered to have acted rationally all
that needs to be shown is that they were prompted to action as the result
of a felt frustration of their objectives. On such a showing, the entire
question of the 'objective', supracontextual soundness of their beliefs is
wholly beside the point.

However, Skinner's concessions to Hollis, I would further insist, are not
merely unnecessary[46] but to a large extent misconceived. For one thing,
I seriously doubt that all thoughtful forming of propositions can arguably
be said a priori to premise even the minimal set of logical presumptions
enumerated by Skinner. In Hegelian dialectics and in any number of mys-
tical texts, one frequently encounters knowing dismissals of even the most
basic canons of logical discourse. And regardless of the historian's own
discursive standards, it would be impertinent to dismiss them as the up-
shot of sheer irrational caprice. Some of these texts were undoubtedly
worked upon, modified, and improved by their authors cool headedly and
reflectively, in accord, at times, with discursive and argumentative rules
and strictures decidedly different from our own. To analyze them as such,
our theory allows us to read them as the outcomes of troubleshooting
exercised from the admittedly exotic vantage point of the aims and discursive standards of their authors, which, of course, may be extremely
difficult to reconstruct—a feat Skinner and certainly Hollis seem incapable of performing.

But the trouble I have with the notion of 'rational belief' is far more
basic and, in an important respect, relevant to the concerns of the present
study. I have insisted at the outset that for an action to be considered
susceptible to judgments of rationality, it must be voluntary. There seems
to me no sense at all in deeming involuntary mental or bodily functions
as rational regardless of their value or worth. The question is whether
the *act* of believing (as opposed to belief-content) can at all be considered
voluntary. Can one intentionally *form* a belief as Skinner's and Hollis's
texts seem at times to imply? We may, of course, treat our beliefs rationally.
We may question them or willingly expose them to the criticism of our
colleagues. We may also consciously desire to believe certain knowledge
claims we find convincing. But, and this is the crucial point, we cannot,
it seems to me, *decide* to believe or disbelieve an assertion any more than
we can decide to fall in love. The very act of adding an assertion to or

removing one from the body of our held beliefs, I submit, is as involuntary as that of digesting our food. The process of digestion is a good example. We can affect and control it indirectly, to a large extent, by attending to our eating habits or conversely by ignoring our doctor's advice. In fact, we can intentionally improve or damage the process dramatically. Nonetheless it would be absurd to rank the digestive process itself a rational activity. The same, I believe, goes for belief formation. By testing propositions, by seeking out and paying attention to arguments for and against them, and by acting on them—all of which are bona fide rational activities—we are capable of partially determining what we will end up believing or disbelieving. Nonetheless, we do not, in my opinion, directly control the very act of belief or disbelief. We simply awake to it, again, like the act of falling in or out of love. We may rationally court, attempt to entice, charm, or, conversely, do our best to discourage a partner—all of which, again, are potentially rational actions. But whether or not we end up loving or ceasing to love the person in question is not something we can arguably claim to have *done*. It simply happens to us. One may rationally examine one's feelings, just as one should rationally examine one's beliefs from time to time. But loving, and more importantly, believing, I concede, cannot themselves be considered feats of rational endeavor.[47] If this is granted, the entire question of 'rational belief' is rendered a non-issue. Assuming, then, that beliefs are given and are not themselves susceptible to judgments of rationality, we can only speak of assessing the rationality of deliberate actions determined, prompted, constrained, or motivated by beliefs—which, of course, almost all human activity is! Such an endeavor, I submit, may be judged to have been rational or irrational solely on the basis of its being or not being the outcome of critical reflection. But neither the content of the beliefs it premises, nor the process by which they were initially formed should, in principle, enter our judgments.

To summarize, I have suggested, following Popper's lead, that we equate rationality with systematic criticism, with a profound awareness of the limits of knowledge, and that to act rationally on a humanly designed or humanly applied goal-directed system is first and foremost to prudently consider it as potentially imperfect regardless of past successes. It is imperative, therefore, that all who are capable, especially persons directly responsible for its design, maintenance, and further development, test it continuously and as thoroughly as possible in relation to its intended objectives and constantly pit it against alternative systems, being willing at all times to modify or replace it as the case requires. To a great extent, modern science is indeed a model of rational inquiry. Scientific knowledge is passed on from one generation of researchers to the next, in order to be carefully studied, constantly tried and criticized, hopefully to be modified and further expanded. But the task of maintaining such scientific continuity is not an easy one. In order to be in a position to effectively criticize or test a theory one needs first thoroughly to understand it. Hence, as Popper himself has stressed,[48] two different yet complementary traditions are required to ensure and perpetuate an ongoing rational inquiry. First, and obviously, the theories themselves need to be transmitted efficiently from one generation to the next, otherwise scientific work would be trivially terminated. To serve this purpose institutions of learning are required in which current theories are taught and understood as thoroughly as possible. But to avoid mere indoctrination, a second tradition—a critical tradition—is equally pertinent. Scientific theories are typically handed on, accompanied, as Popper puts it, by "a silent . . . text of a second-order character" which says in effect: "this is the very best we have, study it carefully, and if possible improve it." It would be better still if theories were taught in their original historical context, as solutions to the problems that originally prompted them. The two traditions that nourish the logic of scientific discourse combine in this manner to toe the delicate line between dogma and mere fancy. Scientific theories need to be taught as tentative but not frivolous; to be taken seriously yet not considered infallible; to be viewed as the very best we have though not as the best possible. If Einstein's scientific revolution and Popper's philosophical revolution have taught us anything about science, it is that all, but all human knowledge, even the most seemingly sound, is fallible and that the essence of rationality is forever to treat it as such.

The existence of *theories* of rationality, however, convincing as they may be, will not suffice to sustain rational discourse. What are needed are well-established institutions geared to value, stimulate, promote, and reward debate, criticism, and reform. In this respect science is today perhaps better equipped than before. Most scientific journals now welcome and encourage the challenge and refutation of accepted views. Scientific conferences are now normally set up with a view to confront rival schools of thought. Even the Ph.D. dissertation, by which novices are supposed to prove their mettle, is now expected to be original, i.e., to call currently held positions to task by devising new ways of testing them. There is undoubtedly much more that can be done, for much conservatism still prevails. Young dissenters still have difficulties securing jobs, and many established researchers still prefer to join the bandwagon rather than to call established views to task. But in science, as in art and industry, especially since 1905, the well-informed, skeptical, speculative avant-garde is becoming more and more the rule rather than the exception. It seems that not only in science itself, but also in philosophy, history, and scientific management, the lesson to be learned from Einstein's science and Popper's novel interpretation of it is gradually taking root.

The lesson, I have argued, extends far beyond science. Science, it is now clear, is best understood when the impulse to demarcate it from other creative human endeavors is resisted rather than emphasized. Science is best conceived of, taught, and cultivated as a human endeavor, employing the same sort of rationality or rational behavior as many other realms of reflective human activity. This idea has taken root enough so that it no longer seems strange to imagine that a baseball player working methodically to improve his swing, or a cellist to improve her fingering, or a poet to improve a verse, or a mechanic to fix a car are all doing the same sort of thing as the researcher mixing batch after batch of experimental vaccine. It no longer seems strange to think that when scientists act rationally to disclose and solve their problems, they act much like people in other fields acting rationally to lay bare and solve their own very different problems. It still seems strange, however, to conceive of those who framed the central formative texts of the Judaic tradition as acting the same way. This is because this literature is often taken for a collection of authoritative dicta. A close reading of these texts will frequently show that their own portrayal of their protagonists was as rabbis acting rationally in precisely the sense outlined, as seeking self-consciously to expose and solve the sorts of problems that they faced. Such a close reading is the subject of the remainder of the book.

PART 2

THE JEWISH COVENANT OF LEARNING

I have so far described and argued for a view of knowledge and how it advances that is implicit in the way that scientists, and with them all self-doubting endeavoring persons, rationally act to achieve their goals. The same notion of rational action, I shall argue in upcoming chapters, is not only implicit in the modus operandi of the rabbis whose disputes and deliberations are reported in the talmudic texts, but can be shown to have been explicitly and self-consciously adhered to by many of the framers of these documents. To some of my readers this claim may seem at first surprising and to others even absurd. If the study of Torah is construed, as it frequently is, as rote treatment of a body of scriptural exegesis and especially halakhic rulings handed down to Moses at Sinai, then it indeed leaves little room for the type of self-doubting troubleshooting I have been describing. But although these texts are often regarded as formative of a culture uncritically bound by its traditions, I wish to show that they are framed to a significant extent as long and sustained arguments against blindly following tradition. My aim in what follows is to locate and retrieve this undogmatic, reflective, self-doubting voice of talmudic culture.

Antitraditionalism is the (lamentably inelegant) label I have reserved for the position for which this voice speaks. Antitraditionalists take the teachings of their forebears in utmost seriousness, but do so, contrary to their traditionalist adversaries, with a view not to following them indiscriminately, so much as to seriously putting them to the test. The antitraditionalist voice is, in other words, the voice of a rabbinic elite aspiring to teach future rabbinic elites to reason rationally about the content of their legacies in the same way open-minded and self-doubting agents were shown, in the previous section, to act rationally, when striving knowingly to improve upon the systems on which they work.

But to repeat a point mentioned earlier: the affinity between the theory of rationality developed in Part 1 and the rabbinic view of Torah-study I shall expose in upcoming chapters is *not* the result of applying the former to the latter. There is an all-important difference between what I have said about science in previous chapters and what I shall be saying about talmudic discourse in those that come. The former builds upon a construction of science that is advanced and developed by a contemporary school of philosophy to which I subscribe, while the latter is an attempt to retrieve the way in which an important group of late antique Jewish

sages constructed *their* world of learning. And although the two construc-
tions turn out to be surprisingly similar, they truly represent the conclusions of two quite different and independent research efforts motivated,
on my part, by two quite different and independent sets of questions. It
follows, therefore, that one may agree with one without necessarily having to agree with the other.[1]

Having said that, however, I cannot deny that my interest in this particular voice of talmudic culture owes much to my own moral and philosophical identifications. To paraphrase Daniel Boyarin (upon whose constant intellectual self-reflection this work is consciously modeled[2]), the
following pages are, in this respect, openly tendentious. They seek to
retrieve, isolate, and amplify one voice in a talmudic polyphony, a polyphony that readily extends to the rabbis' second-order attempts at self-
understanding, in which the part of it that bears upon matters of epistemology is central.

Still, I shall not ignore other talmudic voices, especially those diametrically opposed to antitraditionalism. Interestingly, it is especially here,
in the opponents they are constructed to combat, that the theory of rationality I have proposed and the antitraditionalism contained in the talmudic texts differ most greatly. While the former is forged today in defiance of the relativistic abandonment of standards frequently associated
with postmodernity,[3] the latter's natural opponent, then as today, is, by
contrast, a steadfast, reactionary traditionalism that preaches rigid and
dogmatic adherence to a set of inherited norms for no more reason than
that they are inherited. In a sense, they can thus be said to combine to
occupy a philosophically viable and potent middle ground between two
equally dangerous extremities of uncritical discourse: at the one extreme,
the type of unchecked, standardless free-for-all one frequently experi-
ences in contemporary Western modes of thought, and, at the other, the
sort of equally unchecked dogmatism that characterizes the way many
orthodox Jews have come to construct their religious world views. Needless to say, uncritical tolerance is no more exclusively postmodern and
Western than uncritical intolerance is a uniquely orthodox Jewish way
of thinking. Suitably generalized, they mark the outer boundaries between which and in contrast to which constructive skeptics in general
fashion their systems of thought. But this book is less about constructive
skepticism as such than it is about the particular brand of constructive
skepticism found in the talmudic literature. The fact that it suggestively
resonates with a position developed by latter-day thinkers for entirely different reasons is interesting, though this in itself explains very little about
the rabbis' mind-set. To appreciate their way of thinking, we are required
to set aside the comparative aspects of this study, to leave latter-day philo-

sophical deliberation behind, and to enter as far as possible their world of discourse.

The case I aspire to make about this world of discourse is admittedly tendentious. But it is not essentialist. This book does not pretend to be about the unique spirit, the one mind, the logic, or any supposedly characterizing feature of the Jewish texts of late antiquity. It is about the epistemological discourse and deliberation contained in and evidenced by these writings. In studying these deliberations, it emphasizes and amplifies one of several voices, a voice that is, in the humble opinion of its author, unique and historically intriguing. It is also a voice that seems to me to possess both a special relevance and an urgency for contemporary Jewish orthodoxy, but that, of course, is beside the point.

In at least one major tannaitic corpus, antitraditionalism, we shall see, is not merely presented as one viable option among many but is dramatically confronted with the opposing view and hailed victorious in the most explicit terms. Still, even that is not to say anything about talmudic culture as a whole. The rabbis did not speak in one voice about anything. But speak they did, and one of the voices they spoke in—the antitraditionalist—this book will argue is a voice that is central to any understanding of how the founder-fathers of Torah Judaism conceived of themselves and of their great intellectual undertaking.

But crucial as the antitraditionalist voice is to talmudic culture, it turns out also to be curiously toned down by the framers of the very texts in which one would have expected it to be the most audible. At one primordial foundational level of the rabbis' writings, it is a voice that is valorized and one that speaks loudly and openly, only to be rendered, seemingly for no apparent reason, almost inaudible at the more advanced level of their discursive practices. This point, dealt with in considerable detail in what follows, is one worthy of some elaboration even at this early stage.

The talmudic writings present themselves as comprising two chronologically ordered literatures, generated by two clearly distinguished groups of sages—tannaitic and amoraic. The tannaitic era is the era of the tannaitic sages, the *tannaim,* who flourished prior to the redaction of the Mishna, traditionally ascribed to the *tanna* R. Yehuda "the Patriarch" early in the third century c.e. The amoraic period is that of the amoraic sages, the *amoraim* who flourished during the post-mishnaic era until the generation of R. Ashi who is traditionally deemed responsible for the "signing" of the Bavli early in the sixth century c.e. Roughly speaking, there are four generations of *tannaim* and seven of *amoraim.* Each of the two eras produced a substantive literature of its own. But while the two bodies of texts, the tannaitic and amoraic, possess much in common, they differ considerably in one important respect. Only the latter presents systematic reactions and responses to earlier, authoritative rabbinic material. The two great amoraic compilations, the Talmud of the Land of Israel, the Yerushalmi, and the Babylonian Talmud, the Bavli, are framed as systematic responses to the great tannaitic halakhic compilation, the Mishna. They are presented as commentaries on the Mishna, but make reference to much additional tannaitic material in the course of discussion. However, it is at the tannaitic level, where the question of the proper attitude toward the teachings of one's forebears hardly affects the redactory and narratory level of the text, that the antitraditionalist voice can be heard most audibly, while in amoraic treatments of tannaitic writings, where one would expect the antitraditionalist approach to be felt most conspicuously, it is present, but curiously and vexingly muted—especially in the Bavli. For this reason, we must first learn to listen for, and recognize, the voice of antitraditionalism before turning to the amoraic texts and attempting to witness it in action.

We shall approach the amoraic texts indirectly at first, working our way up in pseudodiachronic fashion from what they themselves set forth as their foundational, tannaitic, second-order grounding. I say pseudodiachronic because I do not intend for the terms *tannaitic texts, tannaitic view,* or *tannaitic level,* as I shall use them in what follows, to be taken as necessarily making historical claims. By *tannaitic* I shall be referring to that which is *presented by the rabbis as tannaitic* even when modern scholarship can prove that it is of amoraic origin. Many of the alleged quotations of tannaitic sources in later documents and a lot of the ascription of halakhic positions to tannaitic authorities, especially in the Bavli, were edited and modified to suit the objectives of their latter-day amoraic discussants and conveyors. My aim is not to set the historical or philological record straight so much as to enter the world of discourse intended and constructed by the latter-day framers of the documents we read. Hence, throughout this book, such terms such as *tannaitic text, the Jabne debate, the disputes between the Houses* and so on—even the names of particular sages—are intended, unless indicated otherwise, as stand-ins for the more cumbersome: "tannaitic text according to the framers of the Bavli's the literary conventions," "the Jabne debate as depicted by the talmudic corpus of Jabne stories," and so forth. The foundational tannaitic texts I shall be dealing with will be privileged not because I believe they were foundational, but because they are set forth by their later amoraic and savoraic redactors as such. These tannaitic texts comprise mostly metahalakhic and legendary material which will prepare the ground more fully for the close examination of the discursive amoraic texts in subsequent chapters. But beforehand, a brief introduction to the theological background and context of the rabbis' enterprise.

Time after time "the tablets of stone, and the Torah, and the commandments" (Exod. 24:12) are referred to in the Hebrew Scriptures as constituting a covenant.

> And the Lord said to Moshe, write thou these words: for after the tenor of these words I have made a covenant with thee and with Yisra'el. (. . .)
> And he wrote upon the tablets the words of the covenant, the ten Words.[4]

Unlike the Calvinist Covenant of Grace, talmudic culture constructs the Hebrew covenant primarily as a Covenant of Works,[5] a matter of righteousness related to action, of practical performance and reward. And for the rabbis *talmud-Torah*—Torah-study—was deemed the very highest of works, an obligation to the Almighty comparable to all other religious obligations put together. In being a prerequisite to all deeds, Torah-study, they state, outweighs all deeds.[6] In the canonical texts of Judaism, Torah-study undoubtedly came to form a class of its own among humankind's obligations and responsibilities, constituting a covenant in its own right—a Covenant of Learning.

It appears, however, that the rabbis themselves might have viewed this as not always having been the case and to have considered the unparalleled pre-eminence that had come to be ascribed to Torah-study as the outcome of historical development. The reception of the Torah at Sinai was undoubtedly a climactic point in the Exodus story. Still, the way it is related in Deuteronomy 1:6–7, and especially the manner in which this passage was rendered by the rabbis, suggests that in their view the reception and study of the Torah might have been originally regarded as but an intermediate step en route to a higher purpose:

> The Lord our God spoke to us in H̱orev, saying, You have dwelt long enough in this mountain: turn, and take your journey, and go to the mountain of the Emori, and to all the places near it, in the plain, in the hills, and in the lowland, and in the Negev, and by the sea side, to the land of the Kena'ani, and the Levanon, as far as the great river Perat.

Having just received the Torah at Sinai (H̱orev), the Children Israel are reprimanded: "You have dwelt enough in this mountain: turn, and take your journey" away from Sinai to the prime task at hand: the possession of the land. And the *Sifri*[7] pointedly interprets verse 7 as denoting more than a mere geographical description of the promised land. "and the Levanon," the *Sifri* elaborates, "He said unto them, when you enter the land thou shalt appoint thyselves a monarch and build the Temple."[8] This particular midrashic reading would imply that the reception of the Torah was thus originally regarded as but a stage, albeit an important one, in the larger process geared to the establishment of a sovereign political entity in the Land of Israel.[9] In contrast to the *Sifri's* rendition of the Bible's implied ordering of the nation's duties and responsibilities, Mishna, *Abot* 6:5 states explicitly: "Greater is Torah than both the Priesthood and than the Monarchy."[10]

The reasons for this significant re-ordering of religious priorities are debatable. It is tempting to explain it historically, as a simple and understandable response to the harsh political reality at the eve of the tannaitic period. Close to the destruction of the Second Temple, the prospects of political and religious independence for the Jews of Palestine had become extremely bleak. Under the prolonged Roman occupation the monarchy was swiftly losing its political standing and had long lost its original religious significance; and after Herod it was virtually abolished. Although we have little firsthand evidence of rabbinic attitude and opinion at the time, one may assume that few would have considered the idea of reestablishing Israel as a politically sovereign "kingdom of priests and a holy nation" (Exod. 19:6) in the conceivable future as less than pure fantasy. It is tempting, therefore, to interpret the rabbis' decision to extol Torah-study, to the (transitory?) devaluation of political independence and the Temple cult, as an essentially pragmatic move designed to set the people an achievable religious goal in view of the circumstances.

There is much in the talmudic literature to suggest, however, that this move might have been taken, or at least viewed so in later years, far less reluctantly than such an historical explanation would imply.[11] R. Yehuda *Ha-Nasi*, the Patriarch, the leading authority during the last generation of *tannaim* and traditionally named as the moving force behind the final compilation of the Mishna, is a case in point. R. Yehuda's patriarchy is justly regarded by historians as the golden age of Jewish life in Palestine during the mishnaic and talmudic periods. During his term in office the Jews of Palestine enjoyed an unprecedented period of prosperity and near independence. The bilateral relationship between the Jewish community and the Roman administration flourished at the time to the extent that the local courts were authorized by the Romans, for example, to administer capital punishment—a right seldom granted to the Provinces.[12] And to judge from their fantastic legendary accounts of R. Yehuda's close personal relationship with Emperor Antoninus Caracallus[13] (Marcus Aurelius, emperor from 211 to 217), the rabbis evidently thought so too. And yet their writings contain no suggestion that the legendary or real R. Yehuda ever considered taking advantage of the situation and requesting the restoration of the Temple—a request that the mythical emperor of their legends might well have granted. On the contrary, not only is he nowhere reported as having sought to restore the Temple cult, but, according to an opinion cited in both *talmudim*, he is reputed to have actually recommended that the fast of the ninth of *Ab* be abolished.[14] It is possible, therefore, that the halakhic discussions and the legends recorded by the talmudic literature attest to the fact that at least some of the rabbis may have regarded, if even in retrospect, the shifting of the prime focus of religious performance from the Temple to the academy, from the altar to the synagogue, as a welcome development, as a step forward more than as a necessary evil.[15]

In all events, whether as a sorry result of the circumstances or as in part a matter of ideology, with the fall of Jerusalem rabbinic Judaism of late antiquity was in fact significantly transformed from a largely ritualistic religion dominated and regulated by the Temple rituals to a "community of learners" whose religious life was both structured and informed by the talmudic academy. The sage had come to fully replace both the monarch and the priest at the culmination point of the social pyramid, and the intellectual challenges and objectives of Torah-study and the meticulous rituals associated with the Temple as the focal point of religious performance.[16]

The high point of the talmudic revolution, if one may call it so, was undoubtedly the establishment of the Jabne center, the first great talmudic academy, by Rabban Yoḥanan b. Zakai and his disciples in 70 A.D. With the establishment of the great Babylonian academies during the third and fourth centuries A.D.,[17] talmudic Judaism was further rendered a community of learners for which even the Holy Land can be seen to have gradually surrendered its centrality.

In the course of these revolutionary developments, the texts comprising the talmudic canon were written, collated, anthologized, and edited. This massive intellectual undertaking, which stands as living proof of the rabbis' ruling that "*talmud torah* is superior to all," is the subject matter of the following chapters.

The fascinating story of the revolution itself, arguably set in motion by Ezra the Scribe[18] and culminating, as noted, in the great Babylonian academies of Nehardea, Sura, Pumbedita, and Meḥuza, lies beyond the modest scope of the present study and the limited expertise of its author. The interests of this study lie primarily in characterizing philosophically rather than historically the epistemological presuppositions that arguably inform the rabbis' self-image(s) of their learning as evidenced by their writings.

We shall focus attention in the first instance on the rabbis' own description, in the form of a series of tannaitic texts, of one crucial moment in the course of one crucial phase of the process. The crucial phase begins with the destruction of the Second Temple and the foundation of the new center of Jewish life and learning at Jabne. The establishment of Jabne marks the beginning of the tannaitic era that will eventually produce the Mishna, Tosefta, and the tannaitic *midrashei halakha*—the body of writing that will later be taken to serve as the inherited subject matter and point of departure for the two amoraic *talmudim*—the Bavli and Yerushalmi. From this point on, according to talmudic lore, hampered no longer by the Sadducees and Boethusians, who had virtually vanished after losing their Jerusalem Temple-centered power base, the rabbis set about in earnest to define and deliberate the order of the day among themselves.[19] This essentially in-house Jabne debate is said to have developed into a dramatic and dramatically resolved controversy that centered, as one might expect, on the very notion of *talmud-Torah*—its goals, sources, methods, and modes of resolution. This foundational dispute will be shown to constitute the literary-philosophical focal point of at least three important corpora of tannaitic material. Chapter 1 is an attempt to retrieve the initial theological and epistemological points of contention by examining two of these corpora in some detail, in particular the interrelated group of legendary accounts known as the Jabne stories.

1

The Great Tannaitic Dispute

The Jabne Legends and Their Context

Of the texts comprising the formative canon of talmudic Judaism, the Babylonian Talmud, the Bavli, is by all accounts the most prominent and important. Compiled, like the earlier Yerushalmi, as a wide-reaching and detailed commentary to large portions of the Mishna,[1] the Bavli represents the culmination and high point of the talmudic era, and since its completion in the sixth century A.D. has come to serve almost single-handedly as the paradigm for and prime subject-matter of Torah-study in most religious institutions of Jewish learning. It is also in and through the Bavli that the antitraditionalist voice of talmudic Judaism receives its clearest articulation as well as its most effective stifling. It is in and by the Bavli that the Jabne stories are woven into a self-consciously related literary whole. And it is in the Bavli that the narration and negotiation of amoraic response to tannaitic legacy, so crucial for the epistemological concerns of the present study, acquires its most potent and developed form. Other talmudic texts will be discussed in upcoming chapters, some in considerable detail, but it is the Bavli that will mainly occupy our attention. As previously noted, to understand the talmudic world of learning requires us to read the talmudic texts from the possible vantage point of their authors and redactors, rather than from that of their later readerships. Of the Bavli we should therefore begin by asking: why was it written and what may it have been intended to accomplish and convey?

The Bavli was undoubtedly intended by its framers to serve as a summary of halakhic deliberation until their time and as a source of legal precedent thereon. But it is doubtful that they ever considered such codificatory functions as its only purpose or even as its main purpose. Had the redactors of the Bavli wished to compile no more than an up-to-date, ordered, and appropriately synthesized halakhic codex, they would probably have produced a very different kind of text. Had that been their intention, the resulting text would have resembled far more than does the Bavli such halakhic codifications as Maimonides's *Mishne Torah* or R. Joseph Karo's *Shulḥan Arukh*—namely, a systematic and annotated compendium of halakhic norms and the many obligations, restrictions, and precautions they entail.[2] Among the talmudic texts only the Mishna comes anywhere close to resembling such a format—a point I shall return to in some detail at a later stage. But although the Bavli takes the Mishna as its subject matter and ultimate point of departure, the Bavli itself does not resemble a halakhic compendium in the least. It is not written as a summary of rabbinic conclusions, but takes the form of an extended *protocol* of rabbinic statements, debates, and disputes. Its authors and redactors offer their readers a view from the inside, as it were, of the complex give and take conducted behind the doors of the academy. Rather than ponder a point in monologue form, units of discussion are narrated in the form of running commentaries of reconstructed or imaginary dialogues. The anonymous narrator, the *s'tam*, will frequently raise a question, intercept an exchange with a comment of his own, or muse aloud on the viability or consistency of this move or that, but for the most part, the main parties to these reconstructed debates are named authorities who, in many cases, functioned in different places and in different times. Not unlike the imaginary classroom of Imre Lakatos's remarkable *Proofs and Refutations*, the fabricated world of talmudic discourse plays itself out in an endless series of fictitious study sessions, in which a variety of opinions, voiced over periods of up to three hundred years, are made to meet in lively discussion. The Bavli reconstructs in great detail *talmud-Torah* in action as it were, aggada and halakha allegedly in the course of their formation. The obvious and immediate conclusion is that over and above their wish to summarize the *products* of their learning, the compilers of the Bavli sought primarily to document its *process*. In doing so their intention appears to have been *to teach their future readerships how to study Torah as they did*. Viewed thus, the Bavli, in its finished form, is first and foremost a didactic text: a text designed to convey a vision and a methodology of Torah-study, rather than merely to summarize the state of halakhic knowledge at the time of its framing. And it does so by extensive example rather than by offering a systematic handbook, or manual, of rules of talmudic reasoning—a genre that would flourish during the High Middle Ages especially in the *sephardic* tradition. In this sense the framers of the Bavli seem to have preferred to transmit the art and practices of *talmud-Torah* mimetically rather than textually, in the form of a mediated "live recording" of talmudic disputation rather than in that of a second-order distillation.[3] And with the possible exception of the Mishna, to which we shall return, the same applies in varying respects and degrees to the entire body of tannaitic and amoraic writing. It is as such that I propose to examine these texts in what follows, with a view to laying bare and to rendering explicit the ontological and epistemic presuppositions that arguably inform the various, and at times conflicting, theories of knowing and learning that appear to have guided their composition.

Another, less formal reason for regarding the rabbis' writings as being concerned with the process more than the product of Torah-study concerns a point I mentioned briefly at the outset. It is that despite the highly polemical nature of their writings—their tendency to entertain different positions, to pit them against each other, and to weigh their respective pros and cons—these texts, seem to refrain as far as possible from closure. In fact, throughout the talmudic literature rulings are made and decrees are issued only and exclusively on matters of halakha—that is to say, on issues bearing directly on the performative aspects of the ritual, on everyday conduct, and on the various aspects of civil and criminal law. The need to resolve debates this way or that is limited in the talmudic literature to halakha and to halakha alone—and even there, most is ultimately left undecided. All other questions—theological, moral, metaphysical, exegetical—issues regarded as doctrinal by other religions, are as energetically pursued and disputed, but, as a matter of course, or, as I prefer to argue, as a matter of principle, are ultimately left open. The overall impression conveyed by their writings is that the rabbis seem to have been unwilling to decide an issue except when absolutely necessary, and in their view absolute necessity in this respect is never exegetical, moral, or theological—only halakhic! The only binding aspect of the rabbis' theological and exegetical deliberations, keen and fierce as they may be, is the very process of deliberation itself. Their countless debates and disputes concerning these matters seem not to have been conducted with a view to arriving at a final and obligating conclusion. They are invitations to join in the pondering and not attempts to somehow put an end to it. This is true even of *midrash halakha*—the monumental exegetical project of grounding the halakha in Scripture. Even for the most avid traditionalist, for whom the halakha itself is considered irreversibly and absolutely binding, these midrashic 'proofs', offered to explain from whence various *halakhot* are learned, are never granted the same epistemic status. One may readily dispute a proof even if what is thereof proven is held to be undebatable. And the same is also true of the most fundamental theological and metaphysical questions. Apart from one's basic duty to contribute to the discussion as best one can, nowhere, except in halakha, is a question bindingly decided.

An illuminating if anecdotal example of the type of undecided exegetical and theological pluralism I have in mind is the following debate
recorded in part in Bavli, *Sanhedrin* 108b, and related fully in *Genesis Rabbah*. In Gen. 6:18 we read: "and thou shalt come into the ark, thou, and
thy sons, and thy wife, and thy sons' wives with thee." Extremely sensitive to any textual irregularity, the midrash takes note of the rather
artificial separation of husbands and wives in the wording of God's instructions to Noah, concluding that "whence Noah entered the ark copu-
lation was forbidden," hence, "thou and thy sons to themselves, and thy
wife and thy sons' wives to themselves."[4] And Noah apparently obeyed:
"So Noah did according to all that God commanded him (. . .) And Noah
went in, and his sons, and his wife, and his sons' wives with him into
the ark" (Gen. 7:5–7). Upon leaving the ark a year or so later family life,
it seems, was allowed to return to normal, for God's instructions no
longer imply a separation of sexes: "Go out of the ark, thou, and thy wife,
and thy sons, and thy sons' wives with thee" (Gen.8:16). This time, however, Noah appears to have decided not to comply, choosing, seemingly
on his own initiative, in the words of the midrash "to extend the commandment": "And Noah went out, and his sons, and his wife, and his
sons' wives with him" (Gen. 8:18). Whereby, the following debate between R. Yehuda and R. Nehemiah is recorded,[5] apropos Gen. 9:8–9:
"And God spoke to Noah, *and to his sons with him*, saying, And behold, I
establish my covenant with you etc." The former is of the opinion that
since Noah "transgressed the commandment" he was disgraced and was
no longer personally addressed by God. The latter concludes, by contrast,
that "since Noah extended the commandment and elected to conduct
himself in holiness, he and his sons were rewarded by God's word." The
controversy is fundamental. Does the Torah teach, as R. Nehemia would
have it, that one achieves sanctification by suppressing the flesh, or rather,
as R. Yehuda opined, by appropriately acknowledging and fulfilling one's
sexual needs? Are human beings considered by the Torah to be immutable souls seeking as far as possible to escape and transcend their bodily
constraints and confinements, or as a well-balanced and constructive
combination of body *and* spirit? The two profoundly conflicting philosophies and subsequent readings of Scripture are sharply stated and played
off against each other, but no attempt whatsoever is made, here or elsewhere, to decide the issue.

What are we to make of such exegetical pluralism? What can it teach
us about the conceptions of learning and intellectual accomplishment that
guided the redactors of such passages. And why is the realm of halakha
constructed so differently? Or perhaps, in the last analysis, it is not.

Traditionalism and Its Discontents

Central to and constitutive of the entire talmudic enterprise is the well-known rabbinic distinction between the Written and the Oral Torah, between the sealed canon of the Hebrew Scriptures and the understanding of its meanings and implications, between the subject matter and the fruits of Torah-study. With regard to the Written Torah, *Torah she-bikhtav*, the naked, as-of-yet uninterpreted text of the Hebrew Scriptures, talmudic Judaism grants human beings no say whatsoever. The bare written word of God is unanimously held by the sages to lie entirely beyond the reach of human discretion. Although even the rabbis acknowledged human contributions to the finalization and sealing of the biblical canon—such as those of Ezra and the later so-called "scribal emendations"—*tikunei sofrim*—they themselves treated it as primordial. Notwithstanding the process by which the Written Torah acquired its final form, for talmudic culture it is not within human power to authorize the amendment of even a single letter in the holy text. There exists no such halakhic procedure. The Written Torah, the unread text itself, is regarded by talmudic culture as unchanged and unchangeable, as invariant to the passage of time, as a primordial Given imposed upon its learners "as a mountain pressed down upon their heads."[6]

According to one early midrash, the Torah is said to have come into existence together with the Throne of God, *kise ha-kavod*, prior to the creation of the heavens and the earth.[7] Another midrash, echoing Plato's *Timaeus* perhaps,[8] states that:

> The Torah declares (of itself): I was the Almighty's tool of art! (For) [I]t is common-practice that a king of flesh and blood does not build (himself) a palace on his own accord [on the basis of his own personal knowledge]. Rather, he consults an expert builder. And neither does the expert builder build it on his own accord, but consults charts and notebooks in order to discern how to make (the various) rooms and windows. In similar fashion the Almighty consulted [looked at] the Torah and created the world.[9]

While both *midrashim* strongly emphasize the ontological primacy of the Written Torah, only the second implies the existence of a primordial divine *reading* of the holy text prior to it being handed over at Sinai as it were, to be read and interpreted by its recipients. Surprisingly perhaps, this latter point was evidently disputed by the rabbis. In a number of aggadic passages God is astonishingly reported to be studying, learning from, and even consulting human renditions of Scripture.[10] We shall return to this point shortly.

With respect to the Written Torah, however, no dispute is in evidence, and the rabbis' writings appear to register a solid consensus. The fundamental priority, transcendence, and immutability of the naked holy text is challenged by no one. Indeed, prior to some act of interpretation, the text of the Written Torah is in itself not only meaningless to its human recipients, it cannot even be read. Devoid of vowel points, punctuation, cantillation marks, and accents—the insertion of which renders a Torah scroll ritually unfit for religious purposes—the strings of consonants that constitute the bare and partial syntax of the Hebrew Scriptures are unreadable.[11] In other words, some initial interpretive act of elucidation is required in order to render the primordial text even *potentially* meaningful. Viewed thus, and considered in its role as an object of study for talmudic Judaism, the Written Torah is found to bear an interesting resemblance to the phenomena in their role as the object of scientific investigation. Scientists do not, of course, normally regard their task as a holy duty, and do not necessarily treat their subject matter as the work of a covenanting Deity. Still, the way they treat the phenomena resembles the way the rabbis treat the Written Torah more than the way an exegete would approach a text written by a human author. Scientists, like students of Torah, will never dismiss data as ornamental, as merely a rhetorical device, as a slip of the tongue. Facts may be deemed irrelevant to the phenomena under consideration, but no fact is ever deemed of itself frivolous, feigned, superfluous to the main theme, or redundant. Scientists and rabbis treat their respective subject matters as codes composed entirely of meaningful data. Neither of God's two books is approached with the sort of empathy with which an exegete would normally approach a work of human origin. Scientists regard the phenomena they study as do the rabbis the words of the Written Torah: as a given, imposed upon them "as a mountain pressed down upon their heads."

But what of the Oral Torah, of the interpretation of the text? What are the origin(s), source(s), status, authority, and methods of acquisition of rabbinic interpretations? From what do they derive? To what extent are they considered binding? Who is authorized to propose an interpretation? Who has the authority to arbitrate in the event of rival interpretations? On these and similar questions concerning the aims, status, and nature of the study and students of Torah, the tannaitic literature is sorely divided. But before examining how the tannaitic texts bear witness to this debate and describe the parties to it, it is useful to first map and analyze the possible approaches to these questions in theory. Such a typology is generated by attending to the principal ontological and epistemological considerations at play.

Considering the status of the Oral Torah from a strictly ontological perspective and utilizing a terminology borrowed from current philosophy of science, one may distinguish in theory between two possible points of view I shall term for convenience: *realist* and *conventionalist*. The realist asserts that there exists, independently of and unaffected by the interpretive and investigatory endeavors of human learners, an a priori, objectively true interpretation of the Written Torah—in the mind of God for instance. The conventionalist, by contrast, presupposes no such "authorial intent." The conventionalist's claim is ontological, however, rather than epistemic: it is not merely the claim that we have no way of *knowing* God's intentions—to which many realists would readily agree—but that there is *in reality* no such thing as an a priori, God-intended, true reading of the Written Torah.

For the realist, the ultimate cognitive aim and purpose of Torah-study is to approximate as closely as possible the one true, originally intended meaning of the holy Scriptures. And regardless of whether or not it is believed that human learners are capable of *recognizing* the truth or truth-nearness of their interpretive and halakhic efforts, their efforts will nonetheless be considered by the realist as possessing truth values in principle. The conventionalist denies this to be the case. For the conventionalist—by far the more theologically radical of the two—the development of the Oral Torah is considered a project assigned by God exclusively and solely to humans, who are obligated to interpret the Written Torah according to whatever standards and criteria of meaningfulness they see fit. Viewed thus, the musings and rulings of human midrashists and halakhists cannot, in principle, be spoken of as true or untrue (other than in the purely pragmatist sense of the term). They correspond to nothing outside of them. Such rulings may be considered authoritative, obligating, meaningless or vacuous, useful, inspiring or uninspiring, but never true or false, for there is nothing *of which* they may be held true or false. Surprising as it might seem in the markedly theological context of the talmudic literature, such an option appears to be clearly voiced in the talmudic texts, despite the constraints it would seem to impose upon God's omniscience.[12]

Among the realists—according to whom each human rendering of the Written Torah may, in principle, be assigned a truth value—one may further distinguish between two possible points of view according to their different epistemological premises: between those who maintain that the truth of the Oral Torah, or certain portions of it, may, in practice, be recognized or in some way ensured, and those who believe it cannot. Finally, among those who maintain that truths of the Oral Torah are knowable as such, a further division of realists may be made according to their placement of the source and warrant of such knowledge. All this leaves us with the following three major classes of realist positions in addition to and in contradistinction from that of the conventionalist:

(i) A realist could maintain that humans are at any time in possession of at least part of the one primordial Truth—interpretive or halakhic— by virtue of divine revelation. There are two basic variants of such a position. According to one view, the Written Torah was revealed and handed down to Moses and the prophets fully interpreted and explained to be carefully studied and observed by each generation and faithfully passed on to the next. On such a showing Torah students do not innovate. The history of *talmud-Torah* exhibits no discovery of new knowledge, only the careful reception and transmission of the old. A less conservative variant of this type of realism maintains that the Written Torah was initially revealed accompanied by the foundations of the halakhic system and a divinely vouched-for 'study kit' of exegetical and hermeneutic rules of inference, the famous "thirteen *midot* (principles) through which the Torah is interpreted." According to this view, the proper application of the *midot* guarantees the validity of their exegetical and judiciary renderings. According to this view, the history of *talmud-Torah* resembles that of certain branches of mathematics in comprising an ever-expanding body of creative new discoveries validly inferred from the old. On either showing, each generation of learners is absolutely obligated by the teachings passed down to it from former generations—they differ only with regard to the prospects of further innovation. I shall refer to this brand of talmudic realism in what follows as the *traditionalist* view of Torah-study.[13]

(ii) Conversely, a realist might reject the view that the Scriptures were received from God fully or potentially interpreted, yet still maintain that their one true meaning is humanly attainable by virtue of humankind's cognitive endowments. According to this view, humans are considered capable, in principle, of fathoming at least part of the one true meaning of God's 'Book of Words' by reasoning from and on the text unassisted from without.

(iii) Finally, a realist could deny the epistemological premises of both positions (i) and (ii), maintaining that humans are, in principle, incapable of ever vouching for the validity of the Oral Torah. On such a showing, the Oral Torah, in both its midrashic and halakhic modes, is considered at all times the tentative, hesitant, conjectured product of

ongoing human exegetical and judicial reasoning—invariably pluralistic, forever open to objection, and decidable whenever thought necessary on the basis of publicly accepted procedures such as some form
of institutionalized authority or majority vote. In both its ontological
and epistemological presuppositions this view of Torah-study closely
resembles the theory of knowledge outlined in Part 1 of the present
study. For reasons that will become clear immediately, I shall term this
version of talmudic realism the *antitraditionalist* approach to Torah-study.

Although a viable option in theory, position (ii) appears in fact not to
have been entertained at all in the talmudic texts. The Platonic, or Aris-
totelian, or Reformation, or Cartesian, or enlightenment idea that, left to
their own resources, human beings are somehow capable of seeking out,
arriving at, and recognizing the one God-intended truth has no talmudic
parallel. The talmudic texts seem not to partake in this epistemic *hubris*
so central to ancient, early modern, and enlightenment Western philosophy, as if people, unassisted from without, are sufficiently endowed to
fully fathom the true meaning of the Word or the Works of God.[14] For
the rabbis, in short, the one true meaning, to paraphrase the famous Lu-
theran dictum, is *not* in the text—if it exists at all, it is either God-given
or humanly unknowable.

This leaves us with positions (i) and (iii) that together represent three
realist talmudic understandings of the aims and nature of Torah-study—
two traditionalist and one antitraditionalist.

The conventionalist is, of course, as equally opposed to traditionalism
as the antitraditionalist realist. However, their very different ontologies
notwithstanding, with regard to the purely legalistic project of halakhic
development (as opposed to the exegetical, midrashic project of interpreting halakhic biblical passages), there will be no discernible difference between the logics, heuristics, and decision-making procedures practiced
by the two schools. In terms of the terminology proposed in Part 1, the
various portions of the law will be similarly treated by proponents of the
two schools as tentative goal-directed halakhic systems that require constant troubleshooting in the light of their perceived goals and changing
circumstances. In the context of pure halakhic amelioration (again, as
opposed to biblical exegesis), truth per se will seldom, if ever, be an issue.
Bodies of halakha will be typically troubleshot, on either showing, for
their consistency, fairness, practical applicability, deterring force, social
function, etc.—all legalistic objectives on which both schools of thought
readily converge. The two schools will diverge, both epistemologically and
methodologically, only with respect to biblical exegesis proper in ways
resembling the realist/antirealist debate in philosophy of science. The con-

ventionalist exegete will conduct his midrashic activity—and in this respect there will be no difference between *midrash halakha* and *aggada*—in much the same way as the instrumentalist goes about science, seeking no more than to efficiently and economically 'save the phenomena'.

Here, in the realm of biblical interpretation, the realist, by contrast, seeks truth although he knows that he is incapable of ever recognizing the truth as such. In his midrashic undertakings, the antitraditionalist realist hence resembles the scientific Popperian. Contrary to all manner of conventionalism, the antitraditionalist exegete and Popperian scientist assign truth values to their interpretative conjectures, considering them explanatory of and refutable by the 'facts' each of them studies.[15]

But to return to the traditionalist. Both brands of traditionalism, to repeat, entail a rigid commitment to the teachings passed on by former generations. Whether it is because the truth of these teachings is attributed directly to divine revelation or to the proper application of divinely approved canons of proof, for traditionalists of either kind they are absolutely binding. And it is precisely this unquestioned assent to one's halakhic traditions that is sorely contested by the antitraditionalist.

Still there are important differences between the two types of traditionalist. One concerns their attitude to halakhic questions for which the available halakhic tradition possesses no answer. If the Oral Torah is considered for any reason incapable of further development, then such questions will be considered unanswerable. However, if it is viewed, as are certain areas of mathematics, as a continuously growing body of validly derived halakhic conclusions, then halakhic innovation is always possible. What both brands of traditionalism disallow the antitraditionalist cherishes most: halakhic *criticism* and subsequently halakhic *revision*. This is the defining feature of traditionalism: in the hands of no generation will the body of received halakhic rulings ever be troubleshot for correctness, suitability, or relevance. At the very most it is liable to be critically scrutinized for lacunae.

None of this, however, is presented as such or analyzed by the rabbis themselves in any detail. Their texts contain no systematic taxonomy of positions and certainly no general theoretical discussion of what each of them premises or entails. But the lack of an abstract mode of philosophical exposition does not necessarily signify a lack of philosophical awareness. On the contrary, in its highly literary and keenly nuanced attempts at self-understanding, talmudic culture can be as profound as any other. But to appreciate these texts for their philosophical import, it is important, before approaching them, to acquaint ourselves as far as possible not only with the theoretical and theological presuppositions on which each of the schools is predicated, but also with the main visible features they are expected to exhibit in their negotiations and rulings. We start with the first and most extreme brand of traditionalist mainly because of the central role he comes to play in the Jabne legends.[16]

For all traditionalists Torah-study is primarily conceived of as a matter of meticulous reception and transmission from one generation to the next of the one allegedly revealed binding meaning of the Written Torah. Whatever halakhic discretion is granted to rabbinic authorities of each generation with respect to new questions, it pales for the traditionalist in comparison to the central and holy task of preserving, synthesizing, and passing on the great legacy of old. "Moses received the Torah at [from] Sinai," states the well-known first Mishna of *Abot*, "and passed it (on) to Joshua, and Joshua to the elders, and the elders to the prophets, and the prophets passed it (on) to the members of the Great Assembly, etc." The term "Torah" is here and elsewhere taken by the traditionalist to denote the combined transmitted legacy of written and oral traditions. The problem of local interpretation is, of course, not entirely avoided by the traditionalist. Even if the Written Torah was received interpreted, each generation of learners is still required to understand for itself the precise meaning of what it is taught. This is especially true in the halakhic domain where understanding is a prerequisite to adequate performance. Paradoxically, the local interpretive effort required of the traditionalist is in a sense greater than that required of the nontraditionalist who is prima facie less obligated by the oral traditions of former generations. While the latter treats all but the biblical text itself as potentially debatable and open to question, the former, by definition, is obliged to treat everything received from his esteemed predecessors as canonical. Both constantly run the risk of unwittingly misrepresenting the views of their teachers. For the traditionalist, however, such mistakes can be fatal. In the necessarily dogmatic atmosphere of traditionalist centers of learning, the chances of successful backtracking, of exposing unwitting distortions and disruptions of the legacy by former generations who will have unwittingly passed on their mistakes as dogmatically and confidently as their unmistaken teachings, are extremely small, and the chances of putting them right long after the event, even smaller. This is because traditionalist study sessions are typically uncritical. The teacher's job is to transmit the entire body of knowledge in his possession as effectively and clearly as possible. Students will normally question their masters only on points of clarification, but once the lesson is learned and thought to be sufficiently understood, they will never challenge the content of their teachings. Knowledge passed on from teacher to disciple will be closely scrutinized, but will be seldom subjected to *critical* scrutiny. A traditionalist study session, and certainly halakhic disagreements between different study groups, may well be polemical, but, as in early forms of medieval *disputatio*, the polemics will normally be directed against rival positions or potential objections to the received view. *Self*-criticism will be applied by the traditionalist only as a justificatory device, as a means to reinforce the received view, never as an expression of genuine self-doubt.[17]

In traditionalist centers of learning of this kind, the only sort of issue ever truly open to question will be the authenticity, reputation, and credibility of the transmitter—especially in cases of conflicting teachings. If, however, it proves impossible univocally to discredit the source of one of two conflicting legacies, the steadfast traditionalist is incapable of deciding the issue. He is thus obliged to either refrain from deciding altogether and to leave the matter inevitably open (until the "return of Elijah"), or reluctantly to allow each member of the community to choose sides—in which case the school will irreversibly be split in two. "Thus," states Tosefta, *Sanhedrin*, "as the disciples of Shammai and Hillel who had not attended upon their Masters sufficiently increased in number, so increased the number of halakhic disputes in Israel," and, adds the version cited in the Bavli, "the Torah is rendered (split into) two (separate) *Torot.*"[18]

Both teachers and students are, therefore, under constant pressure to get things right, whole, and intact while face-to-face dialogue is still possible. For the resolute, extreme traditionalist, disruptions in the process of transmission are virtually irreparable. Portions of the transmitted Oral Torah lost en route are in principle irretrievable until prophecy is restored. Each human link in the "great chain of transmission"—the student of today, the teacher of tomorrow—will, therefore, be carefully selected and constantly scrutinized. The holy task of preserving, promulgating, and transmitting the one revealed truth will normally be charged to a small and dedicated group of devoted initiates.[19] Still, reflective members of the community are well aware that no human learner is perfect, and no human transmitter infallible. Much knowledge, extreme traditionalists know, will inevitably be lost, owing to imperfect reception and faulty transmission. Their own image of Torah-study will inevitably be one of an ideally stagnate, but in practice an almost necessarily degenerative enterprise. As a rule, time or history is regarded by such traditionalists, as in certain quarters of premodern Western culture, as an essentially decaying and corrupting factor.[20]

To summarize, one would expect thoroughly traditionalist centers of learning—if such institutions ever existed—to be (a) unflinchingly dogmatic in preserving their legacies; (b) highly selective; (c) genuinely critical only of the credentials and authenticity of the bearers of tradition, but never of the content of their teachings; and (d) to view themselves as fighting a hopelessly losing battle to preserve and transmit an inevitably dwindling body of revealed truth. All four features are explicitly associated with the traditionalist opposition envisaged by the Jabne stories, and all four are reputed to have been markedly reversed by the triumphant antitraditionalist Jabne reformers.

The Jabne Reforms

In the many stories and anecdotes concerning second-generation Jabne scattered throughout the Bavli, R. Eliezer b. Hyrqanus "the Great" is frequently described as the leading, stern, and uncompromising traditionalist.[21] Hailed by his great teacher Rabban Yoḥanan b. Zakai as "a lime (plastered) cistern that loses not a drop,"[22] Eliezer is reputed to have testified of himself that "if all the seas were ink, all the reeds quills and all humankind scribes, they would not suffice to record all my reciting and teaching nor my attendance upon the sages of the School,"[23] and to have added: "and I have never spoken a word (of halakha) that I had not (previously) received from [the mouths of] my Masters."[24] These descriptions suffice to prove that the character of the great sage is used in these stories as a personification of the most extreme form of traditionalism. It is for historians to determine to what extent R. Eliezer's or for that matter any other sage's literary image remains faithful to the real man. It is obvious that the R. Eliezer b. Hyrqanus that one encounters elsewhere—especially as party to numerous halakhic debates recorded in the Mishna and Tosefta—is very different from the intractable traditionalist he plays in the Jabne stories. But which of the two portrayals comes closer to the truth I cannot and need not say.

R. Eliezer's tenacious traditionalism is forcefully portrayed in the following story related in Bavli, *Hagiga* 3b. Having been visited at his hometown Lyyda by his disciple R. Yossi b. Dormaskit who came to pay his respects, R. Eliezer is said to have asked to know what had transpired at Jabne during his absence. Yossi proudly reported that the halakhic problem of determining the type of tithe appropriate in the territories of Amon and Moab during the seventh year of *sh'mita* was debated, voted upon, and decided by the assembly. "It was decided by majority vote that Amon and Moab (areas not included in the traditional boundaries of the Holy Land, in which agricultural work is, therefore, permitted during the seventh, Sabbatical year) are obliged to distribute the tithe among the poor." Upon hearing this, Eliezer, we are told, became so furious that he caused Yossi to lose his sight. "R. Eliezer wept and told him (. . .) go tell them do not heed to your vote[25] (. . .) for I have received it from Rabban

Yoḥanan b. Zakai, who heard it from his Master, and him from his Master—a halakha (given) to Moses from Sinai—that Amon and Moab are to take a tithe for the poor on the seventh, Sabbatical year"! In other words, Eliezer did not object to the content of their ruling, but only to their impertinence in supposedly taking it upon themselves to decide the issue. Eliezer, we are told, eventually recovered his calm and agreed to restore his disciple's eyesight.

For the legendary R. Eliezer and his fellow traditionalists, the essence of Torah-study is, to paraphrase Bavli, *Rosh Ha-Shana* 20a, "Such thou shalt see, study, recite and sanctify," and pass on to your disciples whole and intact without losing a drop! But even Eliezer of these stories, on any count the perfect recipient, "a lime plastered cistern that loses not a drop," seems to have been fully aware that the traditionalist's ideal is humanly unachievable: "Much Torah have I studied, and much have I taught," the archtraditionalist is told to have lamented on his deathbed, "yet have I but skimmed from the knowledge of my teachers as much as a dog lapping from the sea; and my disciples from mine, as much as the tip of a quill dipped in an ink pot."[26] But nowhere does R. Eliezer's traditionalism emerge more vividly than in the Bavli's account of the dramatic dispute with his colleagues following the equally dramatic removal from office of Jabne's second President, Eliezer's brother-in-law Rabban Gamliel *de-Jabne*.

Following the death of the founder of the Jabne center, Rabban Yoḥanan b. Zakai (c. 90 A.D.), Jabne came under the leadership of three sages: the President, Head of the academy, Rabban Gamliel; his above mentioned brother-in-law R. Eliezer b. Hyrqanus; and R. Yehoshua b. Ḥanania who, according to one source, served as Head of the Jabne *Bet din*.[27] Rabban Gamliel's term in office extended from the last decade of the first century A.D. to the first decade of the second. He is described by both the Bavli and Yerushalmi as a stern and imperious Head of House as it were, who claimed for himself the authority to decide new halakhic issues alone, ex cathedra, on the strength of his official capacity. On three separate occasions he is reported not only to have publicly repudiated the learned opinion of his colleague R. Yehoshua, but to have gone on to humiliate him in the presence of the entire assembly.[28] On one such occasion, the two sages are said to have disagreed on the credibility of the eyewitnesses who claimed to have spotted the new moon of *Tishrei*. Consequently, each of them calculated *Rosh Ha-Shana*, the New Year, to have fallen on a different day. Rabban Gamliel not only overruled R. Yehoshua's judgment, but ordered him to report to him at Jabne on the day that was to be the Day of Atonement according to Yehoshua's reckoning, donned in his workday attire, "with his purse and staff."[29] According to the Jabne stories, Rabban Gamliel's high-handed dealings with R. Yehoshua were met with such agitation at Jabne that after the third incident the assembly rebelled and decided to remove the President from office.

> And it is related that a certain disciple came before R. Yehoshua and asked him, Is the evening prayer optional or compulsory? He replied: It is optional. He then presented himself before Rabban Gamliel and asked him: Is the evening prayer optional or compulsory? He replied: It is compulsory. But, he said, did not R. Yehoshua tell me it is optional? He said: Wait until the sages enter the *Bet midrash*. When they had assembled the questioner rose and inquired, Is the evening prayer optional or compulsory? It is compulsory, answered Rabban Gamliel. He said: Is there a anyone (here) who disputes this matter, and R. Yehoshua said No. But, said Rabban Gamliel, were you not reported to me as saying that it is optional? Stand up and they will testify against you. And R. Yehoshua rose to his feet and (laughing the matter off[30]) said: Were I alive and he [the witness] dead it would have been possible for the living to contradict the dead, but since I and he are both alive how can the living contradict the living!? And Rabban Gamliel remained sitting and expounding, and R. Yehoshua remained standing, until the entire congregation began to grumble and ordered Ḥutzpit the interpreter (whose job it was to recite out loud and expound upon the President's teachings) to be silent. And he was silent. They said, How much longer can he [Rabban Gamliel] go on insulting him [R. Yehoshua]? (. . .) Come, let us remove him (from office)![31]

Following a brief consultation, during which the candidacies of R. Yehoshua and R. Akiva were proposed and rejected—the former for being a side to the dispute, the latter for lacking "ancestral merit"—it was decided to appoint R. Yehoshua's young disciple R. Elazar b. Azaria as successor to Rabban Gamliel.[32]

Despite his tender age, a mere eighteen according to the Bavli and sixteen according to the Yerushalmi,[33] the new President, evidently backed and encouraged by his teacher R. Yehoshua,[34] took the opportunity to introduce three major reforms at Jabne, which in light of our previous analysis of the traditionalist approach are highly significant. Nor does their significance go unnoticed by the talmudic redactor of these legends, who describes them as having been instituted in the course of the one, obviously compressed, mythical day of Elazar b. Azaria's election to office.

First, counter to traditionalist elitism and in line with the stated policy of the House of Hillel,[35] "on that very day"—*bo ba-yom*—we are told, Elazar dismissed the doorkeeper evidently hired by his predecessor,

> and (all) disciples were permitted to enter (the academy). For Rabban Gamliel had issued a proclamation [saying]: No disciple whose character does

not correspond to his exterior may enter the *Bet midrash*. How many benches were added (on) that day? (. . .) One said four hundred benches were added, and one said seven hundred.[36]

Second, Elazar b. Azaria appears to have repealed the right of the President to decide new halakhic issues *ex positio*—a prerogative, as we have seen, frequently exercised by Rabban Gamliel. From "that day" on, first among equals, the President's hand was again[37] to be raised and counted along with those of all his colleagues, and all new halakhic questions presented to the assembly would be openly debated and decided when necessary by majority vote. In doing so the opinion of the majority was granted exclusive sovereignty with respect to all standing matters.

R. Elazar's third decree was perhaps the most consequential, certainly in light of our present concerns. But to appreciate it as such a short detour is required.

According to the Tosefta in *Sanhedrin* and *Hagiga*, cited by both the Bavli and Yerushalmi,[38] prior to the destruction of the Temple and the establishment of Jabne, the existence of a reliable halakhic tradition was granted absolute precedence over majority opinion at all levels of the court system. The passage is worth quoting at length:[39]

> R. Yossi said: Originally, there was no (halakhic) disagreement in Israel. Rather, a *Bet din* of seventy-one members sat in the Hall of Hewn Stones[40] and courts of twenty-three sat in each of the townships of *Eretz Yisrael*. And two other courts of twenty-three sat in Jerusalem, one at the entrance to the Temple Mount, and one at the entrance of the *Azara* (the Temple Court). If a matter of inquiry arose, the local *Bet din* was consulted. If they had a tradition [thereon] they stated it; if not, they went to the nearest *Bet din*. If they had a tradition, they stated it, if not, they went to the *Bet din* situated at the entrance to the Temple Mount; if they had a tradition thereon, they stated it; if not, they went to the one situated at the entrance of the Court. If they had a tradition thereon, they stated it, and if not, they all proceeded to the *Bet din* in the Hall of the Hewn Stones (. . .) where they [i.e., the Great Sanhedrin] sat from the morning daily offering until the evening daily offering. (. . .) The question was then put before them: if they had a tradition thereon, they stated it; if not, they took a vote: if the majority voted 'unclean' they declared it so; if 'clean' they ruled even so. And from there the halakha issued forth to all Israel. But when those disciples of Shammai and Hillel who had not attended upon [their Masters] sufficiently increased [in number], disputes multiplied in Israel, and the Torah became as two Torot.[41]

According to these tannaitic sources, all emanating from the Tosefta,[42] prior to the establishment of Jabne a halakhic tradition would always take absolute precedence over personal opinion. Whenever a reliable halakhic tradition was available, nowhere along the line would the issue be debated or put to vote. Only the members of the Great *Bet din* were granted the authority to debate and decide matters of halakha, but even their discretion was wholly limited to issues that were entirely new to the system, issues to which no existing halakhic tradition was known to apply. Halakhic disagreement could thus be kept to a well-controlled and manageable minimum, wholly confined to the new questions which the system effectively bracketed off and funneled up to the Sanhedrin. And since these were duly decided by majority vote, no halakhic dispute would have remained unresolved for long. The system, however, is alleged to have entered a period of crisis owing to the negligence of the latter-day disciples of Shammai and Hillel who apparently caused the chain of transmission to cleave and diverge.[43]

The traditionalist bias of this repeatedly cited tannaitic passage is unmistakable. It does not voice the most extreme form of traditionalism the talmudic texts have to offer—for it allows for halakhic innovation in the absence of an applicable tradition—but the way in which the system is immunized to halakhic revision preserves the traditionalist's defining feature. And in such a traditionalist culture contradictions between different and equally reliable traditions are, in principle, irresolvable. Once such a split occurs, the Torah is inevitably and permanently rendered two separate dispensations!

The Mishna, however, appears to take a rather different view of the limits of court discretion. The Mishna's version of the hierarchical relationships between the three Temple-based *batei dinim* (Mishna, *Sanhedrin* 11:2) differs subtly from that of the Tosefta, although, as the reader may judge, it seems clear that the tannaitic redactors of the two texts were working from the same material.

> Three courts of law were there, one situated at the entrance to the Temple-Mount, another at the door to the Temple Court, and the third in the Hall of the Hewn Stones. They first went to the *Bet din* which is at the entrance to the Temple Mount. And he stated, Thus have I expounded and thus have my colleagues expounded; thus have I taught, and thus have my colleagues taught. If they (the first *Bet din*) had a tradition, they stated it; if not, they went to the one situated at the entrance of the Temple Court, and he declares, Thus have I expounded and thus have my colleagues expounded; thus have I taught, and thus have my colleagues taught. If they (the second *Bet din*) had a tradition, they stated it; if not, they all proceed to the Great *Bet din* of the Hall of the Hewn Stones whence Torah issued forth to all Israel.

There are at least three significant differences between the two tannaitic accounts. First, by omitting all mention of tradition with regard to the

Great *Bet din's* ruling, the Mishna could be read as decreeing that the highest legal instance is itself quite free to rule as its majority sees fit *despite* traditions to the contrary. Second, unlike the Tosefta and in line with what one would expect of an antitraditionalist, the author of the Mishna makes no mention of, let alone complain about, a split between the two Houses, or about the increase in halakhic disagreement as a result. The existence of conflicting traditions is simply not a problem for the antitraditionalist.

Finally, the very setting of the Mishna's description is different. It is not presented as a general description of the way things used to be, but is brought up apropos the special case of the rebellious elder—a sage of high standing who disputes a ruling of the Great *Bet din* and is accused of issuing a ruling of his own to the contrary—an offense punishable by death (Deut. 17:13). Now, if the Mishna's omission is not accidental and can be said to be ruling that even in the case of a rebellious elder the Great *Bet din* is *not* bound by halakhic tradition, then we are facing a rather special pronouncement of antitraditionalism. The guilt of a rebellious elder, one would think, has nothing, in principle, to do with whether he is right or wrong. His crime is that of taking the law into his own hands as it were. Every person—especially a sage of such standing—is entitled to an opinion. But once a Sanhedrin has had its say, one may think differently, but no longer rule differently. According to the antitraditionalist, only a Sanhedrin can overthrow the ruling of another Sanhedrin, while the traditionalist will not even allow that. A rebellious elder is not a Sanhedrin and is, therefore, either guilty or not. But if my reading of it is correct, the Mishna seems to imply a third possibility, namely, that the elder be found guilty of contradicting a tradition sanctioned by the Sanhedrin, and yet be deemed right in his conclusion. In this case his Great court judges will reverse the transgressed ruling and pronounce the elder innocent of rebellion after the event.[44]

These and other related discrepancies between the Tosefta and the Mishna, to be examined shortly, strongly suggest, in my opinion, that the framers of the two texts took sides in the traditionalist/antitraditionalist dispute. (Inattentive perhaps to the implied disagreement between the two tannaitic sources, the redactor of Yerushalmi, *Sanhedrin* i, 19a, confusingly combines the two versions.) Also worth noting in passing is that according to both versions the President is not even mentioned, let alone granted special halakhic authority.

But to return to the traditionalist version of the court system and to the state of crisis it is alleged to have entered owing to the existence of conflicting halakhic traditions. Tosefta, *Sanhedrin*, and its three parallels give no indication as to how the problem might have been dealt with, although, as we shall see shortly, other passages do. For the steadfast traditionalist, to repeat, such a quandary is indeed irresolvable. And the clear sense of mournful finality conveyed by the way the Tosefta contrasts the idyllic past with the lamentable present further attests to the traditionalist leanings of its author.

It is beyond the scope of the present study, of course, to attempt to describe the real historical background to the Jabne stories associated with R. Elazar b. Azaria's election. Tempting as it is to relate the antitraditionalist nature of R. Elazar's reforms to a desire to overcome the sort of traditionalist deadlock described above, most historians are of the opinion that by the time R. Elazar is alleged to have assumed office the two Houses had long disbanded, and that the Jabne *Bet din* no longer faced such an impasse,[45] certainly not the one described. Historically speaking, it is also quite reasonable to assume that the Bavli's dramatic attribution of the Jabne reforms to R. Elazar's first day in office is a mythical compression and summation of a prolonged process that in all probability had started earlier and certainly ended later, and had perhaps involved the young Elazar far less than the legend implies. Indeed, some historians now maintain that although the authority of Rabban Gamliel might have been challenged for his high-handed behavior, it is doubtful that he was ever actually removed from office or made to share it with anyone else.[46] However, for the modest purposes of the present study, the historical circumstances and viability of the Bavli's stories of Jabne are largely immaterial. Our aim is to assess the possible theological and philosophical significance of the stories rather than their truth. Read thus in the intertextual rather than the historical contexts of related talmudic accounts,[47] as the literary dramatization of ideological conflict rather than as a questionable historical source, these related anecdotes, I shall argue, align themselves quite clearly with the Mishna's antitraditionalist bias. But first to the third and most significant of R. Elazar's alleged reforms.

On the day of his election, we are told, Elazar demanded that all bearers of halakhic traditions present their teachings for the perusal of the Jabne *Bet din*, where he proceeded to put each of them to the vote. "On that day," states the *beraita* in *Berakhot*,

> (Tractate) *Eduyot* (lit. Attestations, Testimonies) was taught (originated[48]). And wherever [an event] is said (to have occurred) *'bo ba-yom'* it (refers to) that very day. And there was no unresolved halakhic issue left pending at the *Bet midrash* that was not interpreted (decided).[49]

How are we to understand this move? Needless to say, the very subjection of long-standing halakhic traditions to majority vote is a procedure wholly unheard of in traditionalist circles—unless, of course, the issue to be decided is the reliability of the testifiers rather than the viability of their testimonies. For the steadfast traditionalist no person, not even a majority of the Great *Bet din*, has the power to overrule a reliable halakhic legacy. A resolute traditionalist will therefore be inclined to interpret R. Elazar's third decree, in good traditionalist spirit, as intended to filter off undependable testimonies, that is, as a vote of confidence rather than as an act of halakhic arbitration. On such a showing, Jabne was not radically reformed by the new President but merely granted itself the authority of a Great *Bet din* along the lines of Tosefta *Sanhedrin*—filtering off unreliable testimonies, remaining firmly committed to those proved dependable, and ruling by majority opinion only on genuinely new halakhic issues. Viewed thus, Jabne would have remained no better capable of dealing with the traditionalist deadlock described by the Tosefta than the earlier Temple-based Sanhedrin.

A less resolute traditionalist might interpret the Jabne reforms more boldly as a reluctant, minimal, and isolated breaching of the strict traditionalist norm, taken temporarily in order to break the deadlock and to unite Israel under one binding system. According to such a rendering of the *beraita*, the new President's third decree should be understood as a decision not merely to sift out trustworthy bodies of transmitted halakha, but to arbitrate by majority vote between those reliable testimonies found to conflict which each other.

Finally, it is possible to interpret R. Elazar's reforms as a motivated and radical move against traditionalism itself. Thus construed, Jabne is perceived to have broken with the traditionalist norm entirely: to have taken upon itself to thoroughly scrutinize the present halakhic situation, to disregard the very force of tradition as such, and to rule anew upon each and every halakhic issue in dispute, new and old, according to the opinion of the majority of its members.

Taken in isolation, the *beraita* lends itself equally well to each of these three readings. But by reading it in isolation we will have achieved very little. The *beraita* was evidently composed, cited, edited, and embedded in the text for a purpose, by writers and redactors who sought to make a statement. It is impossible to determine, on the basis of the text alone, who were responsible for the *beraita* as it stands or what the statement might have been that they, in particular, had sought to make. But in any case, the aims of this study are different. What we seek as far as possible to determine is the image of Torah-study conveyed by whole portions of the corpus as we find them, rather than to characterize the proven intentions of the particular writers responsible for their various parts. While the *beraita* under consideration provides the only detailed account we have in the Bavli of the Jabne reforms, it remains in itself philosophically ambiguous. Other texts, such as Tosefta and Mishna, *Sanhedrin*, though not related to the Jabne reforms directly, appear, as we have seen, to be less ambiguous in this respect. Still others, equally unambiguous in their affirmation of or opposition to traditionalism appear, as we shall see, to relate to the Jabne reforms quite clearly. In what follows, these latter texts will be shown to comprise two separate and quite distinct philosophical subcorpuses, in the context of which the *beraita* in *Berakhot* will inevitably acquire different meanings along the lines delineated above. Only then will we be in a position to attempt to suggests ways of determining which of the two literary contexts is the more appropriate to the *beraita* in hand.

The first of these texts is tractate *Eduyot* itself, the formulation of which, according to the *beraita*, originated "on that very day." Interestingly, the two versions of *Eduyot* we have, that of the Mishna and that of the Tosefta, seem, as in the case of the court system, clearly to present quite different approaches to the issue of traditionalism.

The Testimony of *Eduyot*

Unlike any other tractate of tannaitic origin, with the possible exception of *Abot*, neither version of *Eduyot* is organized thematically. Nor are all the halakhic rulings they enumerate new. Many of them simply repeat *halakhot* listed by the Mishna and Tosefta elsewhere. True to its title, *Eduyot*, in both of its renditions, serves to document and rule upon a disorganized array of frequently conflicting, halakhic attestations. This seems to be the only organizing principle of the tractate. Most of the testimonies, though none of the final rulings, are ascribed to named sources. When stated explicitly, final decisions are attributed anonymously to "the sages" thereby indicating that they were reached by majority vote.

Apart from one exception,[50] Mishna, *Eduyot* does not explicitly situate itself in time or place. The recurring format is "so and so testified . . . " with no mention as to when and where. Tosefta *Eduyot*, by contrast, opens with the following words:

> When the sages assembled at *Kerem be-Yavne*,[51] they said: a time may come when a person might seek an article of the Torah and not find it, or of the words of the scribes[52] and not find it (. . .) They said: let us begin with (from) Hillel and Shammai.[53]

Nonetheless, it is clear, here and also in the Mishna, that over and above the obvious desire to record the particular halakhic rulings listed throughout the tractate, *Eduyot* is also and perhaps mainly about the process and procedures of issuing such rulings. In this sense both versions of the tractate appear to possess a theme. In fact, by not grounding itself in a particular historical context, the Mishna, more than the Tosefta, implies that such a treatment of halakhic traditions should not be viewed as an isolated episode but as the everlasting norm. (More on this below.) Still, neither version of *Eduyot* can be read wholly as an account, not even as a mythical account, of the happenings on that one "day" at Jabne.[54] Three of the recorded attestations are said to have been testified in the presence of (before) R. Yishma'el and three others in the presence of R. Akiva.[55] Likewise, sporadic mention is made by both versions of testifiers and discussants, such as Akavia b. Mehalal'el, R. Yehuda, and R. Meir, who functioned much earlier or much later than the "day" in question. Even so, one can easily imagine earlier Ur-versions of the tractate, comprising large portions of the two later and clearly supplemented versions in hand, as the *Eduyot* possibly referred to in *Berakhot 28a* as having originated at Jabne. Read thus, as required by the *beraita*, as the proceedings at Jabne during R. Elazar's first day in office, the two appropriately sifted versions of *Eduyot* offer intriguing if clearly conflicting evidence as to the nature of the Jabne reforms.

The opening paragraphs of the two versions will suffice to make this clear. First the Mishna. Mishna, *Eduyot*, opens with three halakhic issues said to have been disputed by Hillel and Shammai,[56] all of which are ruled upon by "the sages"—according to this reading, the Jabne assembly presided over by R. Elazar—differently from either of the two received views! The recurring format is: "But the sages say: neither according to the opinion of the one nor according to the opinion of the other, but (. . .)." This in itself seems seriously to undermine both of the above-mentioned optional quasi-traditionalist interpretations of R. Elazar's initiative. Unlike their alleged attitude to certain of the testimonies of latter-day Hillelites and Shammaites, it is inconceivable that R. Elazar and his colleagues would have considered the authentic teachings of Hillel and Shammai themselves as unreliable testimonies. And in any case, if pressed to decide, a resolute traditionalist, unauthorized to work things out for himself in such cases, would have had to cast his lot with one of the two received views. But Jabne is reputed to have rejected both. Nor can it be said, and for the very same reason, that the sages' rulings in these cases may have reflected a reluctant decision to arbitrate between equally reliable existing traditions for the sake of unifying the people exactly because they jettisoned them both.

When read in conjunction with *Berakhot 28a*, the Mishna hence leaves us with little choice other than to interpret the new President's policy as an attack on traditionalism itself. This is further confirmed by the commentary offered by the Mishna on its first three rulings. The parallel passage in the Tosefta, however, appears again to retain a firm traditionalist frame of reference. Let us briefly compare the two texts.

The last of the first three rulings of the Mishna concerns the minimum amount of drawn water sufficient to render a ritual bath (a *mikve*) unfit. Hillel is said to have set the minimum disqualifying amount at one Hin (three Kabs[57]) whereas Shammai set it at nine Kabs.

> But the sages say: neither according to the opinion of the one nor according to the opinion of the other. Rather, when two weavers from the Dung Gate which is in Jerusalem came and testified in the name of Shema'aya and Avtalyon (that) three Logs (three-quarters of a Kab) of drawn water render a *mikve* unfit. And the sages confirmed (affirmed, maintained) their statement.[58]

"Why," then, asks the anonymous narrator of Mishna, "do they (at all) record the opinions of Shammai and Hillel merely in order to dismiss[59] them?," and answers:

> (It is done) to teach future generations that a person should never persist in his opinion. For behold, the fathers of the world did not persist in their opinion.

> And why do they record the opinion of a single person among the many, when the halakha must be according to the opinion of the many? So that if a *Bet din* prefers the opinion of the single person it may depend on him. For one *Bet din* cannot set aside the decision of another *Bet din* unless it is greater in wisdom and in number. If it be greater than it in wisdom but not in number, in number but not in wisdom, it may not overrule its decision unless it is greater than it (both) in wisdom and in number.
>
> R. Yehuda says: if so why are the words of the one (minority) mentioned together with those of the many (only) to dismiss them? In order that if a person should say such (and such) is my received view, he will (can) be told: you have heard the words of so and so.

According to the Mishna, a received halakhic tradition, in other words, does not acquire its authority by virtue of its being long standing. The sole criteria for its acceptance is the relative quality and size of the *Bet din* that had last ruled to endorse it. Whether or not a halakhic ruling was reliably transmitted from generation to generation is not in itself crucially relevant to its acceptance, let alone its truth. In other words, what a *Bet din* is required to rule upon, according to the Mishna, is the halakha itself, not merely the reliability of its promulgators. In principle, implies the Mishna, any ruling can be overturned by a later generation if it is capable of assembling a *Bet din* "greater in wisdom and number" than the one originally responsible for the decree. And if and when this occurs, one is cautioned not to "insist on one's opinion" but to accept the rule of the majority. How exactly are wisdom and number to be evaluated in this context is inconsequential for our present concerns.[60] The important point for now is to note how, on the Mishna's showing, the traditionalist's fundamental conception of Torah-study was squarely challenged by the newly reformed *Bet din* at Jabne. Under the leadership of R. Elazar b. Azaria, Jabne appears not only to have ruled that a sufficiently qualified *Bet din* has absolute halakhic authority to decide an issue as its majority sees fit, but to have considered themselves sufficiently qualified to overrule opinions attributed to the "fathers of the world"—even those of Hillel *and* Shammai.[61]

The Tosefta's version of the same passages[62] tells a very different story. After relating the tale of the dispute concerning the *mikve*, the testimony of the two weavers, and its subsequent resolution by the sages, the Tosefta goes on to ask not why the opinions of Hillel and Shammai remain on record, as does the Mishna, but rather:

> Why are [the weavers'] abode and occupation expressly mentioned—for there exists no occupation lower than that of a weaver, and no location lower than the Dung Gate in Jerusalem?

The Tosefta's answer is party-line traditionalist:

> To teach that if the fathers of the world did not persist in their opinion *in the face of a received tradition* (*bimkom sh'mua*), how much more so that a person should never persist in his opinion in the face of a received tradition.

The Mishna, to recall, does not mention received traditions at all in this context.

The Tosefta then states, as does the Mishna, that "the halakha is always according to the opinion of the many," and goes on to explain, as does the Mishna, why minority views should remain on record. The two answers provided by the Mishna are replicated by the Tosefta, but with two significant differences. First, they are clearly presented as mutually exclusive. Second, and more importantly, the attributions are reversed. The view attributed by the Mishna to "the sages" is now credited to R. Yehuda, while the view attributed by the Mishna to R. Yehuda's minority opinion is credited by the Tosefta to the majority view of "the sages."

> The halakha is always in accord with the opinion of the many. (If so) why do they record the opinion of a single person among the many just in order to set them aside? R. Yehuda says, they record the opinion of a single person among the many to enable a *Bet din* that prefers the opinion of the single person to depend upon it.[63] *But the sages say,* they record the opinion of a single person among the many but for the sake of cases where this one will

rule (a thing) pure and that one will rule (the thing) impure, as did (for example) R. Eliezer. For he can then be told that he had received the (formerly dismissed) words of R. Eliezer.[64]

According to the Tosefta, a rejected minority view remains on record in order to ensure that it stays rejected forever! Tosefta, *Eduyot*'s analysis of the way the halakha is determined once and forever by majority vote thus not only fares well with its description of the Great *Bet din*'s voting procedures described in *Sanhedrin*, but provides an important addition. Not only are long-standing halakhic traditions deemed absolutely binding, equally binding, we now learn, are all the new rulings decided by the majority vote of former Great *batei dinim*. As far as any one generation is concerned, there is no difference, according to the Tosefta, between a bona fide long-standing halakhic legacy and the recent rulings of a Sanhedrin—both, states the Tosefta, are absolutely binding, and will remain so forever. The only obvious departure implied by Tosefta, *Eduyot* from the resolute traditionalism of its version of *Sanhedrin* is the implication, apropos the story of the weavers, that equally reliable, conflicting traditions may be *arbitrated* by majority vote. In this way, the Tosefta here implies, the type of impasse described in *Sanhedrin* and alluded to in the opening paragraph of *Eduyot* can be (was?) avoided.

Although the Tosefta explicitly sets itself up as a historical account of how Jabne actually elected to deal with the halakhic crisis of the day, it goes to great lengths, as we have seen, not to describe the Jabne reforms as an antitraditionalist initiative. The Tosefta covers considerably less ground than the Mishna, and in doing so fails to mention several cases dealt with by the Mishna that are prima facie incompatible with its own quite pronounced traditionalist frame of reference. Nonetheless, two such cases clearly at variance with the general traditionalist atmosphere created by the Tosefta are curiously retained. The first[65] is lodged between the opening paragraph describing the crisis at Jabne and the above-mentioned dispute about the quantity of drawn water required to disqualify a *mikve*. Here the dispute between the conflicting rulings of Shammai and Hillel (concerning in this case the minimum quantity of dough liable to be subjected to the law of *ḥala*) is decided by the sages "not as the words of this and not as the words of that," not by reference to an earlier tradition, as in the Tosefta's rendition of case of the *mikve*, but by reasoning directly and seemingly independently from the relevant verse in Num.15. The second,[66] equally at odds with the Tosefta's otherwise consistent traditionalist bias, is located immediately after its markedly traditionalist explanation of why minority opinions should remain on record. It is one of the five known cases, of which four are listed by Mishna,[67] in which the House of Hillel is reported to have been convinced by the arguments leveled against them by the Shammaites to renounce their original position and accept that of their adversaries.[68] The Hillelites' evident renunciation of their former position—motivated by counterargument rather than by the disqualification of those responsible for its transmission—is inexplicable from a traditionalist perspective. Both cases are out of character with the Tosefta's otherwise traditionalist leanings.

Out of place as they are, the two cases remain isolated exceptions. *Kerem be-Yavne* is portrayed by the Tosefta as having striven, albeit boldly, to achieve halakhic consensus without rocking the traditionalist boat. Apart from the minority, and hence, implicitly rejected view, according to which *batei dinim* are granted the authority to overrule former majority decisions by relying on formerly dismissed minority opinions, the Tosefta, as one would expect, makes no mention of dissenters, certainly not of traditionalist dissenters, who, on the Tosefta's showing, would not have been dissenters at all.[69] The Mishna, by contrast, does, and with considerable force. Mishna, *Eduyot* 5:6–7 relates the following:

> Akavia b. Mehalal'el testified concerning four matters (of halakha). They said to him: Akavia, withdraw these four things which you say and we shall appoint you Head of the *Bet din* in Israel. He said to them: I would rather be called a fool all my days than to have acted wickedly in the eyes of God for one hour. And let not men ever say: He withdrew his opinions in order to gain the power of office. (. . .) Whereupon they excommunicated him and he died while he was under excommunication, and the *Bet din* stoned his coffin. R. Yehuda said: God forbid (to say) that Akavia was excommunicated, for the Temple Court was never closed in the face of any man in Israel who was equal to Akavia b. Mehalal'el in wisdom and in fear of sin. But whom did they excommunicate? Elazar b. Ḥanokh who demurred against the laws concerning the purifying of the hands. And when he died the *Bet din* sent and laid a stone upon his coffin. From which we learn that whoever is excommunicated and dies while under excommunication, his coffin would be stoned.[70]

> On the hour of his death he (Akavia) said to his son: withdraw the four opinions which I used to declare. And he said to him: Why did not you withdraw them? He said to him: (Because) I heard from the mouths of the many, and they heard (the contrary) from the mouths of the many; I stood fast by the tradition which I had heard, and they stood fast by the tradition which they had heard. But you heard (my opinion) from the mouth of a single person, and (theirs) from the mouths of many. It is better to leave the opinion of the single individual and to hold by that of the many.

Unsurprisingly perhaps, Akavia, resolute traditionalist to the end, interprets the decision, later endorsed at Jabne, always to rule by majority vote as a decision always to follow *the tradition adhered to* by the majority. And since this is precisely the way in which the framer of the Tosefta appears to understand the position of Akavia's adversaries, he is incapable of following the Mishna's line in depicting Akavia as a stubborn reactionary susceptible to excommunication—and indeed, the Tosefta makes no mention of the episode.

To conclude solely on the basis of the two texts examined so far that there exists a general ideological incongruity between the editorial policies employed throughout the Mishna and the Tosefta would be rashly to overgeneralize. And I have no intention of doing so. Still, it is impossible to ignore the consistent and compelling *local* differences between their respective accounts of the decision-making procedures employed both in Jerusalem and Jabne and not to conclude that the redactors responsible for these passages wrote consistently from opposite sides of the traditionalist/antitraditionalist divide. The Tosefta's marked historicist narrative, both in *Sanhedrin* and *Eduyot*, combines to tell a story. It is the story of an initial traditionalist utopia followed by a period of crisis (*Sanhedrin*), which was later confidently met and resolved by an authoritative yet equally traditionalist Jabne (*Eduyot*). The Mishna's decidedly less historical text, in both instances, largely refrains from telling a story at all.[71] The Mishna frequently incorporates short narrations of events, such as that of Akavia's steadfast traditionalism and unwillingness to conform, but does so always as part of a wider text apparently designed timelessly to lay down the law. In doing so the Mishna frequently records disagreements, and less frequently actual debates, between named parties, which serve, at most indirectly, to locate particular rulings in particular place and time.[72] Neither of the two Mishnaic texts under consideration is primarily historical. Mishna, *Sanhedrin*, is less a narration of the past happenings in the Jerusalem court system than a normative depiction of how a rebellious elder should be treated in principle. Similarly Mishna, *Eduyot*, clearly appears to have sought to transcend the inventory of the particular rulings it documents, with a view normatively to dictate how *any* future Great *Bet din* should treat existing halakhic traditions.

When read in conjunction with and in the light of the *beraita* in *Berakhot* 28a, the two Mishnaic texts appear to the latter-day reader as ahistorical embodiments rather than historical descriptions of the reforms instituted by R. Elazar b. Azaria—as if the two texts were compiled by latter-day writers sympathetic to and supportive of his antitraditionalist cause.[73] Viewed thus, they appear to extend and extrapolate the alleged one mythical day at Jabne beyond the boundaries of place and time, transforming Elazar's campaign from a quasi-historical episode into a set of binding metahalakhic principles to be applied wherever and whenever the need for halakhic rethinking arises. Conversely, the Tosefta's parallel passages remain firmly situated in place and time and appear as valiant, if somewhat inconsistent, latter-day attempts to rewrite history from the reactionary vantage point of the defeated traditionalist. The Mishna, in short, reads as if it had taken "that very day" as the grounds for a normative program, the Tosefta, as an incentive for a rearguard redescription.

Be this as it may, the philosophical significance of the *beraita* thus still remains undetermined. The problem, of course, is not that of ascertaining the true historical intentions of second-generation Jabne. Nor is it that of deciding which of the two versions of *Eduyot* and *Sanhedrin* was closer to the mark. It is evident by now that, at least in the texts so far examined, the Tosefta and Mishna represent and embody significantly different philosophies of halakhic development. It is also reasonable to assume that they reflect schools of thought that existed at least at the time of their composition. The question I wish to address is whether it is possible to determine which of the two philosophies, if at all, can be said to inform the Bavli and other portions of the corpus; whether, in other words, the Bavli can be said to have self-consciously decided between the two. As indicated earlier, I believe that in its treatments of the Mishna and in its discursive practices and policies, the Bavli can be shown to have aligned itself with the antitraditionalist approach conveyed by Mishna, *Sanhedrin* and *Eduyot*. But first we need to establish the Bavli's understanding of the Jabne reforms. Two additional *beraitot* depicting anecdotes related to second-generation Jabne will suffice to make the point.

The story, related by the Mishna, of Akavia b. Mehalal'el's refusal to renounce his traditionalist convictions, his apparent inability to even comprehend the nontraditionalist approach of his disputants, the inevitable conflict, and its sorry outcome, are all replicated with exceedingly greater force in another text clearly associated by the talmudic redactor with the reforms introduced by R. Elazar b. Azaria. It is the dramatic, oft-quoted story of the bitter conflict between R. Eliezer b. Hyrqanus and his Jabne colleagues, related in Bavli, *Bava Metzia* 59a-b.[74]

"It Is Not in Heaven" (Deut. 30:12)

The occasion for the Bavli to introduce the story in question arises in the course of a lengthy discussion devoted to various halakhic aspects of deceit and cheating. Among the various issues raised and recorded is a seemingly innocent disagreement between R. Eliezer and the (presumably Jabne) sages about the susceptibility to ritual uncleanliness of a certain type of earthen oven—a debate recorded without further comment in Mishna, *Kelim* 5:10, and again in Mishna, *Eduyot* 7:7. Ovens, like any utensil, are susceptible to ritual impurity as long as they are unimpaired.[75] The oven in question, however, is one that was first cut or sawed horizontally "layer by layer," and then repaired by filling the crevices with sand or mortar. It is thus a borderline case of a technically damaged, yet fully functioning utensil. True to the strict letter of the law, R. Eliezer decrees the sawed oven unliable, in principle, to defilement owing to its technical impairment. The sages disagree, fearing perhaps that since such an oven remains fully functional, people would be tempted to cheat by "immunizing" their ovens to charges of uncleanliness by deliberately cutting and then repairing them with sand or mortar. The Mishna in *Kelim* dubs such an oven: *"tanuro shel akhnai"*—the oven of the snake—referring perhaps to its coiled, serpentine cutting line, or to the snake-like slyness of its deceitful owner. The Bavli, however, in both its mentionings of the dispute, offers a different explanation entirely of the oven's curious name:

> Why (is it called the oven of) *Akhnai*? Said R. Yehuda in the name of Shmuel: to inform (us) that they coiled (looped) words [*halakhot*] around him like a snake, and pronounced it (the oven) unclean.[76]

This provides the cue for the redactor of *Bava Metzia* to enlarge upon the story with the help of an astounding, additional *beraita*.

"Tana,"[77] the narrator begins, "On that (same) day R. Eliezer provided all (the) counterarguments in the world, but they did not accept them from him." By this, the reader is clearly meant to associate the day in question with the day on which the snake-like oven was disputed. But as the better informed reader is well aware, this particular dispute is explicitly listed in both versions of *Eduyot* as one of the many halakhic testimonies presented before and voted down by the Jabne assembly.[78] The day mentioned in the second *beraita* is therefore immediately associated with its namesake: i.e., the "very same day" of R. Elazar's election to office, the now famous *bo ba-yom*.

The second implied "linkup" between the two *beraitot* insinuated by the opening sentence has to do with the nature of R. Eliezer's alleged "counterarguments." The majority of commentators take these to be his attempts to disprove his colleagues' arguments in favor of their opinion regarding *Akhnai's* oven that were "looped around him."[79] But given the day in question and the important fact that we are dealing with an additional *beraita* that is cited apropos and not necessarily as a continuation of the story of the oven, and especially in view of the sweeping generality created by the inexplicit, all-inclusive phrase "all (the) counterarguments in the world," it would appear that more was at stake than the counterarguments marshaled by one sage in the course of a single halakhic disagreement. The temptation to reverse the implied subordination of the second story to the first and to regard the sages' reaction to Eliezer's ruling on the oven as but one instance of their attitude toward the totality of his halakhic positions is increased by the continuation of the story and its climactic conclusion, which speak legions of a deep and fundamental ideological controversy that far transcended the imaginable outcome of any particular low-level halakhic difference of opinions. Read in the mythological, literary context generated by Bavli, *Berakhot*'s account of R. Elazar b. Azaria's momentous first day in office and against the normative backdrop of Mishna *Eduyot*, the story in hand becomes that of R. Eliezer's entire testimony before the Jabne assembly, in the course of which he may well have referred, along with many other things, to the specific question of the snake-like oven.[80]

I prefer, therefore, to read the story as follows. First among the bearers of halakhic traditions who were called upon to testify on that day at Jabne was R. Eliezer b. Hyrqanus "the Great." A recipient and uncompromising guardian of ancient halakha, he is said to have offered "all the possible counterarguments in the world" after the legacies to which he had apparently testified, were, to his utter dismay, voted down by the Jabne assembly.[81] No assembly of sages has the authority to question, let alone dismiss, a reliably transmitted halakhic legacy, and the Jabne stories consistently depict R. Eliezer as a strong-minded traditionalist. From his alleged traditionalist perspective, the embarrassing though inevitable conclusion was that his colleagues' refusal to accept his teachings could only mean that, for some reason or other, they had come to doubt his trustworthiness as a credible testifier. And so, after his counterarguments had also been rejected, rather than marshal arguments to support of his position, R. Eliezer is said to have set about in earnest to provide supernatural "proofs" of his personal credibility.

> He said to them: If the halakha agrees with me, let this carob tree prove it. Whereupon the carob tree was torn a hundred cubits out of its place— and some say: four-hundred cubits. They said to him: From a carob tree no proof can be brought! Again he said to them: If the halakha is as I say, let the aquaduct prove it. Whereupon the water (in the duct) turned and flowed backwards. They said to him: From an aquaduct no proof can be brought! Again he said to them: If the halakha is as I say, let the walls of the academy prove it. Whereupon the walls of the *Bet midrash* inclined to fall.

At this point R. Yehoshua b. Ḥanania is said to have intervened, not to counsel or to counter R. Eliezer, but to scold the walls of the academy.

> He said to them: If *Talmidei ḥakhamim* are engaged in halakhic dispute, what have ye to interfere? Hence they ceased from collapsing in honor of R. Yehoshua, [but] refrained from resuming the upright in honor of R. Eliezer— and they remain thus inclined to this very day!

So far, the exchange between the perplexed Eliezer and his colleagues fares well with the suggestion that the present dispute was about more than determining the ritual 'utensilhood' of the oven of *Akhnai*. Indeed, it seems not to have even been about the authority of the Jabne assembly to rule between existing traditions, for by continuing to offer "proofs" of his credentials, Eliezer clearly appears to have heeded to their objections. The dispute appears to have been about what is and what is not generally acceptable as viable "proof" in halakhic debate. But if that is the case, it resembles less a *debate* about the grounds for halakhic decision than a sheer failure to communicate on the matter. While his anonymous interlocutors appear to have understood and, subsequently, to have dismissed whatever he was up to, Eliezer, it seems, failed to even grasp what was expected of him. Time and again he attempts to provide what he evidently took to be living proof of his position, and time and again he is told that his proofs are no proof at all. His disputants do not explain to him what in their opinion should count as viable proof in favor of the sort of claims he aspired to make, but only state forcefully that, as far as they were concerned, the type of proof he was offering was irrelevant. What he seems to be saying is: "Here are indisputable signs that I am a credible and trustworthy testifier, therefore you must accept my teachings!" What they seem to be replying is: "We do not doubt your credibility, nevertheless we do disagree with your teachings, therefore, if you wish to convince us, do so by offering us arguments, not miracles!"

Having failed to impress his colleagues, or, if you wish, to comprehend the nature of their objections, the frustrated traditionalist finally turned to what he evidently regarded as the ultimate and final proof of halakhic verity, calling upon the Heavens themselves to vouch for his position:

> He then said to them: "If the halakha agrees with me, let them prove it from the heavens!" (whereupon) a heavenly voice issued forth and declared: "What have you against R. Eliezer, for the halakha agrees with him everywhere!"

At this, we are told, R. Yehoshua rose to his feet and declared: "It is not in heaven (Deut., 30:12)"! "What does it mean 'It is not in heaven'?" queries the anonymous narrator, and answers: "Said R. Yermia: 'Since the

Torah was given [at] Mount Sinai, *we pay no heed to a heavenly voice* for you have already written in the Torah at Mount Sinai: 'to incline after a multitude' (Exod. 23:2)." And with a view perhaps to reducing the shock of R. Yehoshua's firm dismissal of the heavenly voice's intervention at Jabne, whereby implicitly placing the halakhic pronouncements of God in heaven on a par with those of the stone walls of academy, the Bavli goes on to give us the Almighty's side of the story as it were. R. Nathan, we are told, later "happening upon" Elijah, asked to know what God had done at that hour. "He smiled," replied Elijah, "and said: My children have vanquished me, My children have vanquished me."

Eliezer, as one might expect, remained stubbornly unconvinced. The fact that he had attempted so persistently to prove his trustworthiness in the first place clearly indicates that he had accepted his colleagues' authority to decide the authenticity of halakhic testimonies.[82] But their refusal to even consider his credentials, even when presented with such a heavenly stamp of approval, seems to have finally convinced him that the issue must have been quite different from what he had thought. R. Eliezer "the Great" appears at last to realize that Jabne had resolved to do the unthinkable—to criticize, vote, and reject the very content of his attestations. To this the archtraditionalist could not agree. Under R. Yehoshua b. Hanania and his young disciple R. Elazar b. Azaria, Jabne was willing to tolerate wide first-order halakhic disagreement. In this respect it was thoroughly pluralistic, and, as we shall see, is portrayed as actually encouraging diversity. But Jabne was unwilling to extend its pluralism to the second-order, metahalakhic, procedural level of halakhic decision making. It had granted itself, and by implication any future Great *Bet din*, the authority prudently to consider all matters of halakha, old and new, and to decide them as the majority of its members saw fit, unconstrained by tradition or even by heavenly intervention—their authenticity notwithstanding. And this point presumably marked the extent of their toleration. For the reformed Jabne the principle of tolerance applied widely, but only to those who accepted it. A steadfast traditionalist may be capable of tolerating the existence of rival traditional systems, especially those of unquestionable heritage, by virtue of their common traditionalist premise. Rival traditionalists may thus agree to disagree. But no traditionalist is capable of accepting the idea that his *teachings* be laid open to criticism and subjected to majority vote. And by the same token the antitraditionalist is incapable of accepting the idea that he be denied the right to do so. Traditionalists and antitraditionalists cannot, therefore, agree to disagree. They share no common premise. Each of them regards the other's position not only as thoroughly unjustified, but as thoroughly *unjustifiable*. The two approaches to Torah-study and to halakhic development are, in short, mutually incompatible. And R. Eliezer's refusal to comply left Jabne no choice.

> On that day they brought out all of the things that R. Eliezer had declared pure and they burned them in the fire, and they voted to "bless"[83] him. And they said: Who will go and tell him? R. Akiva said to them: I will go, for if someone who is not fit should tell him, he could destroy the entire world! What did Akiva do? He donned black and wrapped himself in black, and sat before R. Eliezer at a distance of four cubits. He said to him: Akiva, what is different about today? It seems to me that your colleagues are estranging themselves from you, answered Akiva. [R. Eliezer] then also rent his garments and removed his shoes and slipped (from his chair) and sat on the floor and his eyes poured forth tears.

Even if my interpretation of this astonishing story may seem to some forced or biased, most, I believe, would still have to agree that, under any interpretation, it is exceedingly difficult to bring it in line with the Tosefta's implied rendition of the Jabne reforms. Had Jabne been as inclined toward traditionalism as Tosefta, *Sanhedrin* and *Eduyot* jointly suggest, the point of contention between R. Eliezer and his colleagues would have had to have been either his trustworthiness, or the viability of other rival and equally reliable traditions, or the very authority of the Jabne *Bet din* to arbitrate on such matters. In the first two cases, even if his earlier "proofs" were found lacking for some reason,[84] the testimony of the heavenly voice, whose authenticity was contested by no one, should have been quite sufficient. For committed traditionalists seeking evidence of the halakha passed down from Sinai, such heavenly approval should have been ultimate proof![85] Nor is it possible, as I remarked above in passing, to endorse the third possibility, according to which the controversy boiled down to a second-order disagreement about a traditionalist Jabne's license to arbitrate between conflicting traditions, as the Tosefta seems to hint in its brief mention of the episode.[86] Had that been the case, R. Eliezer would not have attempted to convince them in the first place. The mere fact that he sought to *prove* to his colleagues that the halakha was according to him clearly indicates that he acknowledged their jurisdiction.

Read, as seems to be the Bavli's intention, as continuing and expounding upon the story of the Jabne reforms described in Bavli, *Berakhot*, the present *beraita* appears clearly, if tacitly, to endorse the firm antitraditionalist approach to halakhic development conveyed by the Mishna, and in doing so, to render the Jabne reforms themselves unambiguously antitraditionalist. But it does more than to implicitly align itself with one of the two conflicting tannaitic schools of thought. Writ large, it vigorously dramatizes and forcefully resolves the tannaitic disagreement in ways the tannaitic literature does not. Nowhere else in the tannaitic literature are the conflicting positions represented by the Tosefta and the Mishna played off against each other, and nowhere else is their apparent dispute authoritatively resolved. Indeed, nowhere else in the tannaitic literature are the two positions even attributed to particular sages or schools—at least not explicitly.[87] Seemingly well aware of its existence, the *beraita* in *Bava Metzia* appears to address the question head-on: first by personifying the two conflicting philosophies by ascribing them to the two most prominent representatives of second-generation Jabne, and then by portraying their debate and resolving it in the most dramatic terms conceivable.

Furthermore, by allowing the controversy to run its course, the *beraita* enables, or, rather, forces, the victorious antitraditionalist faction, led by R. Yehoshua, to *explain* its position. To this end the narrational, legendary format is especially effective.[88] In the legalistic context of Mishna, *Eduyot* the antitraditionalist is made to explain his citation policy with regard to dismissed minority opinions by pronouncing a normative ruling about halakhic decision making (as to which *Bet din* is authorized to overturn former rulings), the *beraita*, on the other hand, impels R. Yehoshua and his followers to further explain their halakhic decision making by laying bare their philosophy. And in doing so the narrator doesn't mince his words. The responsibility for developing the Oral Torah, declare R. Yehoshua and his followers, is charged exclusively to the mundane, this-worldly sensibilities of the human learner. Human understanding of the halakha is in principle devoid of external warrant, be it in the form of a reliable transmission or even in that of direct revelation—both of which are firmly rejected although the authenticity of neither of them is called into question. By acknowledging the presence of the heavenly voice they assert their ontological realism, but by knowingly ignoring what it has to say, they declare the truth of their epistemic undertakings, in principle, unconfirmable, and by implication, their efforts, in principle, indeterminable. The covenant of learning, of Torah-study, is thus understood by the Jabne antitraditionalists as founded upon a sharp demarcation of divine and human authority. Where the dividing line runs between Word and meaning, between Text and understanding, between *interpretandum* and *interpretans*. And within the hermeneutic space allocated to humans, not only is the force of tradition ruled uncompelling, but the Almighty himself is refrained from interfering.

Daniel Boyarin takes this interpretation an interesting and radical step further. For R. Yehoshua, he aptly observes,

"Oral Torah" means the Torah expounded orally in the interactive process
of dialectical reading of law. Meaning is not in heaven, not in a voice behind
the text, but in the house of midrash, in the voices in front of the text. The
[W]ritten Torah is the Torah which is written and the Oral Torah is the
Torah which is read.[89]

The only binding constraint on the human reader is seemingly the
scriptural text itself. But even that, according to Boyarin, is rendered in
the hands of R. Yehoshua flexible to the point of interpretive anarchy.

"It is not in heaven" is a citation, the use of which is radically different
from its meaning in its "original context."[90] R. Yehoshua is arguing with
God from God's own text. You gave up Your right as author and even as
divine voice to interpret Your Torah, when You said, "It is not in heaven."
(. . .) R. Yehoshua transforms the verse through his citation into meaning
that the Torah is beyond the reach, as it were, of its divine author. (. . .)
"It is not in heaven" means not only that the Torah is not beyond human
reach, but that it is beyond divine reach, as it were[!][91]

Although the story is clearly set in the context of halakhic deliberation,
Boyarin correctly stresses the exegetical, midrashic context in which R.
Yehoshua's hermeneutic philosophy is elaborated. However, with regard
to midrash, conventionalist and (antitraditionalist) realist approaches to
the Oral Torah no longer converge as readily as they do in the realm of
halakha.[92] While the latter's interpretive efforts are regulated by a desire
for truth, the former seeks to fulfill only such 'phenomena-saving' aes-
thetic desiderata as conciseness, elegance, and consistency. Boyarin takes
his insightful interpretation of the *beraita* to place R. Yehoshua and his
followers squarely among the conventionalists. By rejecting the heavenly
voice, he argues, R. Yehoshua and his colleagues bring themselves

squarely up against the dilemma of any hermeneutic theory that does not
allow appeal to [the] author's intention as a curb on interpretation. Once
that control is gone, it seems that any interpretation is the same as any
other, that anything at all can be said to be the meaning of the text.

However, "[s]uch hermeneutic anarchy," he continues, "is clearly *not* the
way that midrash presents itself."[93] (It is not quite clear whether *this* statement is meant to apply to the way this particular midrash presents itself—in which case it could be true—or to the entire midrashic *genre*—in
which case it is contestable.) The dilemma, evidently felt by the narrator,
is resolved by R. Yermia's additional midrash—itself a bold application of
R. Yehoshua's approach: "What does it mean 'It is not in heaven'? Said
R. Yermia: 'Since the Torah was given [at] Mount Sinai, we pay no heed
to a heavenly voice for you have already written in the Torah at Mount

Sinai: "to incline after a multitude" (Exod. 23:2).' "[94] This answer, submits
Boyarin,

> is surprisingly modern: the majority of the community which holds cul-
> tural hegemony controls interpretation. To put in another way: correctness
> of interpretation is a function of the ideology of the interpretive commu-
> nity.[95]

But by equating the *control* of interpretation with its *correctness*, its obligation with its verity, Boyarin ascribes to Jabne a pragmatist, 'might-is-
right' notion of truth. Interpretive anarchy is supposedly avoided by an
agreement-in-advance on behalf of all involved to follow the interpretive
sensibilities of the majority come what may (to the excommunication of
whoever is unwilling to comply—be it even R. Eliezer "the Great"). Even
if, when taken alone, it is possible to read the *beraita* as Boyarin suggests—the dismissal of the heavenly voice as a dismissal of the idea that
human interpretation is at all geared to approximate God's intended
meaning of His written text, and the decision to follow the rule of the
majority as a conventionalist assertion of a pragmatists theory of truth—
as a general statement about the rabbis' midrashic practices, his reading
is problematic. First, and I'm sure Boyarin will agree, the rule of the majority applies throughout the talmudic corpus *only to the narrow realm of
halakha!* No one is granted hegemony in the realm of midrash. There simply exists no body of "correct" rabbinic interpretations of Scripture—not
even with regard to *midrash halakha*. Majority vote applies only to halakhic rulings proper, not even to their exegetical groundings. It is commonplace for the Gemara to ask of a particular uncontested tannaitic ruling "from where do we know this?" and then to suggest (or cite) and
discuss the plausibility of various possible exegetical proofs. These deliberations are frequently polemical and often result in the refutation of
some.[96] More often than not, however, one is left with more than one
possible midrashic grounding, and in such cases the issue is never decided.[97] The polysemic world of talmudic midrash appears to be utterly
oblivious to the possibility that the type of multiple interpretation they
practiced and encouraged could be considered problematic.[98] Therefore,
even if R. Yehoshua's radical reinterpretation of "it is not in heaven" is
taken as a denial of human access to the Author's intention, R. Yermia's
equally radical reinterpretation of "to incline after a multitude" cannot
be viewed as some sort of a pragmatist remedy to the (Hobbsean?) problem of hermeneutic anarchy, which for the rabbis, I insist, seems clearly
not to have been a problem at all.[99]

Second, and more importantly, although R. Yehoshua and R. Yermia
perform bold midrashic speech acts in the course of presenting their phi-

losophy, I disagree with Boyarin that the issue at stake for them was at all interpretative. They employ midrash as a rhetorical resource and in doing so display the full force of rabbinic hermeneutic autonomy. But the problem raised, dramatized, confronted, and resolved by the story is purely metahalakhic. The question under consideration is not at all that of controlling or curbing interpretation, but that of the rules of *judicial* dispute and the sources of *halakhic*, rather than exegetic, authority. It is not even the question of halakhic correctness, but only that of the *procedural* correctness of halakhic decision making. The correct way to decide a halakhic (not a midrashic) issue, states the *beraita*, is by keen public debate followed eventually by a majority vote, the outcome of which is obligating, until a suitably qualified *Bet din* decides to overturn its ruling, regardless of its objective correctness. R. Yehoshua does not claim that the heavenly oracle's halakhic opinions are wrong or mistaken. It is rejected not because it is thought not to know what it is talking about, but because its intervention is perceived as a violation of the rules of the game. The heavenly voice and R. Eliezer, in other words, are *both* banished from Jabne for precisely the same reason!

To summarize, I understand the philosophical point of the story to be an assertive declaration on behalf of the new Jabne leadership that the halakhic development of the Oral Torah is a task ascribed exclusively to the human learner, whose efforts are considered devoid in principle of external warrant. In the context of Torah-study such validation would have to be some reliable form of revelation of God's intended meaning. R. Eliezer's traditionalism, predicated on the reliable transmission of an originally revealed Oral Torah, and the heavenly voice's attempt to reveal the truth here and now are both firmly dismissed as inadmissible evidence, as irrelevant to the deliberations of the *Bet midrash*. Lacking external warrant, the *truth* of the Oral Torah indeed remains undecidable in both its halakhic and exegetical modes, and, by implication, the enterprise of Torah-study is rendered necessarily pluralistic.

Boyarin, we have seen, equates interpretive pluralism with hermeneutic anarchy, and claiming, wrongly in my opinion, that the *beraita* regards this as a problem, reads R. Yermia's rendition of "to incline after a multitude" as its solution. But even Boyarin's premise is questionable.[100] Not all forms of pluralism are nihilistic or anarchic. The fact that a question may be viably answered in more ways than one does not necessarily mean that all its proposed answers are therefore *equally* viable. The Popperian theory of science presented in Part 1 of the present study, for example, describes science ideally as a rational and progressive enterprise, which, though intrinsically pluralistic, is by no means nihilistic. The idea that humans are fallible and therefore lack the means to ever verify their scientific hypotheses need not entail Paul Feyerabend's anarchic (or as he prefers, Dada-ist) claim that therefore "anything goes."[101] Popper's view of science "does not allow appeal to 'author's intention' as a curb on" its attempts to interpret nature, yet few would conclude, as Boyarin does with respect to scriptural exegesis, that therefore any scientific theory "is the same as any other, that anything at all can be said to be the meaning of the 'text'."[102] I agree with Boyarin that, in their firm dismissal of R. Eliezer's traditionalism and the heavenly voice, the *beraita* attributes to R. Yehoshua and his colleagues the view that in both its halakhic and interpretive undertakings Torah-study is fundamentally indeterminable. The *beraita*, however, does not explain the nature of the problem this might have posed to them, nor the nature of the solution conveyed by their insistence on always following the majority in halakhic matters. For a genuine anarchist, who truly maintains that any one position is as good as any other, a decision to follow the majority is a pure rule of thumb.[103] On such a showing consensus is as arbitrary and contingent as disagreement. For an anarchist, the process of debate and reasoning prior to taking a vote is, therefore, at most an exercise in rhetorical manipulation. There is simply nothing to be gained from seriously pitting one position against another if one sincerely maintains that all possible positions are equally viable.[104] Other forms of pluralism, however, may be far less radical. The Popperian, for instance, lays great stress on rational debate. Even if truth itself is indiscernible some positions, he maintains, are meaningfully and arguably better than others, and their relative pros and cons may and should be deliberated. For the Popperian, the voting itself is the culmination point of a keen and serious process of reasoned elimination of less suitable alternatives, for which, he believes, exist agreed and substantive criteria. But in order to determine the Bavli's position on these questions we need to look farther afield. This brings me first to the third and last of the Jabne stories I wish to discuss in detail in the present context.

Jabne's Antitraditionalist Manifesto

The story in question is related in two consecutive *beraitot* in Bavli, *Hagiga* 3a-b, the second of which was briefly mentioned above.[105] If the story of R. Eliezer's excommunication may be read as a definitive description of the Jabne reforms in terms of what is *inadmissible* in the house of learning, the first of the two stories related in *Hagiga* presents us with a positive and normative declaration of the way the new Jabne leadership believed that things should be. Unlike the other two stories we have discussed so far, here the Bavli narrator relies on firm tannaitic sources.[106] The setting of the story clearly recalls R. Elazar's dramatic election to office, and the way in which the two *beraitot* are conjoined by the Bavli's redactor, again, forcefully contrasts the very different approaches to halakhic authority personified in the Jabne stories by R. Yehoshua and R. Eliezer.

Upon being visited at his hometown Peqi'in (Beqi'in, according to the Yerushalmi) by two of his disciples, R. Yohanan b. Baroka and R. Elazar b. Hisma, who had come to pay their respects, R. Yehoshua is told to have inquired:

> What novel teaching[107] was there at the *Bet midrash* (in my absence) today? They said to him: We are your disciples and (therefore) drink [only] from your water. He said to them: Even so it is impossible for a study session to pass without some novel teaching. Whose Sabbath was it? It was (one of) R. Elazar b. Azaria's Sabbath(s) [they replied].

To recall, according to Bavli, *Berakhot* 28a, after humbly begging R. Yehoshua's forgiveness Rabban Gamliel is said to have been reinstated and thereafter to have shared the Presidency with R. Elazar b. Azaria on a rotational basis.[108] As David Stern notes, this allusion to the tri-weekly rotation (Rabban Gamliel was to preside, according to the Bavli, for three Sabbaths for each of R. Elazar's one) clearly places the story in the immediate context of R. Elazar's reforms.[109]

R. Yehoshua then asked to know the theme and details of Elazar's sermons, and his disciples proceeded to rehearse for their Master the three homilies they allegedly heard at Jabne that day.[110] The third of Elazar's sermons is of utmost importance to the nature of the Jabne reforms and, in the light of our discussion so far, may be regarded as an antitraditionalist manifesto. Significantly, Elazar chose as his text the one verse in *Qohelet* (Ecclesiastes) that, in asserting that "the words of the wise (. . .) are given by one shepherd," seemingly applies the notion of human wisdom deliberated throughout the book also to Torah-study: "The words of the wise are like spurs, and like nails well driven in are the sayings of the Masters of collections; they are given by one shepherd"[111] (12:11).[112] As is common practice in the rabbis' midrashic writings, the homily is constructed as a phrase-by-phrase commentary on the verse, firmly setting *Qohelet's* observations in the context of Torah-study and interpreting the metaphors of the spur and the well-planted nail and the allusion to the one shepherd in marked antitraditionalist idiom.

> Why are the words of the Torah likened to spurs? To teach you that just as the spur directs the heifer along its furrow to bring forth life into the

world, so the words of the Torah direct those who study them from the paths of death to the paths of life. But [should you think that] just as the spur can move [and can be removed], so the words of the Torah can move [and be removed]—therefore the text says: nails [which once pounded in cannot be removed]. But [should you think that] just as the nail only diminishes [as it is driven into the wood] and does not increase, so too the words of the Torah only diminish and do not increase—therefore the text says "well-planted." Just as a plant grows and increases, so the words of the Torah grow and increase.

As David Stern notes, the way in which R. Elazar's homily "atomizes" the verse is one of the most common and effective exegetical techniques of midrash.[113] The idea is normally not merely to deconstruct the verse but to interpret it as a structured and ordered whole by ascribing to each of its claims, metaphors, and images both a constructive and an eliminative role. The spur, the nail, and the plant mentioned in the first part of the verse are each shown by R. Elazar to contribute in turn a positive analogy to Torah-study which serves at the same time also to eliminate an unwanted connotation implied by the previous image. By treating the string of metaphors as a structured disjunctive succession, R. Elazar's reading thus accounts meaningfully not only for each of the phrases in the verse but also for their particular ordering. The spur or goad are responsible for directing the heifer along the proper path. In this respect they are likened to the words of the wise, which in turn are taken prima facie to be those of the student of Torah. But spurs and goads are also removable devices. This unwanted analogy to the fruits of Torah-study is eliminated by the additional metaphor of the irremovable well-pounded nail. But how can the words of Torah be likened to nails that appear to decrease and diminish in order to remain permanently fixed? Hence the additional allusion to plants that grow and increase. The words of the wise, the fruits of their learning, are, according to R. Elazar, at once partly spur-like, partly nail-like, and partly plant-like, but wholly resemble none of them.

In light of our discussion so far, Elazar's insistence on the growth and increase of halakhic knowledge suggests an antitraditionalist bias especially in the context of the Jabne stories. The extreme brand of traditionalism represented by R. Eliezer b. Hyrqanus in these stories necessarily views the study of Torah as a diminishing and degenerate enterprise in which the words of the sages acquire their permanence and fixedness by virtue of an inevitably incomplete process of transmission. "The words of the wise," according to this type of traditionalist, represent a constantly dwindling, if irremovable body of knowledge that, when viewed over long periods of time, is indeed analogous to a repeatedly pounded nail.

For R. Elazar, the young reformer of Jabne, by contrast, the proliferation and growth of the words of the Torah is essential. "The words of the wise" not only direct those who study them to "the paths of life," but they themselves, he insists, evolve and develop like well-tended living saplings.

However, though opposed to the extreme traditionalism of the likes of R. Eliezer, the homily so far still lacks the revisionist aspect of the truly antitraditionalist view of Torah-study. The compounded imagery of the nail and the plant lends itself far too readily to an *accumulative* view of the Oral Torah that easily accommodates a "softer" brand of traditionalism such as that of Tosefta *Sanhedrin* and *Eduyot*. "The words of the wise," one might be led to think by the first part of the homily, contribute to an ever-growing body of *permanently fixed* items of knowledge—each of which, once proposed, remains irremovable. But to judge by the way it continues, this appears not at all to have been R. Elazar's intention. By "irremovable" or "fixed," he seems clearly to have meant "temporarily binding" or "irremovable except by 'the Masters of assemblies'." "[What does the phrase] 'the Masters of assemblies' [mean?]" he goes on to say,

> These are the disciples of the wise (*talmidei ḥakhamim*), who sit in assemblies and study the Torah—some pronouncing (a thing) unclean and others pronouncing (it) clean, some prohibiting and others permitting, some declaring unfit and others declaring fit. Should a person say: Since some pronounce unclean and others pronounce clean, some prohibit and others permit, some declare unfit and others fit—how in these circumstances am I to learn Torah? Therefore Scripture says: All of [these conflicting halakhic opinions] "were given from one shepherd." One God gave them, one leader proclaimed them from the mouth of the Lord of all creation, blessed be He, as it is written: "And God spoke *all* these words" (Exod. 20:1). Therefore make your ear like the hopper and acquire a perceptive heart to understand the words of those who pronounce unclean and the words of those who pronounce clean, the words of those who prohibit and the words of those who permit, the words of those who declare unfit and the words of those who declare fit.

The possibility that *Qohelet* may be taken to assert that the increase and development of the words of the (Oral) Torah is a simple matter of accumulating approved items of permanent knowledge is hence flatly rejected in view of the fact that "the words of the wise" most often contradict one another. At any one time, argues R. Elazar, Torah learners are faced with a baffling profusion of conflicting opinions. The halakhic component of the Oral Torah is described not as a gradually diminishing body of reliably received and transmitted truths, but as a vibrant, and ever-changing aggregate of disparate and contrastive halakhic opinions and rulings issued by different people for different reasons at different times

and in different circumstances.[114] "How then am I to learn Torah?," asks
the bewildered student, "Of what, then, can *the* Torah be said to consist?
What of all this am I supposed to study as a true student of Torah?"
R. Elazar's answer to these questions is one and the same: to view the
perplexing variety of opinion as the state of affairs intended by God. The
student is urged to regard the confluence of viewpoints he encounters as
"given from" or sanctioned by "(the) one shepherd"—as if God granted
"the Masters of assemblies" the license to pursue halakhic questions as
each of them sees fit.

The studious reception and comprehension of the various halakhic
viewpoints is, therefore, merely the first phase of Torah-study. What is
ultimately required of students of Torah is eventually to make their ears
"like the hopper" and to acquire and apply a "perceptive heart," or, as
Rashi explains, to eventually make their own particular decisions in accord with their understanding of their own particular circumstances.[115]
In other words, once issued, the rulings of "the Masters of assemblies"
indeed attain a permanence likened to well-pounded nails: they will be
binding and obligatory and will retain their mandatory status, but only
until they are called into question—in which case they are apt to be replaced.

R. Elazar's homily is not offered as a detached description of halakhic
development, but as an injunction. It is formulated as a manifesto, as an
urgent plea to his present and future disciples beseeching them not to be
passive recipients of past halakhic legacies, but to study them carefully
with a view to putting them to the test of their own time and place. To
study Torah at the most advanced level, to be a true *talmid ḥakham*, is,
according to R. Elazar, first to acknowledge and welcome diversity of
opinion; to perceive it not as an unfortunate symptom of epistemic frailty,
but as the standard of perfection; to realize that a variety of conflicting
positions is a necessary prerequisite for the words of the wise to indeed
multiply and increase and should therefore be encouraged rather than
shunned. Second, at its most advanced level Torah-study involves more
than the mere openness to and encouragement of diversity in halakhic
opinion. At its highest level it encourages active participation in the game
of halakhic innovation and rethinking. It is important, therefore, to distinguish clearly between the obligation fully and wholeheartedly to abide
by the existing halakhic norm and the equally compelling obligation to
constantly and critically rethink it. While conformity and consensus are
of central importance to the *practice* of halakha, they are serious impedi-
ments to the fruitful *study and development* of halakhic knowledge. Within
the walls of the academy one should not seek to harmonize conflicting
views or to play down their differences. What the homily seems to imply

is that "the words of the wise" are capable of directing students of Torah away from "the paths of death" to "the paths of life" *precisely because* they constitute a diverse spectrum of conflicting alternatives. Indeed the metaphor of the growing and multiplying plant is symbolic also of life itself. In order for the words of the Oral Torah to apply appropriately to the perpetual flux of humanly experienced reality, they too must be allowed to grow, multiply, and diverge in similar fashion. And since no one is better situated to assess and evaluate the unforeseeable contingencies of the here and the now than the Masters of today's assemblies, each and every generation of Torah students is charged with the grave responsibility of critically surveying anew the teachings and rulings of their predecessors and of deciding what to uphold, what to reverse, what to modify, and what to add of their own. Such is R. Elazar b. Azaria's understanding of the Covenant of Learning.

The homily, I suggest, should be read as more than merely a midrashic interpretation of *Qohelet* 12:11. The *beraita* is first and foremost a Jabne story and should therefore be read, as it was evidently written— homilies included—as part and parcel of the mythical historiosophy established by the wider context of at least those of the other Jabne stories to which it directly or indirectly refers. As we have seen, the immediate and explicit point of reference is the story of R. Elazar's election to office. Read in conjunction with the *beraita* cited in Bavli, *Berakhot*, to which it so clearly alludes, the homily then reads as Elazar's explanation to the assembly (apropos *Qohelet* 12:11) of the reforms he had previously introduced. It constitutes, one might say, the newly elected President's declaration of intent as to why and how tractate *Eduyot* originated on "that very day." Assuming judicial leadership, the Jabne *Bet din* undertook to scrutinize all testimonies of past and present halakhic opinion and to put each of them to the vote. And the atemporal and normative tone of Elazar's sermon appears clearly to indicate that, in his view, the Jabne reforms should not be perceived as a radical one-time remedy, but should serve as a model for all future halakhic authorities, who are expected, in similar fashion, to "originate" their own tractates of "Testimonies." Read thus, the homily quite clearly, if implicitly, sides itself with the Mishna's distinctly antitraditionalist version of the tractate rather than with that of the Tosefta. Finally, the normative import of Elazar's sermon is further confirmed by R. Yehoshua, who is said to have been so taken by the words of his former disciple that he exclaimed: "A generation in which (the spirit of) R. Elazar b. Azaria dwelleth is no orphan"!

In the light of R. Yehoshua's unconditional approval of R. Elazar's declaration, one is naturally puzzled as to why his two students were initially so reluctant to inform him of what had transpired at Jabne in his absence.

The reason for their hesitation, explains the anonymous narrator, was a similar story, namely, that of R. Yossi b. Durmaskit's calamitous pilgrimage to Lydda to pay his respects to R. Eliezer b. Hyrqanus, during which he nearly lost his eyesight—a story briefly discussed above.[116] Here again, as in the *beraita* cited in Bavli, *Bava Metzia*, R. Yehoshua and R. Eliezer's conflicting philosophies of Torah-study are forcefully contrasted. In this case, however, contrary to that of *Akhnai*'s oven, Jabne's ruling regarding the status of Amon and Moab during the *sh'mita* year happened to coincide with R. Eliezer's received tradition. Consequently, no halakhic dispute was called for, and there arose no need for Eliezer to "prove" his position on the matter. His furious disapproval of the happenings at Jabne was confined, therefore, to the second-order question of halakhic authority. In R. Eliezer's view, to quote Tosefta, *Sanhedrin*, a Great *Bet din* is obliged to deliberate an issue and put it to the vote if, and only if, they had not "heard." A *Bet din*'s judicial discretion is exclusively limited, according to the traditionalist, to matters for which no reliable halakhic tradition is known to exist—which, in Eliezer's opinion, was definitely not the case with regard to Amon and Moab, for which an authentic tradition, received from Rabban Yohanan b. Zakai and dating back to Moshe at Sinai, was at hand. Jabne, in his view, had, therefore, seriously overstepped its bounds and had violated the traditionalist's most cherished norm. For R. Elazar and his mentor R. Yehoshua, by contrast, by reopening the issue and submitting the received view to the test of latter-day opinion, Jabne had done precisely what a Great *Bet din* is supposed to do. To paraphrase Mishna, *Sanhedrin*, this time, according to the antitraditionalist, it is from the Great *Bet din*, rather than from the body of bona fide received legacies, that "the Torah issue[s] forth to all Israel." Even when it eventually decides to uphold the received view, as in the case of Amon and Moab, the decision is theirs and theirs alone.

Here once more, the two mutually exclusive philosophies of halakhic development are forcefully contrasted by the Bavli narrator. In this case, however, he evidently felt no need to intervene on his own accord and to decide the issue for his readers. The two standpoints, again personified by R. Yehoshua b. Hanania and R. Eliezer b. Hyrqanus, are confined to separate *beraitot* without the two sages having to meet, as it were, face to face. The second of the two *beraitot* does confront the archtraditionalist R. Eliezer with the outcome of an evidently antitraditionalist Jabne procedure, but the confrontation is played out with a hopelessly outranked disciple unable to present the traditionalist case. Such a better-balanced confrontation is related apropos the story of the snake-like oven debate recorded in *Bava Metzia* 59b, when the traditionalist/antitraditionalist controversy is reported to have been squarely confronted by the Jabne sages themselves and dramatically resolved by the parties involved with no need for editorial intervention.

To summarize, following a brief theoretical analysis of conventionalist, traditionalist, and antitraditionalist ideals of and attitudes toward Torah-study, two intertextually related corpora of tannaitic texts were examined in some detail. The first set of texts, comprising the Mishna and Tosefta versions of tractate *Eduyot* and of a portion of *Sanhedrin*, were shown to be consistently divided along the lines of the traditionalist/antitradition-alist battle lines. They also give the clear impression of either having been worked up differently from a shared, earlier text, or for the later of the two to have consciously reacted to and modified the former. In their significantly different portrayals of the hierarchical structure of the Jew-ish court system in *Sanhedrin* and their different treatment of halakhic testimonies and controversies in *Eduyot*, the Mishna was shown to adopt a marked antitraditionalist line as opposed to the Tosefta's equally marked traditionalism. In these particular texts the two positions are clearly, if implicitly, voiced but nowhere explicitly acknowledge one another and are nowhere compared or contrasted by a third party. They jointly bear witness to there being two such schools of thought, certainly among those responsible for their finished form, but contain no evidence of the two schools ever disputing the issue. They represent different "takes" on the same material. They speak clearly, in other words, but not to each other.

The second set of texts consists of three of the many Jabne stories re-lated throughout the Bavli, but could have easily been extended to in-clude several others.[117] Here no obvious disagreement is apparent at the editorial level.[118] Those responsible for the composition of the Bavli's ver-sions of these particular stories appear to have been single-mindedly an-titraditionalist. The three clearly interrelated narratives examined above seem to tell their stories for the definite purpose of establishing the su-premacy of antitraditionalism. The first—that of R. Elazar b. Azaria's election to office—sets the stage and provides the context for the latter two. The second—the remarkable account of R. Eliezer's great dispute at Jabne and his eventual excommunication—establishes antitraditionalism by default as it were. This is achieved by forcefully dismissing the drama-tized foundations of traditionalism—represented in the story by R. Eliezer's proven testimonies and the revealed halakhic wisdom of the heavenly voice. Finally, in the third story—that of R. Elazar's Jabne ser-mon—the antitraditionalist approach is expounded upon positively in the form of a normative vision grounded in Scripture of an ideal community of learners, who, to borrow the terminology developed in Part 1 of the present study, undertake to treat all received halakhic knowledge with utmost seriousness as potentially problematic goal-directed halakhic systems to be keenly troubleshot for potential deficiencies and ruled upon anew whenever the case arises.

The aim of this section has been to establish the self-felt presence of the two contrastive philosophies of Torah-study at the tannaitic level of the talmudic corpus and to argue that perhaps the main literary-philosophical aim of the Jabne stories, as they are presented by the Bavli, was to dramatize, detail, and, most importantly, forcefully to resolve the issue, clearly in favor of the antitraditionalist approach.

A Traditionalist Response: Hillel and b'nei Beteira

I have suggested in passing that there exists yet a third corpus of primarily tannaitic texts that arguably also bear on the issue at hand—namely, that of the thirty-three recorded halakhic debates between the Houses of Hillel and Shammai.[119] It is beyond the scope of the present study to develop this point any further, however, and I hope to devote a separate study to the examination of these texts in the detail they undoubtedly deserve.[120] Nonetheless, I wish to conclude this chapter by looking briefly at the story, again of tannaitic origin, that comes closest to presenting explicitly the metahalakhic tension between Hillel and his alleged opponents. I am referring to the significantly different accounts, offered by the Tosefta, Bavli, and Yerushalmi, of the young Hillel's eventful encounter with the priestly family of b'nei Beteira—a story that to some extent summarizes our discussion up to this point and will serve, in addition, to pave the way for some of the issues raised in the chapters that follow.

Three versions of the encounter will be examined.[121] Unlike the Jabne stories, the earliest version in hand, that of the Tosefta, is quite clearly the work of an avid traditionalist. Here, as in the traditionalist passages of the Tosefta examined above, traditionalism not only reigns supreme, but goes unchallenged. All involved, including Hillel himself, appear to fully comply with traditionalist standards. It tells a story of halakhic perplexity and its resolution but by no means, as do the other two versions, does it tell one of metahalakhic controversy. As in Eduyot, the Tosefta's version of the story is written as if traditionalism were the only available option. It is not a polemical text and seems clearly not to have been written in order to score points in a wider debate. The version related by the Yerushalmi, by contrast, contrives to achieve precisely that. It is strongly biased in favor of a more extreme version of traditionalism than that of the Tosefta, but, quite unlike the Tosefta, rather than avoid the antitraditionalist alternative, the Yerushalmi attempts to confront it head-on. The Yerushalmi's account of the young Hillel's encounter with the *Beteira* family is as close as one gets to a traditionalist's rejoinder to the antitraditionalist accounts of Akavia b. Mehalal'el's and Eliezer b. Hyrqanus's equally eventful brushes with their adversaries. In a sense it is even more decisive than its antitraditionalist parallels. Hillel, who, according to the Yerushalmi, starts out as a sharp and outspoken antitraditionalist, does not stick to his horses as would Akavia and Eliezer in their day. He is not banished but is, on the contrary, proven wrong, apparently repents, joins the traditionalist opposition, and eventually achieves prominence (as Akavia was promised should he comply) among their ranks. On the other hand, as we shall see, the antitraditionalist position attributed to the young Hillel by the Yerushalmi narrator is significantly weaker than that of R. Eliezer's opponents at Jabne. Thus, for example, at no point does Hillel suggest that a reliable tradition be called into question—let alone be voted down. What he does initially maintain is only that one may rely on one's reason in such cases in which the line of transmission was disrupted. In this sense the conflict between the two parties is less dramatic here than in its antitraditionalist parallels.

The third and chronologically last variation of the story we shall examine—that of the Bavli—is the least unambiguous of the three but certainly the most interesting. It is interesting because, without commenting at all on the significance of the story, the framer of the *sugya* allows for the presentation of two quite different readings of the original text: one, a not entirely successful attempt to tell the story as one of antitraditionalist triumph (boldly modifying the traditionalist original to the point of distortion), followed by a squarely traditionalist rerendering of the encounter along the lines of the Yerushalmi. The Bavli gives its readers no indication that the two interpretations attest to manifestly different metahalakhic sensibilities. In fact they are presented as if they were supplementary. I shall argue in upcoming chapters that this seeming lack of interest in, and conflation of, the traditionalist/antitraditionalist controversy is typical of Bavli discourse. But first to the story.

Mishna, *Pesaḥim* 6:1, states that, among other things, the slaughtering of the Pascal Lamb "puts off," or supersedes, the Sabbath (on which the act of slaughtering an animal is normally prohibited). "This particular halakha," we are told, "eluded *b'nei Beteira*." The Bavli's version of the story begins as follows:

On one occasion the fourteenth (of *Nissan*) fell on the Sabbath, (but) they forgot and could not say whether (the slaughtering of the) *Pessaḥ* overrides the Sabbath or not. Is there anyone who knows, they asked, whether (the)

Pessah overrides the Sabbath or not? There is a certain man, they were told, who came up from Babylonia, by the name of Hillel the Babylonian, who attended upon (studied under) the generation's two great sages Shema'aya and Avtalyon, and who knows whether *Pessah* overrides the Sabbath or not. (Thereupon) they sent for him. They said to him: Do you indeed know whether *Pessah* overrides the Sabbath or not? He said to them: Have we but one *Pessah* during the year which overrides the Sabbath? Surely we have many more than two-hundred *Pessahim* during the year that override the Sabbath![122] How do you know this, they asked.[123] He said to them, it says "*mo'ado*" ("its appointed season") with regard to the *Pessah* (Num. 9:3) and it says "*mo'ado*" with regard to the continual burnt offerings (Num. 28:2), and just as (the) "*mo'ado*" mentioned with regard to the continual offerings that override the Sabbath, so does (the) "*mo'ado*" mentioned in connection with *Pessah* indicate that it too overrides the Sabbath.[124] Moreover, it may be reasoned a fortiori (by inference from the minor to the major): If (failures to observe) the continual burnt offerings which are not (divinely) punishable by excision, still override the Sabbath, is it not logical that (the) *Pessah*, which *is* (divinely) punishable by excision, should override the Sabbath![125] They immediately sat him at their head and appointed him their President, and he sat and lectured (*darash*) to them on the laws of *Pessah* throughout the day.

So far the story appears to be narrated with the sole intention of strongly opposing the extreme brand of traditionalism practiced by *b'nei Beteira* and elsewhere associated with R. Eliezer b. Hyrqanus. The fact that the chain of transmission had been discontinued, the young Hillel seems to be arguing, does not necessarily mean that extinct halakhic knowledge is irretrievable. One may, and indeed should, apply one's reason and divine forgotten *halakhot* by rational argument without having to rely on direct mouth-to-ear transmission. Not only is Hillel manifestly portrayed as an adversary to traditionalism himself, but as having sufficiently impressed the traditionalist priestly family by his ability to reason around such ruptures for them to appoint him their Head. The story so far appears to have been penned by someone largely sympathetic to the antitraditionalist cause.[126]

However, at this point the Bavli's account takes an unexpected turn away from its initial antitraditionalist vantage point. Upon assuming the Presidency, we are told,

[Hillel] began chiding them, saying: What had caused you to have me come up from Babylonia and become your President? It was your indolence, because of your failure to attend upon Shema'aya and Avtalyon! (Ignoring his rebuke) [t]hey said to him, Master, what (is the halakha) if a person forgot to bring his slaughtering knife in advance of the Sabbath?[127] He said to them, I (recall having) heard that particular halakha, but have forgotten it. However, (he went on to say,) leave it to the people, for if they are not prophets they are the sons of prophets. On the morrow, those whose *Pessaḥ* was a lamb, entangled [their knives] in its wool, while those whose *Pessaḥ* was a kid, placed them between its horns (and thus avoided having to carry the knives themselves). And observing their behavior, he recalled the halakha, and declared: such had I received from the mouths of Shema'aya and Avtalyon!

The rather clear opposition to *b'nei Beteira*'s extreme brand of traditionalism conveyed by the first part of the story is substantially obscured by the second part, in which an embarrassed and humbled Hillel is forced to fall back on the lowest form of transmitted knowledge without making the slightest effort to infer an answer logically. There are only two ways of harmonizing the two parts of the story while retaining the antitraditionalist import of the first. One, by interpreting the second part of the story primarily as a criticism of Hillel's arrogance rather than of his antitraditionalism;[128] the other, by interpreting it as being specifically about such a halakhic question for which no convincing line of reasoning came to mind, in which case even the antitraditionalist will be obliged to revert as best he can to custom or *sh'mua*. But Hillel's alleged response to the question concerning the knives renders both readings unsatisfactory. By admitting to have forgotten what he had earlier "heard," without making the least attempt to marshal an effective argument, Hillel himself seems to suggest, contrary to the first part of the story, that the Patriarchy had been offered to him more for his firsthand received knowledge of the law than for his ability to deduce it logically.

The Bavli's initial version of the story thus remains vexingly inconclusive with regard to the traditionalist/antitraditionalist dichotomy. I do not believe, however, that its ambiguity necessarily indicates editorial confusion or unawareness. On the contrary, it seems to me that the Bavli narrator made a bold effort to cast the story in as antitraditionalist terms as the text he was working with permitted. As mentioned at the outset, this particular story was not composed by the Bavli, but only adapted from earlier sources. And the two earlier versions in hand—those recorded in the Tosefta and Yerushalmi—appear, as we shall see, to have been penned by dedicated traditionalists. Moreover, whoever edited the Bavli's initial version of the *beraita* seems to have been in conflict with a second, anonymous amoraic commentator, whose attempt to resolve the ambiguity exhibits noticeable traditionalist convictions.[129]

The Bavli's latter-day discussion of the story begins, innocently enough, by clarifying Hillel's precise application of *gezera shava* in his first argument (based, to recall, on the recurrence of the word *"be-mo'ado"* in the two biblical passages concerning the continual offerings and the Pascal Lamb respectively). The commentator then goes on to question the validity of Hillel's second argument, the *kal va-homer*:

> Says the sage: (it says in the *beraita*:) "Moreover, it may be reasoned a for-tiori: If (failures to observe) the continual burnt offerings which are not (divinely) punishable by excision, still override the Sabbath, is it not logical that (the) *Pessah*, which *is* (divinely) punishable by excision, should over-ride the Sabbath." (But) this can be refuted: for (unlike the *Pessah* offering) the continual burn offerings are (more) frequent (than the Sabbath[130]) and they are also burnt completely (which are two reasons for regarding them as *more*, rather than less severe than the *Pessah* offering).

Rather than suggest a possible counterargument that Hillel might have considered, as amoraic critics frequently do when they question tannait-ic positions, the anonymous commentator, relying, as it seems, on the Yerushalmi's version of the story, submits that what really happened was that:

> He (indeed) began by stating the *kal va-homer* but [*b'nei Beteira*] refuted it, therefore, (as a rejoinder) he presented the *gezera shava*. But if he was able to argue by means of a *gezera shava*, why (did he) propose the *kal va-homer* in the first place?[131] Rather, he was arguing *from their point of view*: (he said to them,) assuming that you did not have a *gezera shava* at your disposal, because one is not permitted to devise a *gezera shava* on one's own accord, you should have (nonetheless) been able to construct a *kal va-homer*—an argument one *is* permitted to propose on one's own accord. To this they answered, the *kal va-homer* (you propose) is refuted.

By inserting this exchange into the dialogue detailed in the Bavli's initial version of the *beraita* and reversing the order of Hillel's two arguments, the story is considerably altered. Hillel's interlocutors now turn out to have been far less impressed by his halakhic reasoning than the Bavli's initial version would imply. Offering a useless *gezera shava* and a defective *kal va-homer*, he could be of little help to them. And yet he was offered the Presidency. One is inclined to conclude, therefore, although the second Bavli commentator does not actually say so, that Hillel was then asked if he had heard Shema'aya and Avtalyon's verdict on the matter directly from them. And only after assuring them that that was indeed the case was he offered the job—as indeed the Yerushalmi and Tosefta state ex-plicitly! On such a reading Hillel emerges, by the end of the day, as even less the antitraditionalist than we were previously led to believe. But, more importantly, the very *moral* of the story as a whole is now signifi-cantly transformed. Contrary to the apparent antitraditionalist intentions of the first Bavli narrator of the *beraita*, the second commentator's modified rendition of the story takes it to convey the idea that human reason is faulty and its fruits are refutable, and that therefore, whenever possible one should do one's utmost to unearth a reliable long-standing tradition. Rather than rely on their capricious reasoning capacities, *b'nei Beteira*, we now learn (along with a bemused Hillel), were quite right to seek trust-worthy evidence of a steadfast tradition. The latter-day redactor of the *sugya* seems to have presented us with an editorial dispute, comprising an initial attempt to reverse the import of an apparently traditionalist legend, countered by a second rendition of the story designed to restore its origi-nal message. There is more to be said about such a traditionalist rendition of the story, concerning, for instance, the status and sources of *gezera shava*,[132] but I prefer first to look at the other two versions, in which, as noted, the traditionalist point of view is even more pronounced.

The version related by the Yerushalmi opens, as does that of the Bavli, with Hillel being summoned and asked by *b'nei Beteira* if he had "ever heard" whether or not the *Pessaḥ* offering overrides the Sabbath.

> He said to them, is it the case that we have only one *Pessaḥ* each year that overrides the Sabbath? No, there are several [such offerings] that override the Sabbath every year! (. . .) They said to him, we have already decided that you may be useful to us. He then commenced to prove (the point) to them by analogy, by *kal va-ḥomer*, and by *gezera shava*. By analogy: Because the continual burnt offerings and the *Pessaḥ* are both communal sacrifices, therefore, since the communal continual offerings override the Sabbath, it stands to reason that the *Pessaḥ*, which is also communal, should similarly override the Sabbath. By *kal va-ḥomer*: Although the continual burnt offer-ings are not punishable by excision, they still override that Sabbath, all the more reason then that the *Pessaḥ*, which *is* punishable by excision, should override the Sabbath. (And) by *gezera shava*: It says "*be-mo'ado*" with regard to the continual burnt offerings and it says "*be-mo'ado*" with regard to the *Pessaḥ*, and just like the continual offerings, of which it is said "*be-mo'ado*," override the Sabbath, so should the *Pessaḥ*, of which it is (also) said "*be-mo'ado*," also override the Sabbath.

To recall, according to the Bavli's initial version (prior to the amoraic com-mentator's intervention), at this point Hillel was immediately appointed President. This was by no means the way things developed according to the Yerushalmi. Here, the dialogue continues.

> They said to him: Indeed we wondered whether a Babylonian might be of any help. (First) the analogy you propose is questionable. For is it not the case that while the extent of the continual offerings is restricted that of the *Pessaḥ* is not?[133] (Second) the *kal va-ḥomer* you propose is (also) question-able. For is it not the case that the continual burnt offerings are of a higher

degree of holiness than the *Pessah*.[134] (Finally) the *gezera shava* you propose (is immaterial) because a person is not supposed to devise a *gezera shava* on his own accord.

While the Bavli commentator leaves it at that, the Yerushalmi leaves nothing to the imagination of the reader, stating explicitly that,

although he sat and lectured on to them all through the day, they would not accept the halakha from him until he solemnly swore that he had actually heard it from Shema'aya and Avtalyon. Having heard this from him, they then rose and pronounced him President.

Except for minor, insignificant changes the remainder of the story coincides with the Bavli's version.

The Yerushalmi narrator clearly sides with *b'nei Beteira*. He tells a story that is fully aware of there being two fundamentally different approaches to Torah-study and that contrives deliberately to confront them. To this end, Hillel is portrayed as an ardent opponent of simple-minded traditionalism, who, upon being asked whether he had *heard* a certain halakha, does his best to impress upon his interlocutors the idea that one need not have heard it from the mouths of one's Masters as long as one is capable of determining it logically. But his arguments are found wanting and are easily dismissed. And the defeated Hillel is forced to play by the strict traditionalist rules: to testify under oath that he had heard the halakha directly from his Masters in order to become a President whose authority derives solely from his being a reliable recipient of former teachings. The reader is not even allowed to entertain the thought that after assuming office, Hillel somehow managed to convert his employers to his way of thinking. Hence the second part of the story. When the question of carrying the slaughtering knife arises, he makes no attempt to tackle the problem by means of logical inference but solves it by seeking residual evidence of a mimetic tradition among the common people. If any one was converted, the story implies, it was Hillel, who now goes about halakhic investigations as a full-blooded traditionalist.

If the narrator of this version may be said to have at least acknowledged the existence of a serious, if eventually defeated, opposition to *b'nei Beteira*'s extreme form of traditionalism, the version of the story related by the Tosefta tells of no conflict and subsequently of no real debate between the parties. Unlike the other two versions, here the concerned party is not even named. Hillel is undramatically questioned, never challenged, and for its larger part, the story consists of a lengthy monologue delivered, as we shall see, by an unassuming traditionalist, Hillel.

Once the fourteenth (of *Nissan*) fell on a Sabbath. (And) Hillel the Elder was asked whether the *Pessah* (offering) overrode the Sabbath. He said to them, do we have but one *Pessah* that overrides the Sabbath? We have more than three-hundred (such) *Pessahim* during the year that override the Sabbath. All (those present in) the *Azara* converged upon him.[135] He said to them, the *Pessah* is a communal offering, and so are the continual burnt offerings, and like the communal continual offerings override the Sabbath so does the (similarly) communal *Pessah*. Another way of arguing is: it says, concerning *Pessah*, "*be-moado*," and it says, concerning the continual offerings, "*be-moado*," and like the continual offering, of which it is said "*be-moado*," overrides the Sabbath, so the *Pessah*, of which it is (also) said "*be-moado*," does override the Sabbath. Yet another way of arguing is (by) *kal va-homer*: Since the continual offerings, though not punishable by excision, override the Sabbath, all the more reason that the *Pessah*, which is punishable by excision, also overrides the Sabbath. *And in addition I have received from my Masters that the Pessah overrides the Sabbath*—and not merely the first *Pessah*, but even the second, and not only the communal *Pessah* offering, but even private *Pessah* offerings! They said to him, if so, what is [the halakha] with regard to those of the people who neglected to bring their knives and their *Pessah* offerings (before the commencement of the Sabbath)? He said to them, leave the matter to them for the Holy Spirit rests upon them—if they are not prophets, then they are the sons of prophets. What did they do the following day? Those whose *Pessah* was a lamb, entangled it in its wool, and (those whose *Pessah* was) a kid, stuck it between its horns, and they thus brought them to the Temple compound, and slaughtered their *Pessah* offerings. On that very day they pronounced Hillel their President, and he continued to instruct them in the laws of *Pessah*.

The main difference between this and the two versions formerly discussed is that from the start Hillel's logical inferences seem *not* to have been set forth as stand-ins for direct reception. The setting of the story is decidedly not one of a traditionalist crisis. Hillel is not approached by a distressed leadership who should have known but had forgotten. He is by no means described as seeking to make a metahalakhic point, and is, therefore, only challenged (if at all) to explain his position. The logical arguments he provides are put forward in addition to rather than with a view to replacing straightforward mouth-to-ear transmission, of which he reports without having to be prompted to do so; their function apparently is merely to reinforce or to uphold a reliably received verdict. There is, therefore, no attempt made to call them into question, precisely because the reliability of his verdict was not meant to rest on the viability of his reasoning, but only on its being formerly "heard" directly from the mouths of his Masters. In other words, the Hillel of the Tosefta is not even mildly antitraditionalist. He is not converted to traditionalism in the course of a confrontation. According to this version, there was no confrontation, and indeed no confrontation was called for—for the simple reason that Hillel is portrayed as a committed traditionalist from the very start.

This particular Tosefta thus fares well with the Toseftist accounts of *Sanhedrin* and *Eduyot* discussed earlier in this chapter, both of which, in like manner, make no mention of there ever being an antitraditionalist faction. The halakhic crisis described in Tosefta *Sanhedrin* is said to have developed because *both* Houses, in failing properly to "attend upon their Masters," were held jointly responsible for causing the line of transmission to cleave and diverge. And in *Eduyot* both Shammai *and* Hillel are portrayed as traditionalists worthy of praise for not "insisting on their opinion in the face of a received tradition." In all three Tosefta passages, then, the past and present are described not merely in ways that *favor* traditionalism, but in ways that fail to even acknowledge the *presence*, let alone the viability, of an antitraditionalist voice. In this case, the editor of the Yerushalmi, although he appears equally biased in favor of traditionalism, makes ample room for the opposing view. In doing so, however, Hillel paradoxically emerges as an even more pronounced antitraditionalist than he does in the Bavli's less antagonistic initial version.

Still, to repeat yet again, by this I do not mean to say anything about the Tosefta's *general* attitude toward the traditionalist/antitraditionalist question, or, for that matter, about that of any other major unit of rabbinic writing. The evidence I have offered is not meant to be *generalized* but only to testify to the clear and audible presence of traditionalist and antitraditionalist voices sparring at the tannaitic level of the talmudic literatures. On no account should I be taken to be inferring inductively that the Tosefta or Mishna or any other large unit of talmudic writing was consistently redacted as a whole in accord with one or the other metahalakhic point of view. In fact, the Tosefta and Yerushalmi, in which the most traditionalist-biased versions of the story at hand are to be found, each contain what seem to be firm antitraditionalist rejoinders to this very story.

I am referring to two almost identical Jabne legends in which R. Akiva's renowned powers of theoretical reasoning are dramatically pitted against R. Tarfon's firsthand testimonies regarding certain Temple practices. It seems as if the two stories originated in the *Sifra* and *Sifri*, respectively, but that in itself is of little consequence. The important point is that they are told, in all the versions we have, with diametrically opposed conclusions to that of the young Hillel.[136] Although they are located in quite separate compilations, they intriguingly preserve identical structures and employ identical turns of phrase that are quite unique to the small number of legends about halakhic debates between the two sages.[137]

The first of the two occurs in the *Sifra* to Lev. 1:5. "And the priests, the sons of Aaron," states the verse, "shall bring the blood, and sprinkle the blood around upon the altar that is by the door of the Tent of Meeting." This, explains the anonymous *tanna* of the *Sifra*, refers to receiving the blood of the sacrifice by the priest and not to its sprinkling. At this, R. Akiva attempts to determine exegetically the level of priestly attendance required.

> Said R. Akiva: How do we know that receiving must be performed by none but a legitimate (i.e., unblemished) priest [robed in priestly vestments]? It speaks here of priesthood and it speaks there of priesthood.[138] Just as there it refers to legitimate priest[s] [robed] in priestly vestments, so here too it means by a legitimate priest [robed] in priestly vestments. R. Tarfon said to him: Akiva, How long will you rake words together and bring them up against us![139] May I lose my sons if I have not heard a distinction made between receiving and sprinkling, although I cannot explain! Said R. Akiva: Allow me to tell you what you taught us. Tell me, he answered. He said to him: (There are differences between receiving and sprinkling:) In the case of receiving intention was not made tantamount to action, whereas in the case of sprinkling intention was made tantamount to action. [Again] if one received [the blood] without [its proper precincts], he is not liable (to *karet*), whereas if one sprinkles [it] without, he is liable (to *karet*). (And) if unfit men received it, they are not liable on its account, if unfit men sprinkled it, they are liable on its account. Said R. Tarfon to him, By the [Temple] service! You have [not] deviated to the right or the left! I heard [it] yet could not explain it, whereas you investigate it and agree with [my] tradition. In these words he addressed him: Akiva! whoever departs from thee is as though he departed from life!

The story speaks for itself. Although R. Tarfon cannot quite remember what he had heard, he assumes he remembers enough to scold Akiva's idle speculations in the harshest of terms. R. Tarfon's initial complaint and even more than that his enthusiastic endorsement of Akiva's position at the end clearly imply that had *he* been asked the question Akiva has addressed, his answer, based on what he had heard, would have been different. But he was not asked the question, and their initial first-order halakhic differences on the matter remain tacit. Thus, although the story speaks loudly of the great value of Akiva's midrashic abilities for filling in lacunae in his colleague's memorized traditions, the two sources of halakha are not directly confronted. This is made to occur, and quite dramatically, in the second of the two legends.

The question this time concerns the ritual status of the priests assigned to blow the trumpets "over" the various sacrifices on feast days and New Moons (Num. 10:10). Num. 10:8 states: "And the sons of Aharon, the priests, shall blow with the trumpets," and the *Sifri* to this verse reports the following exchange:

> (By) "the priests" (it means) both undeformed and deformed—(these are) the words R. Tarfon. Said R. Akiva: It says here "priests" and it says there "priests"; Just as there it refers to unblemished and excludes deformed priests, so here too it means by an unblemished priest and not deformed one. R. Tarfon said to him: Till when will Akiva keep raking (words together) and bringing them up against us! I cannot stand it! May I lose my sons if I have not seen my maternal uncle Simeon, standing and blowing the priestly trumpets although he lacked a leg! Said Akiva to him: Yes, Master, (of course you saw him) but perhaps it was at the hak'hel, or on the Day of Atonement, or (at the announcement of) the Jubilee year that you witnessed (trumpets blown by) legitimate priests who are deformed.[140] Said R. Tarfon to him, By the [Temple] service! You have not made this up! Rejoice O Avraham our patriarch for that Akiva emerged from your loins! Tarfon witnessed and forgot, (while) Akiva reasons of his own accord and (yet) agrees with the halakha. Akiva! whoever departs from thee is as though he departed from life!

If the central purpose of the legend of Hillel and *b'nei Beteira*, as related by the Yerushalmi and the second version of the Bavli, is to dramatically convey the essential frailty and unreliability of human reason in comparison to firsthand knowledge of the halakhic tradition, these exchanges between R. Tarfon and R. Akiva, particularly the second one, seem clearly to be designed to forcefully convey the very opposite. It is Akiva's ability to reason about what the halakha *should be* that, despite the imperious and impatient Tarfon's confidential pooh-poohing, is proven far superior to firsthand memory of what the halakha *was*. Contrary to the story of Hillel, the process of Torah-study, of determining the halakha, is associated primarily with the application of one's reasoning capacities rather than one's memory. And it is not confined to cases in which memory fails. On the contrary, the whole point of the second story is to show just how misleading the memory of even the most conscientious witness can be. The lesson to be learned is that recollections of halakhic practice should be regularly troubleshot. And the way to do so is to measure them up to the conclusions of pure, theoretical reasoning. It is here, in the implied epistemological premise and the similarly implied methodological conclusion that the strong antitraditionalist bias of these stories asserts itself. Like that of the young Hillel, they are not stories of halakhic revision. They are stories of epistemic and methodological preference.

As noted above, the vast majority of post-talmudic writers go to great lengths to preserve an even more stringent form of traditionalism than those we have so far encountered in the talmudic texts themselves. In many of these works, even "the (thirteen) rules by which the Torah is studied" are attributed divine origin.[141] According to this view, which is not without its opponents,[142] even in areas in which no direct *sh'mua* is known to apply and the Great *Bet din* is required to rule on its own accord, rabbinic deliberation remains limited to a divinely dictated discursive framework. As we have seen, the little leeway granted to the rabbis according to most of these writers was limited even further with regard to the *mida* of *gezera shava*. The idea that the human halakhist was at liberty to infer halakhic information from one matter to another merely on the basis of some verbal congruity between the two relevant biblical passages was perceived by many as a serious threat to the system. Relying on the dictum that recurs in both the Bavli and Yerushalmi versions of the story, according to which "one is not permitted to devise a *gezera shava* on one's own accord," they conclude that all *gezerot shavot* recorded in the talmudic literature are one of two kinds: they are either themselves received or at most devised merely to buttress a formerly received teaching.[143] This is certainly the position stated by the traditionalist *b'nei Beteira*. But what can be said of Hillel's own opinion on the matter? Regardless of where the redactors' sympathies lie, in both the Bavli and Yerushalmi Hillel is portrayed as an antitraditionalist. How then did he allegedly stand with regard to *gezera shava*?

Curiously, the Bavli is unclear on this point although its overall treatment of the story is less obviously biased in favor of traditionalism than that of the Yerushalmi. To recall, according to the unnamed Bavli commentator, Hillel is reputed to have been arguing *from the point of view of his adversaries*: "(he said to them,) assuming that you did not have a *gezera shava* at your disposal, because one is not permitted to devise a *gezera shava* on one's own accord, you should have (nonetheless) been able to construct a *kal va-ḥomer*" (to which they answered, the *kal va-ḥomer* [you propose] is refuted). According to this reconstruction, it is not clear whether Hillel himself believed that one should not devise a *gezera shava* on one's own accord, or whether he took this principle itself to be part of his opponents' point of view. Despite the Yerushalmi narrator's far more pronounced traditionalist account of the story, *his* Hillel appears quite clearly *not* to have subscribed to such a view of *gezera shava*. According to the Yerushalmi, he was not arguing at all from the point of view of *b'nei Beteira*, but doing his best to prove his halakhic point by means of three arguments. As the Yerushalmi has it, all three arguments were shot down by his interlocutors—the *gezera shava*, on the grounds that one may not devise such an argument on one's own accord. The point is that if Hillel

himself had believed that one is allowed to draw only on reliably transmitted *gezerot shavot*, then the particular *gezera shava* he had offered them
could not have been of his own! In other words, the fact that he was
directly accused of violating this particular traditionalist norm—and, according to the Yerushalmi, justly so—clearly proves that in the narrator's
opinion Hillel could not have subscribed to it in the first place.Whether
or not *gezera shava* is consistently employed by the rabbis as a real rule of
inference or merely as a catchword for an appropriately transmitted item
of revealed knowledge is, of course, a question that cannot be resolved
solely on the basis of this or that version of the story of Hillel and *b'nei
Beteira*. Even if one is as convinced as I am that the particular type of
antitraditionalist, personified by Hillel in the Yerushalmi's account of the
story, does treat *gezera shava* as a rule of inference given to human discretion, this in itself has little bearing on the question of the rabbis' general attitude(s?) to *gezera shava* thereafter—especially since Hillel's position is flatly rejected in the version that gives it its most clear expression.
Surely this question can only be properly answered by examining all of
the very many *gezerot shavot* resorted to and debated throughout the talmudic literature. Several such studies exist, many in the form of commentaries to the *beraita de-Rabbi Yishma'el* appended to the *Sifra* to Leviticus, in which all thirteen *midot* are detailed and exemplified.[144]
Interestingly, the vast majority of these works were forced to disagree
with Rashi (and with *b'nei Beteira* for that matter) that the rabbis would
admit only such *gezerot shavot* that were either mere rhetorical devices or
demonstrably received at Sinai lock, stock, and barrel. If that had been
the case, they argue, one would not expect to find, as one frequently does,
instances of rabbinic disagreement with regard to the particular halakhic
content of *gezerot shavot* (e.g., Bavli, <u>Hulin</u> 85a and *Sanhedrin* 83b) or the
particular biblical phrases on which to ground them (e.g., Bavli, *Ketubot*
38b). Neither would one expect the Talmud to express praise for the successful construction of a *gezera shava* as it sometimes does (e.g., Bavli,
Shevuot 7a). And perhaps most importantly, one would not have expected
gezera shava to be at all counted among the *midot* by which the Torah is
interpreted. This is because, on such a showing, *gezera shava* is not a *rule*
of interpretation at all. In view of these, and other problems,[145] a middle
way was proposed, according to which the dictum stated by *b'nei Beteira*
should be understood as requiring that only a *part* of a *gezera shava* be
received. As a result, two basic types of *gezera shava* are defined: those in
which the congruent terms are given while the specific halakha to be
inferred on their basis is left open to rabbinic debate and those in which
the halakha itself is given, while the appropriate biblical phrases on
which to ground it is left open to question. But, and this is the important

point, all agree that in all talmudic employments of *gezera shava* at least one component of the argument is always given in advance and is as such indisputable. In other words, although such a view of *gezera shava* fares better with some of the talmudic texts than that of Rashi,[146] it remains strictly traditionalist.

To recall, except for the most extreme, almost mythical, form traditionalism associated in the Jabne stories with R. Eliezer b. Hyrqanus, traditionalists do not oppose the idea that where no reliable tradition is known to apply the rabbis are granted discretion and are required to work things out for themselves. The main point of contention between traditionalists and antitraditionalists concerns the disputability in principle of reliably transmitted knowledge. While the latter perceive their long-standing traditions as the legitimate objects of serious critical rethinking, the former regard reliably received knowledge as absolutely binding and immune to criticism. From this it follows that any general depiction of *gezera shava* that insists on always preserving a received, indisputable element is, by definition, *wholly* traditionalist in this respect. The point to stress, however, is that, to the best of my knowledge, in none of these post-talmudic studies is the very possibility of a nontraditionalist approach to *gezera shava* on behalf of the rabbis ever raised, let alone considered a prima facie viable option. One might say that virtually all the post-talmudic literature on *gezera shava*, and with it the vast majority of all post-talmudic rabbinic writing, follow the three Tosefta texts examined above in failing to even acknowledge the presence of an outspoken antitraditionalist point of view. However, among the tannaitic texts, as this chapter has argued, the approach adopted by the Tosefta is by no means representative. In all other instances in which questions concerning halakhic authority and halakhic development were found to arise, not only are both positions entertained, but in three important sets of texts— Mishna *Sanhedrin* and *Eduyot*, the disputes between the Houses, and the Jabne stories—the issue, I have argued, is clearly resolved in favor of the antitraditionalists.

This short digression suggests that at some point in history traditionalism appears to have achieved a supremacy among students of Torah as the "official" philosophy of *talmud-Torah* to the extent of all but silencing its opponents. The rich and multifaceted post-talmudic discussion related to these questions lies beyond the modest scope of the present study, and I can pursue it here no further.[147] However, it does raise a major question pertinent to the concerns of this book, namely, that of the amoraic texts' attitude toward the traditionalist/antitraditionalist dichotomy. The Bavli's treatment of the story of Hillel and *b'nei Beteira* disappointingly borders on the ambiguous, certainly on the noncommittal. Still, the question is an interesting and pressing one if only for the fact that the most conspicuous assertions of tannaitic antitraditionalism are found, on the one hand, in the Mishna, which is taken by both *talmudim* as their ultimate point of departure and, on the other hand, in the Jabne stories, all of which are adapted, interrelated, and integrated in amoraic writings. More importantly, the amoraic texts comprise the first body of rabbinic writings we have that originated and were intentionally compiled in direct response to an extensive body of authoritative texts received from preceding generations. They offer, therefore, ample opportunity to study the various ways in which whole corpora of canonized halakhic texts were received and treated by subsequent generations. Such is our aim in the chapters that follow.

2

The Changing of the Guard

Amoraic Texts and Tannaitic Legacies

We now turn to the amoraic level of talmudic halakhic discourse. This is found in its most developed form in the Bavli, the monumental composition of sixth-century Jewish Babylonia, a work in which all agree talmudic Judaism reaches its definitive and most complete statement. Like the tannaitic material examined in the previous section, the thirty-seven tractates of the Babylonian Talmud also contain many declarative and aggadic passages of amoraic origin that in similar fashion explore, dramatize, and meditate upon issues pertaining to the nature of Torah-study and halakhic development. Such second-order reflections, as we have seen, can be valuable indicators of their author's self-image and self-understanding, and we shall take the opportunity of studying several of them in the pages that follow. But the Bavli has much more to offer that is relevant to the concerns of the present study. Like its amoraic predecessor, the Yerushalmi, but unlike any antecedent tannaitic document, the Bavli is fashioned as an ordered series of detailed, blow-by-blow responses to an earlier, authoritative rabbinic text, the Mishna.[1] Each paragraph of the Mishna dealt with by the Bavli[2] is discussed in great detail, keenly confronted with a plethora of other tannaitic sources, and deliberated in the light of a broad array of social, political, religious, and scientific data thought relevant to the issues it raises. The amoraic portion of the talmudic text—the Gemara—is both occasioned and structured by the Mishna and related tannaitic material.

The only tannaitic text that comes remotely close to explicitly responding to former documented *rabbinic* opinions is tractate *Eduyot*, the book of halakhic attestations and their reception by the Jabne Sages. But *Eduyot*, in both of its mishnaic and toseftist versions, differs greatly from the amoraic genre of the Bavli and Yerushalmi. *Eduyot* focuses almost entirely on final rulings and exhibits little concern for the process of deliberation by which they might have been reached. Written with apparently no didactic aim in mind, *Eduyot* is little more than a summary report of the conclusions allegedly reached by the Jabne assembly at the end of "that very day." It documents the initial positions and the way the majority ended up voting on each of them, but does not, as a rule, report or speculate on the reasons the majority might have had for doing so or on their debates and deliberations that preceded their final verdict. The Bavli also frequently states its conclusions, but even when it does it is anything but a summary report. Its framers present us with a daunting range of "live" rabbinic debates—actual and imaginary—fashioned as a series of meticulously reconstructed study sessions of received tannaitic material, spanning the best of eight generations of named post-tannaitic authorities.[3] The Bavli appears to have been intentionally framed with a view primarily to laying bare and elaborating upon the *process*, far more than the *product*, of Torah-study, and to do so in great detail. The Gemara[4] ponders, reconstructs, and renders explicit the complex lines of reasoning that supposedly led its predecessors to their conclusions and continues vigorously from the point the prior documents leave off. These are the two most conspicuous and novel features of the amoraic talmudic text: its detailed engagement with all prior documents and the detailed image it conveys of Torah-study as an ongoing *activity*, as a dynamic process of debate and deliberation, and decision only when required.

Discerning the Bavli's Point of View: A Schematic Overview

The language of the Bavli is descriptive. The narrative forms adopted in its discursive portions resemble those of an elaborate, at times laborious, protocol, normally accompanied by a clear, dispassionate running commentary. Even if its framers had intended to produce no more than an unreflective chronicle of amoraic learning, the Bavli would still be considered a rich source for the study of early rabbinic epistemology. Philosophers of science are fond of remarking that all reports of facts are theory laden; that even the most innocent and seemingly unbiased factual description inevitably premises a series of general, theoretical presuppositions about the types of things described. And the same is certainly true of historical narratives. There is no such thing as an innocent unbiased chronicle. However impartial and unassuming a story may appear to be, it will always reflect the particular, if at times tacit, point of view of its narrator. The Bavli, however, hardly poses as an innocent, unassuming chronicle, nor can it be assumed to speak in one voice. It impresses one as being first and foremost a didactic work, and, therefore, an extraordinarily normative one, designed to convey a carefully crafted vision of *talmud-Torah*. The framers of the Bavli offer their reader punctilious descriptions of exemplary *batei midrash* in and out of session. They assume the roles of anonymous narrator and moderator, patiently indicating who and what was asked, and who and what was answered, who cited whom, whose opinion was accepted and whose rejected, and all with a compact and extraordinarily uniform vocabulary of logical connectives and inventory of polemic moves. The reader is implicitly urged to study and ultimately to imitate the way they describe themselves and their forerunners keenly negotiating with their libraries and with each other. If we seek evidence of a talmudic theory or theories of Torah-study *in operation*, it is here, in the Bavli's narratives of reconstructed deliberation and dispute, that we should first look. We shall concentrate, therefore, less on what the amoraic Sages had to say *about* their undertakings and direct our efforts, at least for the time being, to an attempt to retrieve the metahalakhic voices arguably embedded in the first-order discursive practices and strategies documented by the Bavli so thoroughly.

Given the way the Bavli presents itself—as a detailed study of so vast a body of received tannaitic writings—and given the fierce dispute concerning the epistemic and halakhic status of such received legacies contained, as we have seen, within those prior writings, our natural point of departure shall be first to ask how the Bavli stands with regard to the traditionalist/antitraditionalist dichotomy. The fact, for example, that virtually all of the Jabne stories appear in the Bavli—and undoubtedly owe to their latter-day amoraic redactors not only their placement but also much of their finished form—strongly suggests that those who were responsible for the framing and inclusion of those passages were well aware of the dispute. And judging by their unambiguous outcomes, one is tempted to conclude that, at least in theory, they were not merely aware of the dispute, but were significantly biased in favor of the antitraditionalist position they so valorize.

But this is precisely *not* the type of evidence we should be looking for at this juncture, and for at least two reasons. First, although all the Jabne stories appear to have been redacted favorably to the antitraditionalist cause, the Bavli's version and subsequent discussion of the story of Hillel and *b'nei Beteira* (Bavli, *Pesaḥim* 66a), I have argued, attests quite clearly to serious disagreement at the editorial level with regard to this very question. If the Bavli can at all be said philosophically to speak in anything like one voice, it will be that of those responsible for its remarkably uniform language of talmudic negotiation rather than that of those responsible for this or that of its aggadic sections. But, as we shall see shortly, to determine even that is far less easy than it sounds.[5]

The second reason for seeking the Bavli's own voices beyond such redactory feats as its versions of the Jabne stories is that if we wish to expose the Bavli's self-understanding of its own project,[6] we are required to read it by its own terms; to reconstruct, rather than deconstruct, the historiosophy, mythology, and world of discourse it aspires to establish. Within the world of discourse so carefully constructed by the framers of the Bavli—for instance, by using Hebrew only when citing tannaitic material in that language and reserving Aramaic almost exclusively for the entire amoraic level—the Jabne stories (all in Hebrew) are clearly presented as part of the Bavli's tannaitic heritage rather than as part of its own latter-day reflective statement. This is why, despite their inclusion and evident redaction by the framers of the Bavli, I have classified the Jabne stories as a tannaitic corpus throughout this study. To ask how the framing of the Bavli might have been influenced by the traditionalist/antitraditionalist dispute is to ask, first and foremost, whether and to what extent its first-order narratives of amoraic response to *what is described therein* as received tannaitic opinion can be said to arguably premise one of the two antithetical positions.

Clearly, not all forms of first-order halakhic deliberation contained in the Bavli are equally relevant to this particular dichotomy. Thus, for example, all manner of amoraic debate and dispute concerning new questions to which no prior document is known to apply have little bearing on the question at hand. This type of amoraic give and take, primarily geared to locating and filling lacunae within the received tradition, has little to do with the Bavli's attitude(s) toward the tradition itself. As we have seen in the previous section, all but the most extreme traditionalists readily agree that in the event of a new query, unattended to by former generations, it is up to the latter-day authorities to debate and decide the issue as they best see fit.[7] Advocates of the two antithetical positions would be indistinguishable either as participants in or as narrators of such amoraic debates.

But even if we limit ourselves to the realm of direct amoraic response to existing tannaitic opinion, not all forms of talmudic discourse are relevant to the traditionalist/antitraditionalist dispute. Alongside the great effort invested by the Bavli to expose and attend to halakhic hiatuses in the tannaitic materials it discusses, their writings aim at superseding these materials in a different and important way. As Neusner argues in great length and detail,[8] the Bavli aims not merely to preserve and fill the gaps in a body of inherited knowledge but to (re)build it into a new and comprehensive system of its own design. Even in the eyes of the most avid traditionalist, the Bavli (and Yerushalmi for that matter), is not a traditional document in the simple sense of the word. Even he will be forced to admit that it is far from being merely "the sedimentary increment of centuries of Mishna study," to quote Neusner, but is "set forth [as] a system that is essentially autonomous of the past," a system that forms "a judgment upon the received heritage of learning in the Torah, inclusive of the Mishna".[9] Metaphorically, he contends,

> [w]e may compare the framers of the Talmud (. . .) to a weaver of a tapestry. (. . .) The weaver uses yarn that she has not made, yarn that is received from somewhere else. But the weaver uses the yarn to execute a vision of her own. The threads of the tapestry serve the artist's vision; the artist does weave so that the threads show up one by one. The weavers of a tractate of the Bavli (. . .) make ample use of available yarn. But they weave their own tapestry of thought. And it is their vision, not the character of the threads in hand, that dictates the proportions and message of the tapestry. (. . .) [W]here reason governs, it reigns supreme and alone, revising the received materials and, through its own powerful and rigorous logic, restating into a compelling statement the entirety of the prior heritage of information and thought.[10]

Neusner's weaving metaphor is not only powerful and useful, but also partly true. It is useful in the way it contrives intentionally to lure our attention away from the cacophonous "haphazard, episodic, sedimentary process of agglutination and conglomeration" by which, as he colorfully puts it, much of the materials comprising the Talmud reached the hands of its framers, to the harmonious, systematic, "well-considered and orderly composition" they then produced.[11] In this respect it also serves well the purposes of the present study. But it is also misleading in unwittingly blurring the distinction we are, at this point, most interested in making. Neusner's weaver uses different types of yarn to produce a tapestry of her own design. She employs materials by which she constructs new systems that aspire to achieve practical and artistic goals that are quite different from those that had formerly motivated the work of the yarn spinners. In real-life situations yarn manufacturers and weavers normally negotiate directly with each other. Yarn is normally spun with the weavers in mind. Weavers frequently have their yarn made to order and engage in critical dialogue with their suppliers. But that is not the way Neusner intends his metaphor to work. Neusner's talmudic weaver inherits yarn which was not custom made to his own specifications. Not being a yarn spinner himself, he has to make do with whatever he finds. In other words, the weaving metaphor, as employed by Neusner, unwittingly premises a point of view that manages completely to miss the distinction between traditionalism and antitraditionalism. To cast this distinction in terms of Neusner's metaphor: traditionalist and antitraditionalist alike inherit ready-made tapestries woven by former generations rather than mere yarn. The former is obliged to preserve the received tapestry in its entirety, but may add to it new portions of his own design or at most recombine whole sections of the old into a newly conceived patchwork or quilt. The latter, by contrast, considers himself free to *unravel* such parts of the inherited work he finds lacking and to use the yarn to weave new patterns of his own. In the language of goal-directed systems developed in Part 1 we could say that a traditionalist considers the systems he receives as immune to internal troubleshooting and hence capable of improvement only by adding on new components—never by modifying existing ones. The antitraditionalist, on the other hand, is comparably unconstrained in both his critical and constructive functions. He is at liberty to question and test the system with respect to its assigned objectives and in comparison with alternative systems and is free to modify the inherited system or, if necessary, to replace it completely. The image of the talmudic Sages merely inheriting the raw materials of a system rather than a ready-made one inevitably obliterates the very distinction between the two schools of thought. Neusner's metaphor suggests that despite the great originality and autonomy displayed in the scheme and structure of the tapestry they wove, the framers of the Talmud did not engage critically with the teachings they inherited. The merely inherited yarn, according to Neusner, and although they may have woven ingeniously to produce a tapestry which was entirely their own, the yarn itself, his metaphor implies, was not troubleshot by its users and, consequently, not modified in the course of its employment. If, as Neusner puts it, the Bavli represents "a judgment upon the received heritage of learning in the Torah," his weaver metaphor suggests that it may have been a creative judgment, but hardly a critical one.

I do not mean to accuse Neusner of endorsing traditionalism, of course. He may well be sympathetic to the idea of seeking out antitraditionalist elements and tendencies in the Bavli's enterprise and discourse. All I am claiming is that the metaphor he proposed is formative of his theory of the Talmud as presented in *Close Encounters* to the extent of rendering him incapable of even considering such an option. It determines his choice of texts and examples and focuses his discussion of the Talmud on only two central areas of amoraic undertaking: determining the precise meaning of each passage of the Mishna they discuss (closely surveying the yarn at their disposal) and building them into a sustained and comprehensive halakhic system of their own design (the weaving thereafter). To this end the metaphor is indeed illuminating. But at the same time it inevitably blinkers our view of the texts to the exclusion of the very possibility of amoraic criticism of prior documents. "If we want to understand the Bavli," writes Neusner, "we have to ask how the document relates to earlier ones." "This question," he adds, "is an important one. For the answer will dictate how we classify the Judaism that the Bavli defines: traditional or systematic."[12] These are the two antithetical options that for Neusner define the field for understanding the Bavli. For the sake of the classification he aspires to make, the weaver's metaphor is effective, but it renders irrelevant the distinction between traditionalism and antitraditionalism. An enterprising traditionalist can as easily build new innovative, halakhic systems out of uncritically received tannaitic teachings as an uninspired antitraditionalist can eventually end up with little more in hand than the teachings he had formerly received and with which he found nothing wrong. Still, regardless of the outcomes of their efforts, the two schools of thought pertain to two diametrically opposed theories of Torah-study that, in turn, premise and entail significant theological, epistemological, and methodological differences, all crucial to understanding the Bavli. To paraphrase Neusner, the question of how to relate the Talmud to earlier documents is an important one for our present purposes precisely because the answer will dictate how we classify the Judaism that the Bavli defines: traditionalist or antitraditionalist, critical or uncritical of one's inherited learning.

To wit, Neusner's metaphor and theory apply unimpaired to large portions of the Bavli, as his detailed readings of Bavli, *Bava Metzia* 2a–5b, and *Suka* 2a–9b clearly prove. In *sugyot* such as these, one does arguably encounter novel amoraic system building which is prima facie uncritical of the tannaitic building blocks it employs. Of course an antitraditionalist will interpret such employments of unmodified tannaitic materials as the outcomes of failed trouble shooting, for he has no objection to uphold a provenly unproblematic inherited text, as long as he is convinced that it is provenly unproblematic. Traditionalists uphold such texts because they are *inherited*. But their final verdicts, in these instances, will be indistinguishable. In what follows I shall try, therefore, to avoid such *sugyot* that set out merely to extrapolate uncritically from given tannaitic positions and to focus for the main part on cases in which the very content of tannaitic materials are allegedly confronted.

Although traditionalists and antitraditionalists are certain to conduct such confrontations very differently, we shall see that they are frequently narrated by the framers of the Bavli in ways that confusingly lend themselves equally well to either kind of reading. Much of the Bavli's formal surface rhetoric is cast, for reasons I shall discuss in the following section, in traditionalist terms. Even so, it is possible, I believe, to dispel the am-

biguity to a certain degree by looking beyond rhetoric to the inner logic
of such confrontations. But before illustrating this point with an example,
let us first recall the main point of contention between the two schools.

For the traditionalist, to recall, a halakhic ruling, reliably received from
previous generations, is absolutely binding. As Tosefta, *Eduyot* states forcefully, the only reason for preserving rejected minority opinions on the
record is to ensure that they remain rejected forever. They are left on record, according to the Tosefta, as living testimony to the fact that they were
indeed formerly considered and appropriately voted down. For once a
vote is taken on such questions for which no reliably received answer
was known at the time to exist, the matter is to be regarded as settled
once and for all and can never again be reversed. It follows, therefore,
that on a traditionalist showing, an amora will never question an uncontested tannaitic tradition. He may, of course, doubt its authenticity or ponder its meaning, and, subsequently, once satisfied, may even try to work
it into a broader system, but, once it is understood, he will never ever
call it into question. Given a tannaitic statement *A*, an amora inclined
towards traditionalism may ask "What is meant by *A*?", "To which cases
was *A* meant to apply?", "To which additional cases can it be said to be
relevant?", "To which broader issues may it be said to relate?"; he may
ponder the consistency of *A* with other tannaitic statements, but, unlike
the antitraditionalist, it is unthinkable that he will ever ask "Is *A* a true
or viable edict?" An amora so inclined will frequently contest a colleague's understanding of a reliable tannaitic source but never the source
itself.[13] A less stringent version of traditionalism we have encountered enables later generations to escape halakhic deadlocks by allowing them to
adjudicate on their own accord between two equally reliable but conflicting traditions. This happens frequently in the amoraic literature in cases
of conflict between single tannaitic opinions. But even such decisions can
still be interpreted along narrower traditionalists lines, namely, as being
subject to second-order rules which are themselves received on tannaitic
authority.[14]

The kind of *sugyot* around which one would expect traditionalist and
antitraditionalist redactors to differ most conspicuously are those devoted
to negotiating apparent inconsistencies between tannaitic and amoraic
opinions. These are cases, commonplace throughout the Bavli and Yerushalmi, in which a well-formed, articulated amoraic position is cited
and then challenged by reports of seemingly contrary tannaitic opinion.
For a traditionalist, the latter-day amoraic view will always be the one
rendered problematic when faced with such a challenge. On such a showing, amoraim are always committed to tannaitic teachings. Tannaim may
disagree with one another (as can amoraim among themselves), but the

only way, according to the traditionalist, that an amora can be said to contest a tanna is by aligning himself with another tanna. Hence, whenever a traditionalist cites a tannaitic source with reference to an amoraic declaration, it will always be with a view to confirming or troubleshooting the latter.[15] For the antitraditionalist, by contrast, confirmation or criticism may, in principle, go either way. At any one time, antitraditionalists will be as respectfully skeptical of former documents as they are hesitant and self-doubting with regard to their own opinions and those of their peers. Still, in the course of a prudently antitraditionalist study session, the direction of questioning will normally be from new to old, from amoraic to tannaitic rather than vice versa. The antitraditionalist forms his own halakhic opinions by seriously troubleshooting those of his predecessors. Once he has tentatively made up his mind and hesitantly boasts a system of his own, he will naturally regard it as superior to whatever earlier systems he had critically considered in the course of its formation. From his point of view, when the two are confronted, it will normally be for the earlier system to answer to the later one. One would, therefore, expect members of the two schools of thought to conduct and narrate such transgenerational confrontations very differently. However, even here, the talmudic text is far less unequivocal than we would have liked it to be. Even in the case of frontal confrontations between amoraic declaration and tannaitic verdict, it is not always easy to decide which of the two texts is the framer of the *sugya* questioning by means of the other—certainly for those who, like the present author, were brought up to interpret them along traditionalist lines almost instinctively.

Introducing the Bavli's Paradigm: *Berakhot* 19b[16]

The Bavli normally introduces such transgenerational confrontations by means of the words *meitivi* (a response was offered; an objection was raised) or *eitivei* (he offered in response; he raised the following objection).[17] At first blush, both the meaning of the terms and the way they are found to function in the text seem to favor a traditionalist approach. The very terms more than suggest that whenever a reliable tannaitic source appears incongruous with an amoraic ruling, the incongruity be considered a challenge, an objection to the latter rather than the former. Objections to amoraic opinions from amoraic quarters, by comparison, will always take the form of a counter*argument*. But the mere fact that two amoraim happen to differ is not considered a problem to either of them. The terms *meitivi* and *eitivei* imply, however, that in order to render a latter-day amoraic opinion problematic, it is enough to cite a tannaitic source that allegedly claims, implies, or premises the opposite. The Bavli contains over twelve hundred transgenerational objections thus introduced.[18] In more than sixty of them the confrontation results in the Bavli declaring the amoraic position in question "seriously problematic" (*teyuvta*) and in about half of these, downright refuted! In other words, in the Bavli's world of discourse, an amoraic statement is in real jeopardy if found to be discordant with a tannaitic text. The opposite, however, seems never to be the case. Nowhere in the Bavli is a tannaitic statement ever outwardly declared refuted or rejected as the result of a *meitivi*-type confrontation with an amoraic text. When discordant amoraic and tannaitic statements are confronted, the Bavli appears to adopt a clear traditionalist stance, always deeming the former potentially at fault and liable to be rejected. Indeed, the traditionalist atmosphere created by the Bavli's rhetoric in these confrontations is duly reflected in the way many Talmud dictionaries render the word *meitivi* itself. Rabbi Adin Steinzaltz's popular *Guide to the Talmud*, for instance, explains that the word is used when "the talmudic rabbis [the anonymous *s'tam*[19]] call an amora's words into question".[20] But does the inner logic of such negotiations always lend itself as readily to the marked traditionalist idiom employed so readily in the way they are narrated? Are all apparent inconsistencies between tannaitic and amoraic statements dealt with *in practice* as if it is obvious that the latter are always at fault? A reasonably reliable method for determining a person's understanding of a problem is to look at the way he or she elects to solve it. It stands to reason that in the case of two apparently contradictory texts, the one thought to be called into question by the contradiction will normally be the one that is eventually modified, limited, or rejected. (It is generally the case that the remedy is normally applied to the wound.) And it is here that one can frequently detect a disparity between what the Bavli says it does and what it seems actually to be doing. Notwithstanding the traditionalist rhetoric of *meitivi*-type negotiations, the ways in which these objections are met and dealt with in practice often attest quite clearly to a boldly antitraditionalist mindset.

In the remainder of the present chapter, I shall argue even further that the discrepancies between the form and content of many of the Bavli's narratives of transgenerational discourse are arguably contrived and deliberate; that their framers' basic attitude to these confrontations is antitraditionalist, but that, for some reason, they sought not to make this immediately apparent. Its surface rhetoric notwithstanding, the Bavli, I shall argue, contains a clear and central, if disguised, antitraditionalist voice. This shall be done by looking closely at what I believe is the Bavli's own paradigmatic model of such a confrontation, its textbook example as it were of transgenerational give and take.[21] In the next and final chapter of this book I shall hazard a guess as to why it was so important for the antitraditionalist framers of the Bavli to so thoroughly obscure their discourse.

The confrontation in question is related in Bavli, *Berakhot* 19b–20a as part of the amoraic discussion of Mishna, *Berakhot* 3:1 which rules as follows:

> One whose dead (relative) lies before him[22] is exempt from the recital of the *shema* and from the *tefillah* and from *tefillin* and from all precepts laid down in the Torah. With regard to the bearers of the bier and those who relieve them (in carrying the bier to the grave) and those who relieve them again, whether in front of the bier (having yet to carry it) or behind the bier (and having carried it already)—those in front of the bier, if they are still required, are exempt; but those behind the bier, even if still required, are not exempt. Both, however, are exempt from saying the *tefillah*. When they have buried the dead and returned (from the grave), if they have time to begin and finish (the *shema*) before forming a row,[23] they should begin, but if not they should not begin. As for those who stand in the row, those on the inside[24] are exempt, but those on the outside are not exempt.

The Mishna does not explain *why* active participation at a funeral entails a suspension of other religious duties. It simply states that it does, and goes on to detail precisely to whom it applies. Now, when one religious duty takes precedence over another it is usually for a reason. What metahalakhic principle is at play here? A possible candidate is the talmudic principle that a person engaged in the performance of one *mitzva* is exempt from others.[25] The problem is that unlike the mourner's halakhic duty to attend to his dead relative, one's obligation to carry the bier and to form the row (or to listen to the funeral speeches mentioned by the Tosefta) are not *mitzvot* of the Torah. They are at most rabbinic edicts that, of themselves, can hardly be thought to merit temporary exemption from major religious duties, such as *tefillin*, that are grounded in the Torah itself. The reason for the exemption seems, in this case, to have to do with a general obligation to pay respect for the dead—*mipnei kevod ha-met*. And although Mishna, *Berakhot* gives no reason for its ruling, in a parallel passage, one of the Minor Tractates does suggest quite clearly that *kevod ha-met* is the principle at play.[26]

But if that be the case, what of respect for the living? To what extent, one is inclined to ask, is a person supposed to defer or disobey "precepts laid down in the Torah" for fear of disgracing, embarrassing, or hurting the feelings of a living, fellow human? Does the halakha consider the

dead and the living to be analogous in this respect? Possibly motivated
by such questions, the Gemara begins its discussion by citing the following, slightly modified rendition of the Tosefta's version of the Mishna under consideration:

> Our Rabbis taught:[27] The row(s) that can see inside [are] exempt, but
> [those] which cannot see inside [are] not exempt. R. Yehuda said: Those
> who come on account *of the mourner* are exempt, but those who come for
> their own purposes[28] are not exempt.

This additional source establishes a subtle difference between the rituals
performed before and after the actual burial. Carrying the bier and attending the funeral speeches (mentioned by the Tosefta) postpone and
supersede other religious duties because in performing them respect is
being paid to the deceased. The beraita, however, implies that the exemption granted to participants in the rows formed after the burial is granted,
at least according to R. Yehuda, by virtue of the respect they are paying
to the living mourner. By citing the Tosefta—for which there seems to
me to be no other reason—the framer of the *sugya* subtly and surreptitiously suggests that a general tannaitic principle may be at work here,
according to which the obligation to respect any human being—dead or
alive—overrides other religious duties. But he does so inaudibly, sheerly
by implication, without even hinting at it openly. The two tannaitic texts
are set side by side for the readers to reach their own conclusions as it
were.

This is not to say that the notion of respect is not discussed. On the
contrary, the Gemara's discussion of the Mishna focuses almost entirely
on questions related to respect. But although various of aspects of human
dignity are discussed in some detail, the word "respect" itself is never specifically associated with the Mishna or the above-quoted *beraita*. By the
time the discussion reaches its climax on fol. 19a, the issue of human
dignity is already "in the air," but in a manner seemingly unrelated to
the two main tannaitic documents under consideration. One has to be
well acquainted with the Mishna's two other parallels to realize that both
the Mishna and *beraita* bear decisively on the question of the relative
weight of the obligation of respecting human dignity in comparison to
other religious duties (if for the tannaitic authorities it was indeed ever
a *question*). If one is not aware of the parallels, the following statement,
attributed to the first-generation amora Rav, will not appear to be at all
related to the Mishna. The less advanced talmudic student will have
passed it by without making the connection at all and associated its appearance, most probably, with the Gemara's prior, seemingly unrelated
discussion of one's obligation to respect one's Masters—a point we shall

return to shortly. But for those who are aware of them, the statement attributed to Rav does more than merely relate to the Mishna. Read in conjunction with the *beraita*, as the framer of the *sugya* would have us do, the statement attributed to Rav appears to assert the very opposite: namely, that one should always attend to one's ritual duties without delay, even at the cost of publicly disgracing oneself, or at the cost of having to act disrespectfully even towards one's Master! In other words, if the Mishna is taken to assert that respect for the dead is to override all religious duties, and if the *beraita* is supposed to imply that in this respect the dead and the living are analogous, then Rav's ruling squarely contradicts them.

> R. Yehuda said in the name of Rav: If one discovers mixed kinds[29] in his garment, he should take it off even in the street. What is the reason? [it says:] "There is no wisdom nor understanding nor counsel against the Lord" (Prov. 21:30); wherever a profanation of God's name is involved no respect is paid (even) to a teacher.[30]

Because the framer of the *sugya* says nothing about the possible relationship that Rav's statement might bear to the Mishna and/or to the *beraita*, it is not at all clear at the outset how he intends it to be understood. Are we to understand it as not necessarily contesting the Mishna's ruling regarding the dead, and as insisting only that respect for the *living* should under no circumstances take precedence over other religious duties?[31] For those who are aware of the fact that the Mishna *is* about respect for the dead, Rav's stated position would then imply that, if the Mishna is to be maintained, then respect for the dead must be viewed as a unique and limited subcategory of respect that cannot be extended to the living. However, in the course of the Bavli's immediate discussion of Rav's statement, it swiftly becomes apparent (a) that Rav's principle *is* understood to extend to the dead, and therefore, by implication, *does* contradict the Mishna; and (b) that the fact that it contradicts the Mishna is passed over by the framer of the *sugya* in utter silence. In fact, from the moment Rav's principle is cited, all mention of both the Mishna and *beraita* are dropped. But although the discrepancy between Rav's statement and the two tannaitic texts previously under discussion is passed over by the *s'tam* without a word, he proceeds to confront it energetically with a series of five other texts of tannaitic origin, all of which also appear to contradict it quite clearly.[32] And it is here that the series of *meitivi*-type confrontations I wish to discuss are conducted.

We are dealing, then, with a general amoraic assertion, placed by the framer of the *sugya* in the immediate context of two major tannaitic sources: one of them is the very Mishna under discussion, with which it is, though it is not actually said to be, clearly at variance and which is then openly contrasted with a series of five other, equally conflicting, tannaitic sources as follows:

> *Meitivi* (an objection was raised). (Says the *beraita:*) If they have buried the body and are returning, and there are two ways open to them, one (ritually) clean and the other unclean,[33] if [the bereaved heading the procession] goes by the clean one they go with him by the clean one, and if he goes by the unclean one they (and with them even the priests) go with him by the unclean one, out of respect for him.[34] Why (is this so)? Let us say: "There is no wisdom nor understanding nor counsel against the Lord" (and hence: wherever a profanation of God's name is involved no respect is paid to a teacher)? R. Abba explained the statement to refer to a *bet ha-p'ras*,[35] which is declared unclean only by the rabbis (but not by the Torah itself); for R. Yehuda has said in the name of Shmuel: A man may blow in front of him (in order to scatter the small bones) in a *bet ha-p'ras* and proceed (without fear of defilement). And R. Yehuda b. Ashi also said in the name of Rav: A *bet ha-p'ras* which has been well trodden is clean.

Whichever view one takes of the Gemara's initial question, there can be little doubt about the way in which it is answered. The apparent incongruity between the statement cited by R. Yehuda in the name of Rav and the *beraita* leveled against it is harmonized away by R. Abba (a disciple of R. Yehuda) by retaining Rav's latter-day opinion and radically limiting the tannaitic ruling of the *beraita* to the case of a *bet ha-p'ras*—a particular type of burial site which, on the authority of both R. Yehuda and Rav, does not involve the sort of uncleanness prohibited of priests by the Torah itself. We began with an amoraic statement in conflict with a tannaitic text, and end with harmony achieved by a significant modification of the latter's straightforward meaning. And the pattern repeats itself. The four additional tannaitic texts cited in contrast to Rav's statement are met by precisely the same procedure. "Come and hear (another *beraita*)," continues the anonymous *s'tam,*

> For R. Elazar b. Zadok[36] said: We (even the priests!) used to leap over coffins containing bodies to greet Israelite kings (. . .) Why (is this so)? Let us say: "There is no wisdom nor understanding nor counsel against the Lord" (and hence: wherever a profanation of God's name is involved no respect is paid to a teacher)? [It is in accord with the dictum of Raba] For Raba said: It is a rule of the Torah that a 'tent'[37] which has a hollow space of a handbreadth (between its outside and what it contains) forms a partition against uncleanness, but if it has not a hollow space of a handbreadth it forms no partition against uncleanness.[38] Now most coffins do have a space of a handbreadth, but [the rabbis] decreed that those which had such a space [should form no partition] for fear that they should be confused with

those which had no space. Still, where respect to kings was involved they did not enforce the decree.

Come and hear (yet another *beraita*): "Great is human dignity, since it over-rides negative precepts of the Torah".[39] Why should it? Let us apply the rule: "There is no wisdom nor understanding nor counsel against the Lord"? Rav b. Shaba explained the *beraita* in the presence of R. Kahana to refer only to the (one, single) negative precept "thou shalt not deviate [from the sentence which they shall tell thee, to the right hand or to the left]" (Deut. 17:11).[40] They laughed at him: (even so) the negative precept of "thou shalt not deviate" is itself from the Torah (and, therefore, the objection still stands)! Said R. Kahana: If a great man makes a statement, you should not laugh at him. All the ordinances of the rabbis were based by them on the prohibition of "thou shalt not deviate," but where the question of [human] dignity is concerned, the rabbis permitted it.

Once again, the tannaitic text is reinterpreted so as to comply completely with Rav's latter-day principle. (This time by two amoraim who functioned five generations after its alleged author.) The strategy is the same as before: to argue that Rav has really said nothing new. All have always agreed, the *s'tam* implies, that concerns for human dignity are powerless when the actual prohibitions of the Torah are involved. Only with respect to their own latter-day interdictions were the rabbis willing to allow considerations of respect to take precedence. With this, we are led to understand, Rav has no quarrel, and the appearance of perfect transgenerational harmony is duly, if somewhat artificially, preserved.

But at a deeper, more subtle level the *s'tam* seems to be calling his own bluff. Hard as they try, the better-versed students cannot ignore the Mishna and Tosefta. On the contrary, the greater the effort invested in harmonizing Rav's statement with other seemingly conflicting texts, harder felt is the total absence of a similar treatment of the Mishna and Tosefta. For those aware of their origins and original contexts, they present equally clear and authoritative cases of tannaitic rulings that sanction transgressions of "all the precepts laid down in the Torah" in the name of respect, for both the dead (Mishna) and the living (Tosefta). In fact, the conflict between Rav's ruling and the two texts becomes more and more apparent as challenges to it from other tannaitic quarters are presented and easily warded off. I find it extremely hard to believe that the framer of the *sugya* was unaware of his omission and shall suggest further down that both the buildup of the tension and the fact that it is passed over in utter silence are intentional. By the end of the day, I submit, the disturbing presence of the Mishna and Tosefta, persisting ignored in the background, serves, intentionally in my opinion, to subtly lay bare the unseriousness of his entire harmonization project. It is as if the *s'tam* were

declaring: "I do not object to the idea that Rav's ruling was in fact revolutionary, I only object to presenting it as such. And as long as the lower-
level student is unaware of what I'm doing, I don't mind leaving a contradicting tannaitic text unattended"! But back to the *sugya* itself.

The questioning and answering go on. The next case is different, however, because it involves the overriding, in the name of human dignity,
of a negative precept that cannot, on any count, be dismissed as involving
no more than a rabbinic injunction.

> Come and hear, (says the *beraita*): (It says:) "[Thou shalt not see thy
> brother's ox or his sheep go astray] and hide thyself from (ignore) them
> [thy shalt surely bring them back to thy brother]" (Deut. 22: 1, 4)—(From
> the positive phrasing of the negative precept we learn that) there are times
> when you mayest ignore them and times when thou mayest not ignore
> them. How so? If [the person who sees the animal] is a priest and [the
> animal] is in a graveyard, or if he is an elder and it is not in accordance
> with his dignity (to pursue or tend to the animal), or if his own work was
> greater than that of his fellow.[41] Therefore it is said "ignore." But why so?
> Let us apply the rule: "There is no wisdom nor understanding nor counsel
> against the Lord"? The case is different there, because it says expressly: "and
> hide thyself from (ignore) them."

In other words, this apparent contradiction to Rav's principle is avoided
by portraying the *beraita* as describing not a matter of principle but a
unique exception to the rule specifically required by the Torah. Why then,
asks the Gemara, is it regarded an exception? Why don't we take this
particular teaching of the Torah as the basis for deriving the general rule
to the conclusion, contrary to Rav, that showing respect indeed always overrides negative precepts of the Torah? Because, the Gemara answers, of the technical reason that one cannot derive a ruling concerning matters of ritual from one related merely to property. In other words,
the theoretical possibility of using the tannaitic materials to construct
(weave?) a general system rival to the one proposed by Rav is considered
and firmly rejected—rather unceremoniously, one has to admit.

So far the four tannaitic sources brought against Rav have all dealt exclusively with the respect payable to the living. Hence, one could still argue that Rav's principle was perhaps meant to be understood not to apply
to respect for the dead and, therefore, not to contradict the Mishna and
its parallels. But the impression is abruptly dispelled by the fifth and final
ta shema.

> Come and hear, (says the *beraita*): (It says:) "[He shall not make himself
> unclean for his father, or for his mother, for his brother,] or for his sis-
> ter[, when they die]" (Num. 6:7) What does this teach us? Suppose he[42]

was going to kill his Paschal Lamb or to circumcise his son, and he heard that a near relative had died, am I to say that he should go back and defile himself? You say, he should not defile himself.[43] Shall I say, therefore, that just as he does not defile himself for them, so he should not defile himself for a *met mitzva*?[44] It says significantly: "or for his sister": [only] for his sister he does not defile himself but he does defile himself for a *met mitzva*. But why should this be? Let us apply the rule: "There is no wisdom nor understanding nor counsel against the Lord"? The case is different here, because it is written expressly "or for his sister."

Again, the apparent contradiction between Rav's principle and the ruling recorded by the *beraita* is avoided by interpreting the latter as describing not a matter of principle but a singular exception to the rule—Rav's rule of course—specifically required by Scripture. In this way the last two *beraitot* are made to work in favor of Rav's statement. Had a rule different from that of Rav been known to apply, namely, that transgressions of the Torah be permitted in order to avoid disrespect, the Torah would not have needed to issue special rulings with regard to the elderly finder of a misplaced animal or the nazirite priest. The exceptions, as it were, are taken to *prove*, rather than disprove the rule, namely, that deference to others should normally *not* override a person's other ritual duties. This is a typical and standard argument in rabbinic *midrash halakha*. The Torah does not waste its words. It should be read in ways that presuppose as far as possible a complete absence of contradiction, redundancy, or arbitrariness. If the Torah goes to the trouble of explicitly stating or suggesting a particular ruling, it is, therefore, either to state an exception, and in doing so to affirm the rule, or to establish a paradigm. The latter option, as in the previous passage, is here also briefly considered and rejected,[45] and Rav's initial ruling is left intact.

As noted, this last objection raised against Rav's ruling is highly significant because of the conflation it presupposes between acting respectfully towards the dead and the living. The very confrontation of Rav's ruling with this particular tannaitic midrash by the framers of the *sugya* proves that they understood his ruling to apply to all manner of deference, including paying respect to the dead. But if that is the case, it must have been equally clear to the framer of the *sugya* that Rav's principle is equally at variance with both the Mishna and Tosefta under consideration. Of this, as noted, not a word is said. In other words, when considered in terms of its general context, as part of the discussion of the Mishna under whose heading it is introduced, the citation of Rav's principle and its subsequent examination constitute, it seems to me, an intriguing, if typical case of deliberately unannounced transgenerational dissent. While the apparent incongruities it exhibits with relation to five other tannaitic sources are energetically pursued and resolved (more on their resolution immediately), the fact that Rav's position turns out to be comparably incompatible with the Mishna itself is passed over in utter silence.

The Logic and Rhetoric of Transgenerational Negotiation

Let us briefly recall how this is achieved. The Mishna rules, without explaining why, that the mourners and active participants in a burial service are exempt "from all precepts laid down in the Torah." Parallel sources clearly indicate that the reason for the deferment is *kevod ha-met*, respect for the dead. Nowhere throughout its discussion of the Mishna does the Gemara explicitly attribute the notion of respect for the dead specifically to the Mishna's ruling, but it certainly does so implicitly. Following a brief discussion, lasting less than a page, of various halakhic technicalities related to the Mishna's specific rulings, the Gemara, on fol. 18a, launches a lengthy, four-page discussion of various forms and aspects of the notion of respect. The opening statement[46] expressly, if incidentally associates, by means of a clever midrashic wordplay, one's duty to partake in a funeral procession with the notion of *kavod*.[47] The remark is not addressed to the Mishna explicitly. Nowhere does the *s'tam* raise the question of the Mishna's possible reasons for ruling as it does. But for students of the text who have asked themselves this question an answer is certainly insinuated.

Read thus, the implications are obvious: if respect for the dead is the reason for partaking in burial services, and if, according to the Mishna, such participation entails an exemption "from all precepts laid down in the Torah," then the Mishna inevitably premises a principle which is squarely opposed to the one attributed to Rav. Moreover, the fact that Rav's statement is eventually found to survive the *meitivi* confrontations with the five other tannaitic sources almost intact[48] clearly suggests that, in the *s'tam's* opinion, the Mishna is not merely contradicted but is actually superseded by Rav's ruling![49] I realize that this is a bold conclusion that will come as a shock to many students of the Talmud, especially to those of traditional schooling like myself. But it is unavoidable. It seems quite clear to me that the framer of the *sugya* goes to considerable lengths in order to intimate it. The *sugya*, in both its content and inner logic, clearly suggests that a substantially antitraditionalist attitude is adopted toward the Mishna on a significant matter of metahalakhic principle.

Why, then, does the *s'tam* seem so reluctant to discuss or even state his strategy openly? Why do the larger contours of the *sugya* read as if a major point is being made and concealed at one and the same time? It is as if the *sugya* was constructed by a confident, yet covert antitraditionalist. In the Bavli's version of the story of Hillel and *b'nei Beteira* we noticed the presence of two opposed, yet distinct interpreters—the narrator, and the amoraic commentator and modifier of the story. Those responsible for the finished form of that passage of the Bavli enable it to speak in two distinct and well-defined voices, allowing room for a traditionalist to criticize and modify an initial antitraditionalist attempt to render the story. As is the Bavli's custom in all nonhalakhic matters, the anonymous framers of the *sugyot* do not intervene in the "editorial disagreement" they record and keep their own opinions to themselves—at least at the most explicit level.[50] Here, in *Berakhot*, however, the ambiguity resides in the very framing of the *sugya*. Here the undecidedness is not due to an unresolved editorial or redactory difference of opinion. Here one cannot even distinguish different voices. The air of ambiguity is generated by what seems to be a basic failure or reluctance, on behalf of the framers, to articulate the basic drift of the *sugya* they are framing. Its entire content and structure seem to be directed *primarily* and *intentionally* to making a significant point, a point to which everything converges but about which all mention is carefully avoided. It is as if the *sugya* was deliberately fashioned in order to ensure that only the most advanced students should get it.

And the same applies to the *meitivi* confrontations themselves. Here too a closer look reveals a sharp disparity between what is said and what is being done; between the ways in which the discrepancies between Rav's statement and the five tannaitic texts are in fact negotiated, and the rhetoric with which such negotiations are usually related. There is a difference, however. The ambiguity surrounding their wider context notwithstanding, the five *meitivi*-type objections dealt with in this particular *sugya* are managed in a manner that is singularly *less ambiguous* in comparison to others of their kind.

In our preliminary analysis of how one would expect traditionalists and antitraditionalists to conduct *meitivi*-type confrontations, I suggested that, for the traditionalist, an apparent discrepancy between an amoraic and a tannaitic ruling will always, without exception, be interpreted as constituting a problem for the former. It is for the amora, or his discursive representatives, either to succeed in explaining the discrepancy away or to withdraw his statement—which is indeed the outcome in about forty of the almost seven hundred *meitivi* objections recorded throughout the Bavli. For the traditionalist, the very word *meitivi* cannot mean anything but "how is it possible for the amora in question to say what he has said in view of a tannaitic source that quite clearly claims the opposite?" And although a traditionalist may, on occasion, end up resolving such a trans-generational discrepancy by suggesting that the tannaitic source should be interpreted differently, he will never, ever initially view it as *a challenge to*, or *criticism of* the tannaitic source.

The very mark of antitraditionalism, by contrast, is that latter-day (amoraic) positions are frequently acquired as a result of critically re-thinking and subsequently modifying or rejecting earlier (tannaitic) ones. Therefore, in the event of apparent discordance between a tannaitic and an amoraic ruling, an antitraditionalist may very well decide to view the former, the tannaitic ruling, as the more defective of the two. In which case, he will have, in theory, two options: to either declare the tannaitic source superseded by the more recent amoraic edict or to preserve it by proposing ways of harmonizing it with the amoraic ruling. Now, it is a matter of fact that in *meitivi*-type situations the Bavli opts for the latter. Nowhere in the Bavli is a tannaitic source declared refuted, rejected, or superseded by virtue of an amoraic statement to the contrary. If a *mei-tivi*-type confrontation ever culminates in the refutation of one of the two rulings involved, it is always the amoraic one that is declared repealed. This in itself constitutes a serious problem for any antitraditionalist inter-pretation of talmudic discourse. How can anyone who appears never to discard an earlier source be at all considered an antitraditionalist? We shall return to this question in greater detail below. At this point, suffice it to say that there is a difference between being or acting as an antitra-ditionalist, and describing or relating to oneself as one. The passage we are dealing with from Bavli, *Berakhot,* as we have already seen, seems to offer a good example of an *unreported* antitraditionalist move, in which the Gemara clearly sides with an amoraic opinion that appears to con-tradict the very Mishna it is discussing. But if that is the case, it again seems nearly-impossible to distinguish between traditionalist and antitra-ditionalist framers of even *meitivi*, transgenerational confrontations. If the former permit themselves on occasion to reinterpret the tannaitic sources involved, and the latter refrain from ever explicitly rejecting them, and if members of neither school of thought actually state which of the texts involved they consider problematic, how is it ever possible to distinguish between them?

At least with regard to the very last point—that of expressly pointing to the text considered by the framer of the *sugya* to be the one open to question—the *sugya* in *Berakhot* is not entirely less ambiguous than others of its kind, but it does have something quite unique to offer. Unlike any other *meitivi*-type confrontation of the Bavli known to me, the anony-mous *s'tam* repeatedly and explicitly states his difficulty. "Why is this so?" he asks of each of the five tannaitic sources supposedly leveled against

Rav, why isn't the rule attributed to Rav—that "There is no wisdom nor understanding nor counsel against the Lord"—also applied here? Each of the tannaitic texts, in other words, seems to be expressly questioned by the *s'tam* in the light of Rav's ruling, rather than vice versa. The very word *meitivi* seems to function here differently from what we are accustomed to.[51] Here, the *s'tam* can be understood to say it signals a query directed in the first instance against what seems to be a tannaitic program, rather than a series of objections to an amoraic statement leveled from different tannaitic quarters. *Given* Rav's general ruling, he puzzles, what are we to make of these tannaitic texts? Of course, there are ways to render the *s'tam's* question along customary traditionalist lines. Still, when taken in conjunction with the other remarkable features of this *sugya*, such traditionalist renderings of the *s'tam's* wholly unique articulation will sound increasingly unconvincing.[52]

The five tannaitic texts confronted by Rav's principle represent four genres of halakhic literature. They include a specific tannaitic ruling (to accompany the mourner, even when he chooses an unclean way, "out of respect for him"); a reliable tannaitic record of widespread and apparently uncontested custom (that of priests leaping over coffins containing bodies in order to show respect for kings); a tannaitic statement of a general metahalakhic principle ("Great is human dignity, since it overrides negative precepts of the Torah"); and two evidently undisputed tannaitic *midrashei halakha* (that a person need not disgrace himself in order to return another person's lost animal, and that nazirites, even if they are High Priests, are obliged to defile themselves for the sake of attending to a *met mitzva*). Read, as the framer of the *sugya* initially intended us to understand them, and taken together with the Mishna and Tosefta, as the wider reaches of the *sugya* imply that we should, the five texts leveled against Rav's ruling clearly attest to the existence and wide application of a general tannaitic metahalakhic principle quite at odds with that of Rav, according to which considerations related to the respect and dignity of both the dead and the living generally take precedence over other religious duties, including precepts of the Torah itself.

The *sugya*, however, is not about the *formation* of Rav's principle. It is not about the reasons that Rav and his generation might have had for departing from the opinion of their Masters, or for adopting the views they had come to accept. *Meitivi* confrontations are not about the *dynamics* of halakhic development. They function as rearguard mop-ups of transgenerational inconsistencies. That is why they are so important for our present concerns. For the framer of the *sugya*, Rav's principle is a given, as are the five tannaitic sources with which it is confronted (and the Mishna and Tosefta with which confrontation is so cannily avoided). But he appears to have decided to deal with the crisis by a curious combination of a traditionalist and an antitraditionalist strategy. On the one hand, he seems to have decided in advance that none of the tannaitic texts explicitly confronted will be actually discarded or even outwardly modified in the process. On the other hand, he seems also to have decided that Rav's position will be retained as it stands come what may. The only way he can achieve both objectives concurrently is by reinterpreting all the tannaitic materials involved in ways that render them inoffensive to Rav's position. In this manner both of the two extreme 'purist' responses are deliberately avoided. Despite the apparently conclusive tannaitic evidence against it, Rav's opinion is not declared refuted as is often the outcome of *meitivi* confrontations. On the other hand, despite the decision to retain the amoraic ruling, the confrontations and their resolution succeed in carrying no outspoken antitraditionalist overtones. The tannaitic texts are not even *said* to have been modified—although as we shall see shortly this is obviously the case—they are merely reread as they were supposedly intended. The middle way, adopted by the *s'tam*, enables him to enjoy the best of both worlds: to avoid all explicit involvement in the traditionalist/antitraditionalist dispute (and God forbid to actively take sides in it); to be an antitraditionalist but to sound as if he is not.

The *s'tam* manages to perform a bold antitraditionalist move—just how bold we shall see immediately—but to do so with little chance of it ever being comprehended as such, except by the most well-versed and attentive of his readers. This in itself does not yet make this particular *sugya* unique. On the contrary, in this respect it is quite typical of its kind. Its uniqueness resides, in my opinion, in the way it subtly gives the game away while still playing it. Unlike any other *meitivi*-type confrontation I have studied, this particular *sugya* strikes me as being self-consciously and uniquely paradigmatic. It appears to have been deliberately set up with a view to drawing attention to and partly dispelling the confusion created by the type of double-talk it so nicely exemplifies. It does so by a variety of means: substantial, contextual, rhetorical—all cunningly aimed at only the most advanced level of readership. But to appreciate both the extent of its antitraditionalism and the almost didactic quality of its self-exposure we need to look briefly again at the ways the incongruities between Rav's ruling and the five tannaitic texts are dealt with.

To repeat, the objective is to show that the incongruities are in truth but apparent. This is achieved by marginally modifying Rav's ruling by allowing, in the name of human dignity, transgressions of Scribal edicts that lack Scriptural status,[53] while maintaining that the first three tannaitic texts in truth permit transgressions of only such decrees. In this respect, it is surely the third *beraita* that undergoes the most radical re-interpretation. The third *beraita* states quite explicitly that care for human dignity is so great that "it overrides negative precepts of the Torah." No, explains Rav b. Shaba, causing his colleagues to snicker, not "negative precepts of the Torah" in general, but only one such precept—namely, the Torah's negative commandment not to disobey such Scribal decrees of the kind that indeed lack Scriptural status! The radicalism, some might even say the sheer chutzpah, of the Bavli's proposed interpretation of this particular tannaitic principle—indeed the radicalism that permeates the entire *sugya*—becomes even more apparent when one looks at earlier employments of the very same principle by the Yerushalmi. The Bavli, I should add immediately, in keeping with its reading here, does in fact apply it consistently only to Scribal decrees.[54]

Yerushalmi and Bavli Compared

The Yerushalmi[55] cites and discusses slightly different versions of the principle on three separate occasions. On two of them it is introduced in the course of discussing the first of the five *beraitot* confronted with Rav by the Bavli. "What is meant by requiring a priest to defile himself out of respect for the many?"[56] asks the Yerushalmi in *Berakhot* iii 6b and in *Nazir* vii 56a,

> We have learned (in the *beraita*): If there were two (equally) suited ways, one long and clean, and the other shorter but unclean, if the majority take the long one, he should take the long one, and if not, he should take the short one (with them) out of respect for the majority.

Notice, that, unlike the Bavli's version of the *beraita*, the setting here is not necessarily that of a burial service. In fact, to the embarrassment of several commentators, the Yerushalmi's version does not even indicate that the priest and his colleagues were at all on their way to perform a religious duty.[57] In the two *sugyot* devoted to this *beraita* the issue for the Yerushalmi appears simply to be that of permitting a priest to defile himself in order to avoid hurting the feelings of his cotravelers, regardless of the reasons or destination of their joint excursion. But of what level of uncleanliness are we speaking? "So far", the Yerushalmi goes on to ask, "have we spoken only of uncleanliness due to Scribal decree?":

> Does [the same principle] also apply to uncleanliness prohibited by the words of the Torah? It (does, and) follows from the saying of R. Zeira:

> "Great is the dignity of the many, since it temporarily overrides (a) negative precept[s].[58]" Ada said, (therefore, a priest should follow the many along the unclean path) even in the case of uncleanliness prohibited by the words of the Torah.

Although the saying, attributed by the Yerushalmi to R. Zeira, does not ever explicitly say, as does the version cited by the Bavli, that the negative precept(s) in question are indeed those *of the Torah*, it is understood throughout the Yerushalmi, without exception,[59] to mean precisely that! In other words, when Rav b. Shaba is reputed by the Bavli to have "explained the dictum in the presence of R. Kahana" two generations later, he was not merely ignoring the *theoretical possibility* that the text might actually mean what it says but seems to be overlooking the fact, possibly willingly, that, at least in Palestinian halakhic discourse, the dictum was indeed thus read. If anything, the Yerushalmi bears clear and incontestable witness to the existence[60] of a school of thought that was apparently associated among second- and third-generation *amoraim* with R. Zeira and R. Ada, who, as a matter of (literary) fact, read the dictum and apparently acted upon it very differently from what Rav b. Shaba would have us believe.

But the Yerushalmi has more to offer that is relevant to the Bavli's *meitivi*-type negotiation of Rav's principle than its very different reading of this particular dictum. On the very question of the discovery of mixed kinds in public, the case offered by Rav as a paradigmatic example of his principle, Yerushalmi, *Kilayim* ix 32a, records a debate that further emphasizes just how strained are the Bavli's relentless attempts to harmonize Rav's teaching with everything that supposedly went before it. The relevant passage reads as follows:

> One who was walking in the marketplace and found himself wearing mixed kinds. Two *amoraim* (debated the issue). One said that it is prohibited (and therefore the garment must be removed immediately). The other said that it is permitted (for him to go on wearing the garment until he can remove it in the privacy of his home). The one who said that it is prohibited maintains that [wearing mixed kinds] is (a transgression of) a law of the Torah. The one who said it is permitted accords with R. Zeira who said that great is human dignity that temporarily overrides a negative commandment.

Seeking to harmonize this passage with the one in Bavli, *Berakhot*, traditional commentators explain the dispute recorded here as if it were no more than a factual disagreement.[61] The amora who prohibits and demands the removal of the garment, they explain, is of the opinion that the mixed kinds in question are of the sort prohibited by the Torah. His colleague, on the other hand, believes that the mixed kinds in question are of the sort prohibited only by Scribal decree, and therefore does not demand the removal of the garment in the marketplace. This, they add, is in accord with R. Zeira's principle, which, in keeping with its rendition by the Bavli, supposedly allows transgressions of Scribal decrees in order not to offend the public or disgrace oneself. Although such a reading of the debate succeeds in squaring it with the Bavli's version of Rav's allegedly uncontested view on the matter, it does little justice to the text of the Yerushalmi. On such a showing, the debate claimed by the Yerushalmi to have obtained between the two *amoraim* is in truth no debate at all. Both disputants agree that the sort of mixed kinds prohibited by the Torah must be removed the moment they are discovered, regardless of where one happens to be at the time, and that in the case of mixed kinds prohibited only by Scribal decree, one may postpone their removal in order to avoid inflicting, or suffering public disgrace. By reading the Yerushalmi thus, two related objectives are achieved. First, Rav's ruling regarding the discovery of mixed kinds in public—and, subsequently, his view on the entire question of human dignity—remains uncontested in keeping with the Bavli's keen efforts. Second, and again in keeping with the Bavli's program, the realm of application of R. Zeira's principle is conveniently limited to Scribal decrees. But as much as the framer of Bavli, *Berakhot* 19b, might have liked Yerushalmi, *Kilayim*, to comply with his agenda, such a reading of the latter is implausible. For one thing, the Yerushalmi states quite clearly that on the *question* of the discovery of mixed kinds in the marketplace—Rav's question—two Sages *differed*. As is often the case in the Yerushalmi, the text that follows is not as lucidly and as carefully redacted as we would have liked it be. Still, to interpret even the Yerushalmi as announcing a difference of opinion only to mean that in truth the two Sages merely differed in their *understanding of the question* is to do real violence to the text. And in view of the Yerushalmi's other two references of R. Zeira's principle, to limit its application here exclusively to Scribal decrees is to render the Yerushalmi's various appeals to the principle mutually inconsistent.

The inescapable conclusion is that where the Bavli goes to incredible lengths to present a harmonious transgenerational front united around Rav's principle, the Yerushalmi does not display the slightest uneasiness about presenting the field as deeply divided twice over. According to the Yerushalmi, not only was the position attributed by the Bavli to Rav directly disputed in its time, but its adversaries are said to have justified their move by appealing to no other than R. Zeira's dictum. Again, the

Yerushalmi bears witness to the "real" existence of a very different understanding of this dictum than the strained interpretation provided by the Bavli. This time, however, it is not merely read differently than in the Bavli, but is employed as an argument against the very position to which the Bavli would have it conform!

In view of these earlier texts, it seems rather unlikely that Bavli, *Berakhot* 19b, could have been framed in genuine unawareness of there being real opposition to Rav. On the other hand, the *sugya*, especially that part of it devoted to Rav's ruling, by no means gives the impression of being confused or poorly edited. There is a distinctly contrived, structured, even didactic air about the entire text. The *s'tam* appears to know exactly what is at stake, where he is going, and why he is going there. (Such hunches, of course, are hard to demonstrate, let alone prove. But they do make good methodological sense. We are always liable to miss something of importance by attributing ignorance, innocence, sloppiness, or confusion to the texts we seek to understand before serious attempts at more charitable readings have proved to fail.) Assuming, then, that the Bavli is aware of the fact that attitudes toward human dignity different from Rav's, such as those recorded by the Yerushalmi, may well have originally motivated the tannaitic texts it considers and that Rav himself had knowingly formed his views in defiance of these earlier systems of thought, what was to be achieved by such a far-fetched and, to an extent, false, exercise in transgenerational harmonizing? Why is the Bavli so eager to conceal disagreement where the Yerushalmi seems quite content to display it? All the more so if, according to the Yerushalmi, Rav's position remained contested among amoraim!

One obvious difference between the Yerushalmi and the Bavli on the question of the discovery of mixed kinds in public, is the former's apparent disinterest in resolving it and making a ruling. On this particular question the Yerushalmi's objective is no more than to describe and explain the two rival positions. The Bavli, on the other hand, has apparently not only decided to resolve the issue, but to do so in favor of an amora facing substantial tannaitic opposition and apparently enjoying no known tannaitic support. This perhaps explains the great effort invested by the Bavli in making the tannaitic texts *seem* to have anticipated Rav's latter-day point of view. But while the two *talmudim* do so differ in their approaches of the question of mixed kinds, their differences can hardly account for the Bavli's feigned traditionalism. There are at least two good reasons for not accepting the Bavli's obvious desire to resolve the issue as a viable explanation of its equally obvious efforts to conceal the opposition. First, traditionalists and antitraditionalists maintain mutually exclusive positions. Hence one of two: had the *sugya* been framed by a traditionalist, the halakha should not have been ruled in accord with Rav's view in the first place; had it been framed by an antitraditionalist, the feigned traditionalism would have hardly been called for. Why perform what looks like a decidedly bold antitraditionalist move while doing your utmost to sound as if you're not?

Secondly, although, in the case of the mixed kinds, the Yerushalmi does no more than to note and explain the two conflicting views, there are several other cases of transgenerational confrontation in which the Yerushalmi not only decides the issue, but, like the Bavli in *Berakhot*, does so by ruling against the tannaitic sources in favor of amoraic opinions.[62] In all these cases, however, and this is the important point, quite unlike the Bavli, the Yerushalmi makes no effort at all to cover up its tracks. The differences between the ways each of the two *talmudim* presents these negotiations is quite striking.

It is significant perhaps that the most clear and detailed *meitivi*-type negotiation of this sort that the Yerushalmi has to offer is contained in none other than tractate *Kilayim* ix 32a in the passage immediately preceding its account of the case of mixed kinds. Here too, as in Bavli, *Berakhot* 19b, the amoraic position under consideration is one of Rav's oft-quoted rulings, and here too it is confronted by a series tannaitic texts that appear clearly to assert the opposite. But apart from that, the two *sugyot*—each of them absolutely typical of its kind—are worlds apart. Despite the special transitional status granted to Rav by the Bavli (more on this below), the difference between the two amoraic redactions remains stunning.

For one thing the Yerushalmi does not, as a rule, announce the transgenerational incongruities it discusses by a loaded term such as the Bavli's *meitivi*. The introduction of an apparently contradicting tannaitic source is not described as an "objection" to the amoraic position under consideration, thereby creating, as does the Bavli, a traditionalist atmosphere in which one instinctively takes for granted that it is for the amora to meet the tannaitic challenge rather than the other way around. The term used to set forth each of the 230 or so transgenerational confrontations conducted throughout the Yerushalmi is completely neutral in this respect: *"matnita* (or *matnitin*) *pliga de-* . . . "—which is simply to declare that "(the following) tannaitic text ('mishna') disagrees with. . . . " The difference, as we shall see, is more than verbal; the Yerushalmi in truth seems to assume nothing in advance. Still, in this case even the merely verbal differences between the two *talmudim* are telling.

The halakhic principle that is attributed to Rav and discussed in

Kilayim ix 32a is: "All that is prohibited for appearance's sake is also forbidden when done privately".[63] Six different tannaitic texts are then introduced in quick succession:

> (The following) [tannaitic source][64] disagrees with Rav: One should not use linen that is dyed black[65] to make a visible hem, but it is permitted (to use such black tinted linen) in pillows and bed-covers.[66]
>
> A(nother) [tannaitic source][67] (that) disagrees with Rav: Should one's coins fall and scatter in front of an idol, one should not stoop down to repossess them in order not to appear to be bowing to the idol, but, if no one is present, it is permitted.
>
> A(nother) [tannaitic source][68] (that) disagrees with Rav: One should not touch one's mouth to a public fountain fashioned as a face (of a false deity) in order not to appear to be kissing it, but if no one is present, it is permitted.
>
> (Yet) a(nother) [tannaitic source][69] (that) disagrees with Rav: One should not prepare a hole in the ground to receive the blood of a slaughtered beast, but in confines of one's home it is permitted. He should not do so in public (in the marketplace) in order not to appear a heretic.[70]
>
> A(nother) [tannaitic source][71] (that) disagrees with Rav: It is permitted (on the Sabbath) to lay out (in one's yard, garments that were drenched by the rain in order to dry) in the sun, but not within view of the passersby.
>
> A(nother) [tannaitic source][72] (that) disagrees with Rav: One should not pour water on the Sabbath (from one's yard) into the street (even) through a four-cubit deep drain[73] (. . .) (However) R. Kapara has taught that if no one is present, it is permitted.

No attempt is made to explain the discrepancies away or to seek for minority views among the tannaitic sources that might conform to Rav's position.[74] Despite the fact that in this *sugya*, unlike that of the mixed kinds, the Yerushalmi is working toward a halakhic decision, it shows not the slightest inclination to harmonize the texts under consideration, gloss their differences, or somehow argue them away. And even if the Yerushalmi does presuppose Rav's special license to contest tannaitic opinion, it makes no mention of it and certainly refrains from rejecting Rav's obviously minority view in favor of the uncontested tannaitic evidence mounted so clearly against it. Rather, the Yerushalmi sums up the situation in one short and potent sentence, declaring simply and unceremoniously that:

> *All of these (beraitot) are in disagreement with Rav, and (therefore?)*[75] *have no standing!*[76]

This is a different matter from the special privilege of *disputing* tannaitic opinion granted to Rav by the Bavli on occasion.[77] It is no longer a question of simply acknowledging Rav's license to think differently from his predecessors, but of the framer of the *sugya* ruling in his favor *in the face of* wide and varied tannaitic opposition. To this end the Yerushalmi appears to have no qualms about declaring bona fide tannaitic sources null and void. Once these earlier documents are found to contradict Rav's ruling it is not the fate of *our understanding of them*[78] that seems to be at stake, but the fate *of the documents themselves!* And this is not the only case. On six other occasions the Yerushalmi concludes such *meitivi*-type confrontations with the same formula, similarly dismissing the tannaitic sources in question in favor of the amoraic positions under consideration.[79]

The Bavli, by contrast, never resolves a transgenerational conflict of views by outwardly rejecting a tannaitic source.[80] It goes to great pains to retain a strict traditionalist frame of discourse while managing to achieve the same results as the Yerushalmi. It does so with the aid of other devices, all of which formally conform with its basic traditionalist vocabulary. We have already encountered the Bavli's method of radically reinterpreting a former text in order to harmonize it with a latter-day position. With regard to transgenerational confrontations, this is probably the most widely employed strategy. But it is not the only one. The Bavli will also often declare the actual *text* of a problematic tannaitic source (as opposed to our understanding of it) warped or misstated and will suggest ways of modifying it. Such moves are usually introduced by the phrase: *hisurei m'hasra ve-hakha ketani*—literally, "words are missing (from the mishna or *beraita*) and it should be read (learned) as follows. . . ."[81] Somewhat less frequently, the Bavli will redescribe an anonymous, seemingly uncontested tannaitic source that has proven itself problematic as the minority opinion of an individual tanna and will then usually produce evidence of it being contested elsewhere by a tannaitic majority or a superior tannaitic authority. In these cases the source is declared "lone" or "singular."[82] Unlike the Yerushalmi's no-nonsense dismissal of problematic tannaitic documents, all these devises are designed to create an atmosphere (illusion?) of prima facie harmony. To declare a prior text garbled and to modify it for no better reason than to bring it in line with a latter-day suggestion could, of course, be the honest work of a prudent traditionalist. If one truly believes that an amora would be incapable of ever contradicting an uncontested, authentic tannaitic ruling, then, if he appears to do so, something has to be wrong. And one possibility is always that the version we have of the earlier text is corrupt and that the very fact that the amora in question ruled differently is proof enough of both the fact that he was working from a different version and of its content.[83] We have already raised serious doubts with regard to the forthrightness of the Bavli's traditionalist rhetoric. There can be no doubt, however, that responding to a problem by tampering with an authentic earlier text, however radically, but with a view to retaining it by all costs, is a different matter entirely from doing so by actually discarding the former document—the latter, if anything, is in principle incompatible with traditionalism.

In order to fully appreciate the differences between the two *talmudim* on this point, it is worth looking briefly at the way the very same principle of Rav's—that "All that is prohibited for appearance's sake is also forbidden when done privately"—is dealt with by the Bavli. The principle, similarly attributed to Rav and similarly confronted with similar versions of the same tannaitic texts, is cited and discussed by the Bavli on four separate occasions.[84] Where the Yerushalmi permits itself to dismiss as groundless the tannaitic texts that contradict Rav, the Bavli manages to retain them, never to overstep the conservative limits of a strictly traditionalist discursive framework, and yet to achieve the very same! In all four cases, just as in *Berakhot* 19b, the tannaitic texts are reinterpreted so as to comply with Rav's view, which, from the Bavli's assumed traditionalist perspective, if it is not contradicted by tannaitic opinion can always be retained. The Bavli's rhetoric here is so different from that of the Yerushalmi that it does not even seem fit to employ the *meitivi* format. Thus in Bavli, *Avoda Zara* 12a, to take but one example, the mention of Rav's principle is initially prompted by a *beraita* that is concerned with prohibitions for appearance's sake per se.

> Our Rabbis taught: It is forbidden to enter a city in which idolatrous worship is taking place therein, or to go from there to another city; this is the opinion of R. Meir. But the Sages say, only when the road leads solely to that city is it forbidden; if, however, the road does not lead exclusively to that place, it is permitted.[85]

Both parties hence agree that it is forbidden to appear as if one is on one's way to participate, or from participating, in idolatrous worship. R. Meir adopts a stricter view than the Sages in cases where there is more than one plausible interpretation of a person's behavior. In the Sages' opinion, prohibition for appearance's sake applies only to cases in which one's actions are unlikely to be interpreted differently. If they are liable to be understood inoffensively they are permitted. But all agree on the principle of the matter. The question of whether or not actions forbidden for appearance's sake are allowed when performed privately is not raised by the *beraita*, nor do the dispute and ruling recorded by it bear on the question in any way. As for the issue of prohibition due to appearances, the Bavli goes on to cite three additional tannaitic rulings. And it is here that Rav's views on the matter appear to be contradicted.

If a splinter has got into his foot while in front of an idol, he should not bend down to extract it, because he may appear as bowing to the idol; but if he is not seen[86] it is permitted.

If his coins got scattered in front of an idol he should not bend to retrieve them, for he may be taken as bowing to the idol; but if he is not seen it is permitted. If there is a spring flowing in front of an idol he should not bend down and drink, because he may appear to be bowing to the idol; but if he is not seen it is permitted.

Not surprisingly, Rav's principle makes its appearance in the *sugya* apropos the second, qualifying clause of the *beraitot*, in which permission is granted if and when one is "not seen." "What is meant by 'not being seen'?" asks the *s'tam*,

Shall we say (that it means) that he is not being observed? Surely (not, because) R. Yehuda stated in the name of Rav that whatever the Sages prohibited for appearance's sake is also forbidden in one's innermost chambers! It can only mean that it is permitted if (by bending) he does not *seem to be bowing* to the idol.

The Bavli's response to the seeming contradiction between the three *beraitot* and Rav's ruling is not to declare any of the sources involved invalid, but, as in *Berakhot* 19b, to harmonize all by rereading the three tannaitic ones. What they *really* mean, explains the *s'tam*, is that a person standing in the vicinity of an idol is permitted to retrieve his fallen coins, kneel to drink, and so on, not when his actions happen to go unobserved, but only when he is able to do so without appearing to be bowing! This feat of harmonization is achieved without the Bavli having to declare Rav's ruling even prima facie problematic. The *meitivi* format is not employed because the tannaitic sources—as the Bavli quotes them—require very little interpretive effort in order to cohere with Rav's latter-day position.

Significantly, the original Tosefta includes *two* rather than one qualifying clause. It is prohibited to appear as if one is bowing to an idol, "but," continues the Tosefta, "if he should crouch down (to retrieve his coins or drink) with his back turned (to the idol), *or* if he is in a place where he is not seen, it is permitted." Indeed the Tosefta permits one to stoop in the presence of an idol in *two* separate cases: if one does so without *seeming* to be bowing, or if one does so without being *seen*. Focusing exclusively on Rav's edict, the Yerushalmi understandably cites only the latter[87] and is therefore forced to make a choice. The Bavli, on the other hand, avoids the confrontation entirely by running together the Tosefta's two qualifications[88] and pooh-poohing the suggestion that Rav might have defied a bona fide, uncontested tannaitic ruling. But by sounding less radical the Bavli does not *achieve* less than the Yerushalmi. As in *Berakhot*, here too Rav's view is retained lock, stock, and barrel; the seemingly discordant tannaitic sources are rendered inoffensive by a stroke of creative reformulation; and, once more, a significant halakhic revision is achieved without the Bavli having to *sound* as if it is.

Antitraditionalism for the Advanced

Such is the framers of the Bavli's way of negotiating across the great amoraic-tannaitic divide: free, as any antitraditionalist would be, to align themselves with whichever position they see fit, yet, at the same time, very careful to harmonize the texts involved whenever ruling in favor of the latter-day position. In this sense Bavli, *Berakhot* 19b, is absolutely typical. What makes it in my opinion absolutely unique is the way in which the Bavli appears to have made a lesson out of it. In *Berakhot* 19b the framer of the *sugya* is engaged in more than the negotiation of yet another *meitivi* confrontation. It is here, I suggest, that the *s'tam* has chosen inaudibly to address his better-versed readerships in order to all but outwardly declare his antitraditionalism, but to do so without ever missing a beat of his carefully crafted traditionalist way of speaking. I would even suggest further, that it is here perhaps that the framers of the Bavli also make their stand in what appears to have been a significant disagreement with their Palestinian counterparts about the appropriate ways of conveying and passing down their shared vision of *talmud-Torah*.

If one is willing to entertain the possibility—as a provisional working hypothesis for which, however, to the best of my knowledge there exists no independent evidence—that the framer of this particular *sugya* was writing with some knowledge of the Yerushalmi's treatment of at least some of the relevant material, then it is even less plausible that he was engaged in a straightforward traditionalist exercise. Thus for example, had he been seeking for no more than to ward off alleged tannaitic challenges to Rav's ruling, it would have made much better sense for him to have cited the Yerushalmi's version of R. Zeira's dictum than the one he does. The version attributed to R. Zeira speaks sufficiently vaguely of consideration for human dignity overriding "negative precept[s]" without actually saying "of the Torah." In this manner it would have been much easier to limit it, as the Bavli labors hard to argue, exclusively to rabbinic edicts. But no. Defiantly, or so it seems, the *s'tam* confronts Rav's affirmation that care for the dignity of even one's own teacher is powerless to override a transgression of God's word (name) with a tannaitic formulation that prima facie could not have been more explicit in stating the very opposite—so great are considerations of human dignity, the tanna asserts, that it overrides the (all?) negative precepts *of the Torah*. Rather than employ a less offensive version such as R. Zeira's—assuming, of course that such a version was available to the Bavli redactor—or fall back on the special license granted by the Bavli to Rav that allows him to contest tannaitic opinion, the *s'tam* chooses to challenge us, his readers, with a bold, almost reckless, display of hermeneutic freedom. "Look how far one can go," he seems to exclaim, "see how much can be achieved without appearing to violate the polite traditionalist rules of transgenerational discourse." "Even as we speak," he seems to whisper inaudibly to all those capable or willing to hear, "most of my readers remain quite oblivious to what we are really up to." So far fetched is Rav b. Shaba's rendition of the *beraita* that the *s'tam*, fully aware of how far he has allowed himself to go, has the members of that academy of old snigger along with the most attentive of his present-day students.

Intriguing as it may seem and as plausible as it might sound, the idea that the Bavli's engagement with its tannaitic heritage was even partly conducted in willful disagreement with that of the Yerushalmi is one for which the present author lacks both the knowledge and training to seriously pursue. The differences in style, temperament, and idiom between the ways in which the transgenerational divide is negotiated by the two *talmudim* speak for themselves and, in my opinion, are worthy of further study. They are certainly relevant to the task of laying bare the Bavli's treatment of *meitivi*-type confrontations in *general*. But in order to appreciate the *uniqueness* of *Berakhot* 19b one need not even be aware of the Yerushalmi's different way of doing things or, if aware, need not necessarily agree with my account of the Yerushalmi, let alone assume that the Bavli was consciously reacting to it. Whether or not the *sugya* in hand was intended to score points in a wider, second-order debate with the *s'tam*'s former Palestinian counterparts is irrelevant to assessing my claim that this particular set of transgenerational confrontations was (also?) constructed with a view to giving away the Bavli's game to its better-trained readers. The evidence for this is, as it should be, wholly internal.

First, as we have already seen, although keenly confronted with five different, seemingly contradicting tannaitic sources, the framer of the *sugya* does much to suggest, though without actually saying so, that Rav's ruling is also contradicted by the Mishna under discussion and the *beraita* apropos of which it was quoted in the first place. Nowhere does the Gemara state explicitly that it understands the Mishna and *beraita*'s exemptions from major religious duties granted to mourners and funeral-goers to be due to considerations of human dignity, but by devoting virtually all of its discussion of the Mishna to various aspects of such considerations, the point, which reference to other sources would have actually proven, is strongly implied. But if that is the case, the better-versed reader gradually realizes, if what the Mishna and *beraita* are claiming is that for consideration of human dignity "all precepts laid down in the Torah" are overridden, then they too are in worrying violation of Rav's later ruling. In fact once one notices that the Mishna and *beraita* should have also been confronted with Rav, the confrontations that are considered and resolved only add to the confusion. This is because the ways in which they are resolved are irrelevant to the Mishna and *beraita*, neither of which can be said to involve transgressions of only rabbinic edicts—both texts exempt the participants from *tefillin* for instance—and neither of which can be said to be required explicitly by the Torah. The increasingly felt tension between the three texts, lingering unmentioned and unresolved in the background, serves, deliberately in my opinion, to invalidate the air of harmony created by the *meitivi* confrontations that are negotiated.

It is difficult to explain to someone who has not studied the Bavli closely just how odd this is. It is highly uncommon for the Gemara to miss or overlook such a clear discrepancy between a text that it has deliberately introduced and is closely scrutinizing and the very Mishna under consideration—especially if the text is attributed to an amora! Had the Mishna (or the *beraita*) explicitly mentioned the notions of respect or dignity, such an oversight would have been inconceivable. On the other hand, the closer one studies the *sugya*, the less and less it appears conceivable that the Gemara could have understood the Mishna's ruling as motivated by anything *but* considerations of respect and human dignity! Given the specific rulings of the Mishna and *beraita*, what was the point of citing Rav's edict in the first place, if not to confront the three texts? And yet their mutual incongruity is passed over in incredible silence. Perhaps, one inevitably finds oneself asking, there are texts, such as the five mentioned, that can be somehow squared with Rav's views, but there are others, among them clearly the Mishna itself, that, by the end of the day, remain in opposition to Rav despite their obvious priority? In which case perhaps the name of the game is not, as the Bavli normally has us believe, harmony at all costs? Perhaps, regardless of the way the five *meitivi* queries were resolved, Rav and other leading amoraic authorities are not strictly required to follow in the footsteps of their predecessors in all matters? Perhaps the halakha admits not only to innovation, but also to revolution? Perhaps the Bavli's rhetoric of transgenerational harmony is *merely* rhetoric? In the very act of harmonizing away *meitivi*-type objections in the normal manner, the *s'tam* seems to be indicating, quite strongly, that he shouldn't be taken seriously. Rav's move *is* revolutionary, he all but says out loud, for it knowingly contradicts the entire tannaitic tradition represented by the Mishna and Tosefta, and if the harmonization of the other five texts seems to you affected and strained, well, you are quite right!

The second unusual feature of the *meitivi* confrontations presented in *Berakhot* 19b is the absolutely unique manner in which they are introduced. As noted previously, nowhere else in the Bavli is a *meitivi*-type objection explained or accompanied by a running commentary. Nowhere else does the *s'tam*, after presenting the two conflicting sources, expressly state which of the two texts constitutes his source of puzzlement and what he is puzzled about.[89] All *meitivi*-type confrontations are introduced by the exact same format:

> amora *A* says *X*; "*meitivi*": tannaitic authority
> (*B*) implies the negation of *X*.

The *s'tam's* only "original" contribution of his own at this stage (apart from actually citing the two texts) is the word "*meitivi*" itself. Given the incongruity between the texts, the problem that it constitutes is apparently considered to be self explanatory—otherwise it would have had to have been explained.

Self-evident or not, what the problem *is*, however, will crucially depend on the point of view of whoever raises it (who, in the case of *meitivi*, as opposed to *eitivei* confrontations, is the *s'tam* himself).[90] For a dedicated traditionalist the difficulty generated by a transgenerational incongruity will always be one and the same: how was it possible for amora *A* to say *X* if an authoritative tannaitic source appears to imply not-*X*? For the traditionalist, halakha accumulates. At any one time the imperative, mandatory legal system will comprise the entire body of authoritative rulings *ever made*. Traditionalism, by definition, admits no change to the body of binding rulings, other than mere augmentation, and acknowledges no procedure for overturning authentic former rulings. Like any legislator he requires the obligatory halakhic system to be at all times consistent. And since the system necessarily preserves all former rulings, troubleshooting it for transgenerational consistency is of paramount importance. It is the first and basic condition of adequacy for any potential latter-day addendum to the system. In order to merit inclusion within the system, latter-day rulings must be shown (a) to fully cohere with all former rulings, and (b) to legislate for cases, situations, or circumstances for which no former ruling is known to apply. In the event of an apparent transgenerational incongruity, it will never be the tannaitic source that comes as a surprise, only the latter-day, amoraic ruling.

But if one's metahalakhic point of departure is antitraditionalist, such texts are allowed to contradict one another, and, in a truly antitradition-alist world of discourse, they frequently will. Here the mere existence of transgenerational incongruity does not of itself necessarily constitute a problem for the system. Since the antitraditionalist allows for halakhic change, his view of halakhic system building is revisionary rather than cumulatory. Like any legislator the antitraditionalist also requires the mandatory halakhic system to be at all times consistent. But antitradi-tionalism, by definition, does *not* require that the demand for consistency extend to all authoritative rulings that were ever made—but only to those still preserved in the system. For the prudent antitraditionalist, casting about for transgenerational contradictions is hardly a form of trouble-shooting the system. It is more a form of bookkeeping, a checklisting of which rulings are in and which are out—normally privileging the most recent ones in the process. Because former rulings are not expected to enjoy special status, and certainly not immunity, within an antitradition-alist's halakhic system, they will not be subjected to special tests of co-herence. For antitraditionalists, a ruling, any ruling, old or new, becomes questionable when they see no good reason for it, or, even worse, when they happen to see good reasons for ruling differently. But even the most avid antitraditionalist does not advocate halakhic change for the sake of sheer novelty. Antitraditionalists are more than willing to accept halakhic reform, even halakhic revolution, but they have to have good reasons for doing so. It follows, therefore, that when an antitraditionalist presents a *meitivi*-type challenge, that is to say, whenever he presents transgenera-tional inconsistencies between two rulings *as a problem*, the ruling he will be challenging will be the one for which he sees relatively less reason *regardless of its formal priority*.

It follows that if asked to fully spell out their questions, traditionalists and antitraditionalists will elaborate very differently on *meitivi*-type con-frontations. For the former, to repeat, the question will always reduce to: how was it possible for amora *A* to say *X* if an authoritative tannaitic source appears to imply the exact opposite? Since it was obviously *possible* for this to happen—because it did!—in cases in which mere oversight is ruled out, a slightly more sophisticated traditionalist formulation of the question would be: because it is unthinkable for amora *A* to have know-ingly contradicted the tannaitic source in hand, we must assume that he had read it differently—if so, how? Unlike the first formulation, this sec-ond traditionalist rendering of a *meitivi*-type challenge, beautifully exem-plified in the Bavli's above-mentioned discussion of Rav's principle in *Avoda Zara* 12a, indeed addresses the question to the tannaitic source rather than to the amoraic ruling. However, it is *not* a plea to understand the tanna's reasons for claiming what he appears to claim, but a plea for a plausible, alternative rendering of his words that will fare better with the amoraic position under consideration. In the context of a *meitivi*-type challenge, a traditionalist will never seriously question the tanna's motives for saying what he said. That will never be the kind of explanation sought for. To do so would be to violate the traditionalist code twice over. First, by seeking for possible reasons for the tannaitic source claiming what it appears to claim, the traditionalist will be tacitly assuming that it could, in principle, be contradicted by an amoraic ruling. Second, in so inquiring he will not only be acknowledging the *possibility* of transgenerational inconsistency, but will in fact be *questioning* a tannaitic source in the light of latter-day opinion.

By contrast, the *only* real difficulties that an antitraditionalist would encounter in the course of a *meitivi*-type confrontation, would be questions concerning motives. (For the antitraditionalist, remember, transgenerational contradictions are not problematic of themselves.) And owing to the fact that in the antitraditionalist's world of discourse latter-day positions are expected to be outcomes of serious troubleshooting of earlier ones, it is only natural that the earlier, tannaitic source will frequently turn out to be the more puzzling of the two. If asked to spell out the difficulty perceived in a *meitivi*-type confrontation, an antitraditionalist will, therefore, often describe it thus: amora *A* claims that the halakha should be *X* for reason *R*. Curiously, the tannaitic text in hand implies the negation of *X*. Why, as seems to be the case, did the tannaitic authors not think of *R* or deem *R* inappropriate? Only an avid antitraditionalist will be able to seriously formulate a *meitivi*-type query in this fashion.[91] *And this is precisely the way the framer of Berakhot 19b scrupulously describes the puzzlement entailed by each of the five* meitivi-*type confrontations he conducts. Of each of the five* beraitot *he asks the same: "Why (is this so)?" Why doesn't the* beraita *say as does Rav that "There is no wisdom nor understanding nor counsel against the Lord" (and hence: wherever a profanation of God's name is involved no respect is paid [even] to a teacher)"?* Unless we interpret the *s'tam*'s formula as a rhetorical, somewhat sarcastic challenge to Rav—a reading I find highly inappropriate—it is hard *not* to read it as a clear if fleeting revelation of antitraditionalist bias.

In addition to the clear antitraditionalist gist and content of the *s'tam*'s comments, it is the insistent, case-by-case, chant-like repetition of identical interrogative formulae that serves to create the distinct impression of a schoolroom example I spoke of above. If one studies *Berakhot* 19b attentively, with some knowledge of similar *sugyot*, one has the distinct feeling of being patiently, very patiently taught to what a *meitivi* question amounts, and how to ask it. The lesson, however, is completely lost on two kinds of reader: the less knowledgeable students who simply fail to register both the uniqueness and significance of the *s'tam's* explicatory remarks, and the quicker, better informed readers who, perhaps because of their former experience with *meitivi*-type queries, no longer bother to pause and think between question and answer despite the *s'tam's* patient prompting. Reluctant to give the game away too obviously, or so it seems, the *s'tam* does not allow the antitraditionalism of his didactic interventions to linger long enough to really sink in. Although the problems generated by each of the *beraitot* are explicated as only an antitraditionalist could describe them, *they are each immediately resolved, or rather harmonized out of existence, in ways only a dedicated traditionalist would tolerate.* In other words, with regard to the *meitivi* confrontations themselves, the framer of *Berakhot* 19b flashes his antitraditionalism repeatedly, but each time only for the briefest of moments between posing the question and proposing an answer, never long enough for it to have a real effect.

The Bavli's repertoire of responses to *meitivi*-type confrontations consists of (a) declaring the amoraic position exceedingly problematic or refuted[92]—which is not the case in *Berakhot* 19b—(b) relegating the two rulings to quite different, unrelated realms of halakhic reality, and (c) rendering the tannaitic source supportive of the amoraic ruling by reinterpretation or reformulation. (b) and (c) amount to the same thing: harmonization at all cost. Once pointed out, contradictions that straddle the great generational divide are simply not allowed.[93] Hence, either the amoraic ruling involved must be dismissed[94] or the tannaitic texts involved support it—at the very least by being shown not to address the issue. Unlike the Yerushalmi, nowhere in the way the Bavli responds to transgenerational incongruities is it even hinted that, halakhically speaking, the *tannaitic* source involved may be inappropriate. Its text may be corrupted, or our understanding of it lacking, but a tannaitic source cannot be *mistaken*. If anyone is to blame, it is us for the inattentive way it was handed down or the slipshod way it was read and understood. The only place in the Bavli in which the *s'tam* appears to at least premise the exclusively antitraditionalist claim that the earlier of the two sources be answerable is here, in the way the *meitivi*-type problematic is repeatedly elaborated. But in response to each of the five confrontations the *s'tam* swiftly returns to the Bavli's well-honed methods of harmonization. Each of the *beraitot*, we are told, either speaks merely of rabbinical edicts that are excluded from Rav's principle anyway, or of precepts actually laid down in the Torah *as exceptions to the rule*—to Rav's rule of course. By the time each of the questions has been answered, all trace of whatever antitraditionalism was allowed to briefly make its appearance in the course of presenting them has been effectively obliterated.

Indeed, had the unusual formulation of the five questions been the only grounds I offered, my case would have been exceedingly unconvincing, if not downright silly. But it is not the only evidence. We have already seen that although Rav's ruling is stubbornly harmonized with the five *beraitot*, it remains conspicuously *un*harmonized with both the Mishna and Tosefta. If it would be out of character for a traditionalist seriously to entertain the thought that, or wonder in all seriousness why, a tannaitic authority had failed to employ a line of reasoning successfully applied by an amora, it would be downright unthinkable for a traditionalist construction of a *sugya* such as this to leave two such major transgenerational contradictions untreated. On the other hand, the way the *sugya* responds to the five questions, nicely corroborates such a reading. From an antitraditionalist perspective, however, the exact opposite is true. It is equally unthinkable that a hard-nosed antitraditionalist would invest such effort in harmonizing away the five transgenerational incongruities. On the other hand, the unresolved tension between Rav's ruling and the two tannaitic texts, not to say the nonironic expression of puzzlement regarding the motivation for the rulings registered in the *beraitot*, would be the natural response for a narrator writing from an unabashed antitraditionalist perspective.

The situation is not symmetrical, however. The assumptions that a traditionalist or an antitraditionalist could have framed the *sugya* are *not* equally improbable. Only the former option is genuinely inconceivable. The only way the *sugya* could be attributed to a prudent traditionalist is by writing off his failure to attend to the inconsistency between Rav's ruling and the two tannaitic sources as a fantastic oversight, and by reading his declaration of surprise at the five tannaitic rulings as expressions of sarcasm addressed to Rav—both of which seem so out of character as to count as almost conclusive grounds to reject them. By contrast, all one needs to assume in order to sustain an antitraditionalist rendition of the *sugya* is that the harmonization of the five *beraitot* with Rav's later ruling is not performed with utmost seriousness and that the *sugya* is contrived in this respect ironically to lay bare the hollowness and sheer formality of the Bavli's feigned traditionalism in this type of confrontation. From an antitraditionalist perspective, the *sugya* under consideration is a masterly crafted attempt subtly to expose, to a small and select readership, the policy and program that ground an entire class of similar units of talmudic discourse. By contrast, those who prefer to take the traditionalist format literally are forced to explain the *sugya*, ad hoc, as a remarkable case of careless editing. Two additional points serve, in my opinion, to tip the balance decisively in favor of the antitraditionalist option. The first concerns the passage preceding, and leading up to the discussion of Rav's ruling, the second concerns the passage immediately succeeding it.

Giving away the Game, or The Gentle Art
of Inaudible Instruction

Rav's dictum that "wherever a profanation of God's name is involved no respect is paid (even) to a teacher" is not the first mention of respect for teachers, *kevod la-rav* or *kevod ha-rav*, that occurs in the *sugya*. The issue is raised a page earlier in a relatively long and rather strange discussion of a claim made by another first generation amora, R. Yehoshua b. Levi who asserts that, "In twenty-four places (it is taught that) the *Bet din* excommunicates a person for (reasons to do with) respect to a teacher—and we learn them all in our Mishna." Despite its vague phrasing, all commentators without exception take R. Yehoshua b. Levi to be claiming that, as Rashi puts it, on twenty-four occasions the Sages are reported by the Mishna to have excommunicated individuals for acting disrespectfully toward their teachers.[95] The matter, however, may not be as simple as it sounds. In all, five incidents of rabbinic excommunication are cited. Taking up R. Yehoshua b. Levi's challenge, R. Elazar is said to have found only three cases in which excommunication was contemplated by the rabbis that are explicitly mentioned by the Mishna, of which in only two was the ban actually enforced. And the *s'tam* adds two more, mentioned only in *beraitot*. The fact that none of the cases mentioned seems to have anything to do with disrespect for teachers is puzzling in itself. But the strangest and, in my opinion, the most interesting aspect of the discussion is the fact that at least two of the cases cited imply the exact opposite—namely, that individuals were excommunicated by Sages for being *too* respectful of their teachers! Alongside the actual excommunication of one Elazar b. Ḥanokh—for opposing the ritual washing of hands decreed by the rabbis—and would-be banning of Ḥoni ha-Ma'agel and Todos of Rome—the former for the petulant manner of his pleading with the Almighty[96] and the latter for permitting his community to eat on Passover a kid roasted in a way that resembled too closely the Temple ritual—we find mentioned the two great antitraditionalist tannaitic legends of excommunication discussed in the previous section—that of Akavia b. Mehalal'el and that of R. Eliezer b. Hyrqanus. Neither of these is told here in any detail. (If anything the former is told misleadingly.) They are briefly alluded to, assuming, so it seems, that the reader is well acquainted with the material.

As in the discussion of Rav's principle that follows almost immediately, here too, I shall argue, one encounters a surface impression that not only contradicts a more sophisticated reading of the text, but seems to have been deliberately constructed in order to conceal it. To begin at the most explicit and seemingly obvious level: what we appear to be told by R. Ye-hoshua b. Levi is of a firm and apparently uncontested tannaitic tradition, widely attested to by the Mishna, that so valorizes the honor and respect due to teachers as to deem those who fail in this respect liable to be banned.[97] The discussion that follows seeks to corroborate both of R. Ye-hoshua b. Levi's claims: (a) that such a tannaitic tradition exists, and (b) that evidence for it is found not only in *beraitot*, but also in the Mishna itself. Still keeping to initial, superficial impressions, the discussion seems to yield whatever it was intended to yield. The details, at this level of read-ing, are unimportant. What matters is that (a) and (b) are not outwardly disputed, and Mishnaic evidence of such a policy seems to be produced to the *s'tam's* satisfaction. The only aspect of R. Yehoshua b. Levi's claim that is contested is the number of relevant examples to be found in the Mishna. Rather than twenty-four there turn out to be only three, or at most not to exceed five. But that is beside the point. The idea that the Mishna *maintains* that disrespectful behavior toward a teacher is suf-ficient reason for excommunication goes unchallenged and, by implica-tion, undisputed. Less than a page later, Rav will, of course, claim that the value laid on the respect due even to one's teachers is nevertheless outranked by the value laid on fully and promptly carrying out one's re-ligious duties. According to Rav, respect for the words of the Torah sur-passes respect for those responsible for teaching them. If mixed kinds are found in a garment, one is obliged, according to Rav, to remove it on the spot regardless of the embarrassment it may cause even to one's teacher. At the cursory level of having to treat one's mentors *politely* the two claims jar somewhat, but do not contradict each other. One stresses the value of respecting one's Master while the other explicates its limits. But in tal-mudic culture, *kevod ha-rav* means more than courtesy.

In talmudic culture, the very term *respect for teachers* refers first and foremost to respect for their *teachings*. A disciple who issues an edict or even pronounces a particular low-level ruling in the vicinity of his still functioning Master, we are told in several places—not surprisingly per-haps, always apropos a story about the archtraditionalist R. Eliezer b. Hyr-qanus—is punishable by death, even if his ruling coincides with his mas-ter's opinion and is perfectly correct.[98] To openly *disagree* with one's teacher is even worse. "To dispute one's teacher" says R. Ḥisda, "is like disputing the Almighty," to which R. Ḥanina b. R. Papa adds: "To even think of doing so, is as if one was thinking of contesting the Almighty."[99] All this is not prima facie as blatantly traditionalist as it sounds. Once a person ends his studentship, we are told elsewhere, and becomes a *talmid-*

ḥaver—a recognized fellow academician in his own right—he is no longer expected to follow his former Master to the letter, or to belittle himself to such an extent in his presence.[100] Still, when applied to the special realm of *kevod ha-rav*, the realm of a person's duties regarding the *teachings* of his master, the incongruity between the position attributed by R. Yehoshua b. Levi to the Mishna and that attributed to Rav becomes far more pronounced. Bearing this connotation of *kevod ha-rav* in mind, Rav's ruling acquires an additional and new aspect: if one is convinced that a profanation of God's name is in the making, Rav's principle now implies, one is expected no longer to pay respect to one's teacher's *teachings*. While R. Yehoshua b. Levi appears to be claiming in the name of the Mishna that those who contradict their mentors are liable to be excommunicated, when duly generalized Rav's principle implies that if one truly believes one's masters to be wrong and liable, as a result, to cause a "profanation of God's name," one is *required* to set aside all considerations of honor and respect and to speak one's mind. Interestingly, elsewhere in the Bavli Rav's ruling is explicitly employed to mean just that.[101]

Read thus, Rav's principle is rendered more than the mere *outcome* of a covert antitraditionalist move, more than a curiously veiled, surreptitious halakhic breaking with the past. It now reads, also, or perhaps even mainly, as an all but open *declaration* of antitraditionalism itself! Not only is one obliged to publicly strip, even in the presence of one's teacher, if mixed kinds are discovered, but, Rav now seems strongly to suggest, one is required outwardly to contest one's teacher's teachings and rulings if one genuinely finds them inappropriate. The implication is that just as one is encouraged constantly to check all garments for mixed kinds with a view to taking action whenever one considers it necessary, one is likewise encouraged critically to scrutinize everything one learns with a view similarly to speaking one's mind whenever one thinks it necessary, even in public. Even great teachers are fallible and capable, therefore, in *both* their conduct and their teaching, of unwittingly causing the Divine Name to be profaned. And if teachers are believed to be fallible, so are their students. Hence, individuals taking action on the grounds of Rav's principle should willy-nilly regard their own views and decisions to act equally liable to be mistaken and should, therefore, act cautiously, keep an open mind, and be willing to hear whatever criticism comes their way. It is most significant, in my opinion, that closely following the discussion of Rav's principle the *s'tam* presents the story of R. Ada b. Ahava who mistook a heathen for an Israelite and tore from her head an improper headgear, only to be humbled by a high fine of four hundred *Zuzim*.[102]

In short, if Rav's principle can be said to sustain such a reading—and Bavli, *Eruvin* 63a, apparently presumes that it does[103]—then it clearly implies, if not actually proclaims, all the main elements of antitraditional-ism. Needless to say, according to R. Yehoshua b. Levi "our Mishna" appears to suggest the opposite. If one is in real danger of excommunication, who would ever even consider publicly stripping or contradicting a teacher! Viewed thus, the Mishna implies that society cannot tolerate challenges to the authority of its teachers. Taking the place of a teacher, let alone actually defying his teachings, even for the best of causes may result in untimely death by the hands of Heaven and/or banishment by the hands of a human *Bet din*. In the context of a tutor's teachings—rather than in that of his attire or conduct—the implications of this assessment of "our Mishna" are centrally traditionalist. At the most superficial level, then, before studying the discussion of R. Yehoshua b. Levi's claim in depth, the passage devoted to it by the *s'tam* would seem to give rise to yet another set of tensions between the ruling attributed to Rav and the alleged policies of the Mishna. While Rav's words are confronted with a series of seemingly contradicting *beraitot*, and the apparent contradiction between his ruling and the specific ruling of the Mishna in relation to which it is cited hovers seemingly unperceived in the background, the *s'tam* appears to be doing his best to prove that the Mishna, *in particular* and as a whole, *beraitot* notwithstanding,[104] is not party to the antitradi-tionalism suggested by Rav's principle. Here, though, the tension is no longer contained within the tolerable limits of a specific incongruity between specific halakhic rulings. If until now Rav's ruling had seemed at most to be in disagreement with the particular edict recorded in the particular mishna under consideration, Rav now is seen to be contesting the Mishna's general metahalakhic principle according to which such discordancy is itself improper!

The way in which a cursory reading of one passage seems to amplify a considerably deeper understanding of the one that follows is interesting, perhaps even elegant, but it is also slightly worrying. It is worrying because by presenting the antitraditionalism residing in Rav's principle and surreptitiously revealed in the *meitivi*-type confrontations it undergoes, as *opposed* to the Mishna's general outlook, the *s'tam*'s position is debili-tated rather than bolstered. If, as I suspect, the framer of the *sugya* con-trives quietly to draw attention to the extent of his own antitraditionalist commitment, it is hard to imagine what he intended to achieve simply by citing and discussing the words of R. Yehoshua b. Levi. If indeed this was his intention, it would have been far more effective, it seems to me, for him to raise and discuss a less obviously traditionalist pronouncement of the Mishna's point of view. As we have seen in the previous section, there are several tannaitic sources that express an antitraditionalist per-spective quite clearly.

But then why should we compare a superficial reading of one unit of discourse with a far more subtle and deeper reading of the one that follows it? It makes far more sense to maintain an even level of reading for both passages. Read superficially, the *meitivi* discussion of Rav's principle, I have argued all along, is as traditionalist as they come. The entire discussion is geared on such a showing to prove that Rav's ruling conveyed absolutely nothing new—or so it seems. At such a cursory level of reading it is almost certain that the tension between Rav's ruling and the Mishna would also go unnoticed. At this level of reading, then, the two passages mesh nicely—with the traditionalism implied by R. Yehoshua b. Levi's claim effectively grounding that of the blissful harmony allegedly sought and found between the ruling of his amoraic colleague and the entire tannaitic tradition that went before it. A closer look, however, not only changes our perception of the discussion of Rav completely as we have seen, but also that of R. Yehoshua b. Levi.

In its discussion of R. Yehoshua b. Levi's assertion, the Gemara cites five examples of actual and would-be bans issued by tannaitic authorities, of which, it claims, only four are mentioned in the Mishna itself. The claim is true. The excommunications of Akavia b. Mehalal'el and Elazar b. Ḥanokh are related in Mishna, *Eduyot* 5:6–7; the threat to excommunicate Ḥoni ha-Ma'agel is noted in Mishna, *Ta'anit* 3:8; and the dispute associated with the banishment of R. Eliezer b. Hyrqanus is mentioned in Mishna, *Kelim* 5:10 and *Eduyot* 7:7. On the other hand, the threat we find in Tosefta *Beitza* 2:15, to banish Todos of Rome for permitting his congregation to eat kids roasted in their entirety on Passover, is nowhere mentioned in the Mishna. Despite this, I believe this fifth case, unmentioned by the Mishna, may be a key to the entire passage.

Although the threat to Todos is not noted by the Mishna, the issue of eating this type of roasted kid on Passover is mentioned twice. Both Mishna, *Beitza* 2:7, and *Eduyot* 3:11 report that Raban Gamliel held to the view that kids prepared in this fashion can be eaten outside the Temple on Passover night. His, we are told, was a minority opinion, however, that was voted down by a majority of his Jabne colleagues. As far as the Mishna is concerned—and the question in point *is* the Mishna's view on these matters—there was nothing prima facie wrong in Raban Gamliel *believing* that a full roast be permitted on Passover, or in his announcing his belief, or in his arguing the case with his colleagues. The Mishna does not give the Sages' reasons for voting against Raban Gamliel's proposal, but we may assume, following the Tosefta's remarks with regard to Todos, that the issue at stake was probably a desire not to emulate the Temple rituals too closely outside the Temple Compound. The point is that Raban Gamliel is not reprimanded for *adopting* the more liberal approach, but, as is commonplace in any Great *Bet din* or *Bet midrash*, is merely voted down. Todos, on the other hand, is not reported to have merely *opined* that such a dish was permitted on Passover but to have issued a ruling to the effect. Because of this he was threatened with excommunication. If dignity and respect can at all be said to be involved here, it is respect for the current halakhic authority, the majority of one's peers, and not that of one's present or former teachers. Had Raban Gamliel refused to accept the ruling of his colleagues, we may safely assume that he would have been called to task. And, as in the case of R. Eliezer b. Hyrqanus or Akavia b. Mehalal'el, the opinions of his former teachers would have had nothing to do with it!

This I believe is the intended import of at least three of the four incidents found by the Gemara to be of the Mishna itself. The case of Honi ha-Ma'agel is the exception. His sin is that of "arrogance against the Most High," as A. Cohen puts it.[105] The other three—Akavia, R. Eliezer and Elazar b. Hanokh—were banished for arrogantly opposing not the Almighty but the halakhic authorities of their day. The point is that the halakhic authorities of their day were *not their teachers or mentors, but the majority of their colleagues*! As I have noted repeatedly, those acquainted with the original stories well know that the banishing of Akavia and R. Eliezer was the result of their having followed the received teachings of their masters *too closely*! How could it be that the two most vividly antitraditionalist legends in the entire talmudic corpus are recruited here as confirmation of the Mishna's alleged traditionalism? The answer, I have indicated and shall now flesh out, is that, again, the *s'tam* has ingeniously constructed a *sugya* that speaks deliberately in two distinct and disharmonious voices, addressed to two distinct readerships—two voices that mesh remarkably well with the two equally discordant voices that, I have suggested, typify the *meitivi*-discussion of Rav's principle that follows closely in its wake.

To summarize, there is no way in which R. Eliezer's excommunication can be seriously presented as a penalty for violating *kevod ha-rav*.[106] He and Akavia and apparently also Elazar b. Hanokh were banished for refusing to submit to the majority ruling of their colleagues who had ruled *against* the traditions to which they were committed. But if they were banished for adhering too strongly to the words of their former teachers, the two stories, if not all three, would tend to disconfirm rather than corroborate R. Yehoshua b. Levi's assertion—unless, of course, he really meant to say something different.

First, as indicated at the outset, the phrasing of his statement is sufficiently vague to enable it (with a pinch) to state the opposite. His precise wording is: "In twenty-four places the *Bet din* excommunicates for respect of a teacher, and we have learned them all in our Mishna." It is possible, though, I admit, not as natural as the alternative, to read R. Yehoshua b. Levi as claiming, in accord with the examples that follow, that the *Bet din* is known to have excommunicated individuals for showing *excessive* respect for the lessons of their former teachers, to the extent of rendering themselves incapable of adjusting to the changing circumstances, of appreciating the advancement of learning, and of submitting to the authority of the current halakhic system. Obviously, the Bavli's slightly ambiguous wording of R. Yehoshua b. Levi's statement is not enough to sustain such a reading. But there is more. Unlike the subsequent discussion of Rav's principle, in which the five sources alluded to are cited in full, that of R. Yehoshua b. Levi's assertion assumes that the reader is well acquainted with the five examples mentioned. It is enough for the *s'tam* to merely *remind* his readers of each of the stories by quoting a relevant phrase. He does not tell the stories anew. Here he clearly seems to have the better-versed reader in mind. And we may safely assume, and even assume that the *s'tam* himself assumes that such a reader, puzzling over the apparent discrepancy between the claim and its alleged manifestations, will most naturally consult R. Yehoshua b. Levi's original statement cited by the Yerushalmi. I would even go further and say that just as the five examples of excommunication are merely pointed to rather than quoted, so is the assertion attributed to R. Yehoshua b. Levi, which appears for the first time in Yerushalmi, *Moed Katan* iii 81d, in the course of a long discussion of the entire issue of rabbinical banning.

"They sought to excommunicate R. Meir," begins the Yerushalmi. No reason for this is given, but R. Meir's reaction serves to prompt the detailed discussion that follows. "I shall not heed to you," he answered them, "until you tell me who are those who are (liable to be) excommunicated, on what basis do they (seek to) excommunicate (me), and for how many causes do they excommunicate" (*loc. cit.* 81c). In response, the Yerushalmi works through the list of known cases of rabbinical banning, presenting each of them as a paradigmatic example of the types of misconduct for which the rabbis are liable to excommunicate. These include the five cases mentioned by Bavli, *Berakhot*. And there, lodged between the Yerushalmi's interesting version of the banning of R. Eliezer b. Hyrqanus and of the would-be banning of Ḥoni ha-Ma'agel, one finds the following:

> R. Yehoshua b. Levi summoned a man (to his court) three times, but he did not come. He sent for him saying: If not for the fact that I have never, in all my life, declared a man banned, I would have banned that person,

because they excommunicate (a person) for twenty-four things (reasons) and this is one of them—(as it says:) "And that whoever would not come within three days, according to the counsel of the princes and the elders, all his property should be forfeited, and himself banned from the congregation of the exiles" (Ezra 10:8).

According to the Yerushalmi, R. Yehoshua b. Levi did not speak of twenty-four different cases of banishment for the same reason (*kevod ha-rav*, respect for teachers), but of twenty-four different reasons for banning people, of which treating an acting halakhic authority disrespectfully is apparently one. The term *kevod ha-rav* is not mentioned by the Yerushalmi at all in this context, let alone actually counted as a reason for banning. In fact, none of the cases mentioned by the Yerushalmi, including that of R. Yehoshua b. Levi himself, have anything specifically to do with the relationship between teachers and students or masters and disciples. The causes for excommunication that *are* listed are by the Yerushalmi: (i) not paying attention to the principle that the law follows the majority (the case of R. Eliezer); (ii) failing to come when summoned by the local halakhic authority (R. Yehoshua b. Levi with reference to Ezra); (iii) bringing the public to profane the Divine Name (Honi ha-Ma'agel); (iv) holding the community back from carrying out a religious duty (Raban Gamliel, R. Akiva);[107] (v) causing the community to eat (dishes that resemble) Holy Things outside of the Temple (Todos of Rome); (vi) doubting and disobeying a ruling, even if it derives from the rulings of Scribes (Elazar b. Hanokh); (vii) disgracing a Sage, even after his death (Akavia b. Mehalal'el); (viii) behaving like Yerov'am son of Nevat—that is to say, to cause the public to sin.[108]

According to the Yerushalmi, then, excommunication, expulsion from the community of Israel, would normally be the penalty for two types of essentially public offenses: for challenging and defying the authority of the community's institutions of power: the *Bet din* and its rulings (i, ii, vi); or for persons in positions of power to lead the community astray (iii, iv, v, viii). The two most pertinent cases are, of course, those of Eliezer and Akavia—two major rabbinic figures who, with the best of intentions, the former even with unquestionable proof of the truth of his halakhic traditions, were tragically unable to accept the fact that views they had good reasons for holding sacred could be voted down.

How offenses against the collective and the authority of its institutions of power came to be subsumed by the Bavli under the heading *kevod ha-rav* is hard to say. There are two possible explanations. The easy way out, and to some perhaps the most tempting, would be to blame it on some form of scribal error. Although I have found no evidence for such an error in any of the manuscript sources of Bavli, *Berakhot*, it is still extremely easy, for example, to mistake 'כבוד הרב—common shorthand for כבוד הרבים, respect for the public or for the many—for כבוד הרב—respect for teachers. And it may be tempting to suppose further that the *s'tam*'s real intention was merely to synthesize the Yerushalmi's initial and rather disorganized discussion of the rabbis' policy toward banishment under the general rubric of "disrespect for the public."

The other option is to assume that the term *kevod ha-rav* was intentional. By this I do *not* mean to say, of course, that the *s'tam* might have seriously considered the five examples he cites to be *examples* of individuals who were excommunicated for being disrespectful of their teachers. To the best of my knowledge, there is nothing to indicate that anywhere in the talmudic literature is disrespect for one's teacher ever deemed a transgression punishable by excommunication or one for which anyone is ever reported to have been banned or threatened with banishment. And since R. Yehoshua b. Levi and his friendly discussants do not aspire to offer their own opinion on the matter, but only to describe and substantiate what they take the position of "our Mishna" to be, it would be very unlikely for them to have diverged so thoroughly from all that is in fact contained in the tannaitic texts. The question, then, cannot be whether the parties to the discussion seriously believed that the five examples prove that the Mishna considers violations of *kevod ha-rav* punishable by excommunication, or whether the Mishna thus rules despite the examples. The only viable option other than scribal error is to assume that the *s'tam*, knowing perfectly well that the Mishna does *not* hold to such a view, still chose to formulate and discuss R. Yehoshua b. Levi's assertion as if it did.

But I shall leave it to the reader to decide between the two options. According to both of them, the *s'tam* knowingly introduces the *meitivi*-challenging of Rav's ruling by means of the two most dramatic legendary accounts of antitraditionalism contained in the tannaitic literature— which, of course, fare very well with any antitraditionalist rendition of the *sugya* as a whole. The "scribal error" option loses some of its appeal, however, when one recalls how hard the *s'tam* works in order to cover up his antitraditionalist tracks in the *meitivi* confrontations that follow. (Why he does so remains, at this juncture, an open question. But the *fact* that he does is now, I hope, reasonably well established.) If one inserts *kevod ha-rabim* for *kevod ha-rav* one is left with a forceful and visibly antitraditionalist description of the social and halakhic power and authority of the current leadership. If, on the other hand, one views the use of *kevod ha-rav* as deliberate, the entire passage is rendered a master-piece of the type of semitransparent double-talk that was seen to characterize the transgenerational negotiation that follow. Talk of a tannaitic policy of banning for reasons of *kevod ha-rav*, as we have seen, introduces the kind of formal yet hollow traditionalist superstructure, similar to the rhetoric of *meitivi* confrontations, through which those acquainted with the material can hardly fail to see. It is precisely the type of self-refuting doublespeak that distinguishes the advanced students from the beginners.

For the last several pages, I have been speaking of the *s'tam's* discussion of Rav's ruling as if it was obviously motivated not only by an antitraditionalist outlook, but by a desire to making it public only to a select, advanced readership. And although there is ample evidence to prefer this reading of *Berakhot* 19b to the traditionalist alternative, I had set out to study the preceding discussion of R. Yehoshua b. Levi's assertion with a view to deciding the issue. It swiftly became evident, however, that although R. Yehoshua b. Levi's claim itself sounds traditionalist, its discussion is even less capable of sustaining a traditionalist reading than that of Rav's ruling. As a result, the traditionalist option for the entire *sugya* was at some point abandoned, and rather than read the *meitivi* confrontation in the light of what comes before it, I ended up arguing in the opposite direction.

The final point I wish to make regarding the antitraditionalist subtext of *Berakhot* 19b concerns the culmination of the *sugya*. Here there is no longer need for subtlety or knowledge of other texts. For a brief moment, the *s'tam* seems to throw all caution to the wind and to announce his metahalakhic preferences for all to hear. He does so in the form of a dialogue between R. Papa and Abaye that constitutes one of the boldest amoraic statements of antitraditionalism of which I am aware. The dialogue, I believe, speaks for itself.

Asked R. Papa of Abaye: In what were our predecessors different from us that miracles occurred for them though for us they do not? Could it be a question of learning? (Surely not.) In the days of R. Yehuda (for example) all they learned (knew) was (the Order of) *Nezikin*, whereas we study (know) all six Orders[109]—(Indeed) when R. Yehuda used to reach (the mishna in) *Uktzin*[110] (concerning) "If a woman presses vegetables in a pot"[111]—and some say (it was when he reached the one concerning) "Olives pressed with their leaves are ritually clean"[112]—he would say 'I see here (a complexity of the same order as) the disputations of (my teachers) Rav and Shmuel' (but was unable to explain it himself). We (by contrast, are able to) study *Uktzin* in thirteen different ways.[113] Yet R. Yehuda would only have to remove one of his shoes for rain to begin to fall,[114] while we can afflict our souls and cry and cry and no one listens! He said to him: (it is not a matter of learning, rather) our predecessors jeopardized their lives for the sanctification of the Name, and we do not.[115]

The previously noted story of R. Ada b. Ahava's unfortunate mistake and four hundred *Zuzim* fine is then presented presumably as an example of how former Sages would risk all for the sanctification of the Name, *kiddush ha-shem*.

On the one hand, R. Papa's statement conveys far more than an antitraditionalist would ever need. In order to justify the adoption of a critical attitude toward the knowledge and teachings of former generations, it is quite sufficient to assume that all humans are fallible, that circumstances can change in ways that are liable to defy the most canny foresight, that even if the quality of learning of later generations is less than that of their predecessors they may still be considered dwarfs standing on the shoulders of giants. More than it may be considered a justification of antitraditionalism, R. Papa's bold assessment of the vast superiority of the knowledge and learning of his own generation in comparison to that of R. Yehuda constitutes a flat refutation of traditionalism. If it is true, it leaves no room at all for the type of commitment to former learning necessarily premised by all forms of traditionalism and seemingly presupposed in the rhetoric of standard *meitivi*-type confrontations.

On the other hand, the *s'tam* still succeeds in remaining ambiguous with regard to one crucial aspect. Forceful as it is, R. Papa's declaration is still not enough to prevent a traditionalist rendition of *meitivi*-type challenges completely, for the simple reason that he cautiously limits his assessment to the amoraic period. Traditionalists can still claim—and I can personally attest to the fact that many of them do!—that even if R. Papa and Abaye were right about the standard of their amoraic predecessors, it is unthinkable to even consider disputing a tannaitic ruling, let alone criticize the quality of *their* learning. Those who failed to take note of the unnoted incongruity between Rav's ruling and the Mishna and Tosefta, who passed by the *s'tam*'s tedious running commentary to the five *meitivi* confrontations, who ignored the discrepancy between R. Yehoshua b. Levi b. Levi's generalization and its alleged instances, will also, in all probability, read R. Papa's exchange with Abaye for its moral conclusion rather than for its epistemological premise. Hence, despite the bold and seemingly unequivocal message conveyed by R. Papa's statement, the antithetical, two-tier meaning of the *sugya* as a whole is retained to the very end.

As I said at the outset, the significance of the *meitivi* confrontations of *Berakhot* 19b is enormous precisely because qua *meitivi* confrontations they are wholly and totally typical. The five tannaitic challenges to Rav's ruling and their subsequent resolutions are so commonplace that, when studied, they are normally skimmed through with ease. They are not a parody of their kind in any obvious sense. They resemble any of the thousands of similar negotiations scattered throughout the Bavli. Even the sniggers earned by Rav b. Shaba's attempt to harmonize Rav's ruling with the tannaitic principle associated by the Yerushalmi with R. Zeira are not immediately taken by the reader as having anything at all to do with the idea of transgenerational harmonizing itself. The ways in which the five *meitivi* challenges are resolved by the *s'tam* do not appear to ridicule such solutions. On the contrary, the force and subtlety of the *s'tam*'s ironical exposition of their real meaning owes much to the fact that, of themselves, they are entirely credible and wholly representative of the Bavli's vast stock of similar negotiations. This is why I suggest that *Berakhot* 19b should be seen as a paradigm rather than a parody, as an extraordinarily constructed *sugya* that contrives to explain rather than satirize the class of *meitivi*-type, transgenerational negotiations. Viewed thus, *Berakhot* 19b is seen as an instructive, explanatory effort on behalf of one of the framers of the Bavli's many other *sugyot* of its kind, rather than that of an antagonistic critic. It is, I urge, the work of an antitraditionalist doing his best to *explain* the Bavli's antitraditionalist project, rather than that of an antitraditionalist aspiring to ridicule a traditionalist one.

All of this, however, is ingenuously concealed. There is little chance that innocent beginners will be deprived of their innocence by studying *Berakhot* 19b in its immediate context, any more than practiced, committed traditionalists are liable to be forced to rethink their former commitments. The *s'tam*, as we have seen, has skillfully provided the former with blinkers and the latter with a convenient safety net, in the form of a largely plausible, if somewhat disconnected, surface rhetoric that serves intriguingly well to mask the accumulating evidence of his own unmistakable antitraditionalism from the eyes of those who should not, or those who can, but desire not to appreciate it. I doubt, however, that he would have mindfully catered to the latter. When it comes to the inheritance of former authoritative documents of the type to which the Bavli and Yerushalmi are almost wholly devoted, traditionalists and antitraditionalist are too bitterly opposed to be expected to write and edit their works with a view considerately to accommodate their adversaries. The *s'tam*'s ambiguity cannot be attributed, in my opinion, to a desire to provide the traditionalist with an honorable way out. The fact that in retrospect it is

seen to do so is beside the point. His ambiguous double-talk, I have suggested all along, seems intentionally to be catering to the beginner rather
than to the opposition.

So far we have seen *how* this was achieved and must now turn to the
inevitable why-question: Why bother? What might have the *s'tam* hoped
to achieve by thus misguiding the novices?

Understanding the Bavli

Problem One: Explaining the Bavli's Double-Talk

Few would disagree that for the most part, the basic rhetorical devices and discursive frameworks the Bavli employs are consistently traditionalist. Traditionalists go further, claiming that the Bavli's surface rhetoric and vocabulary are not *merely* rhetorical, not simply a facade, but faithfully reflect the theological and epistemological commitments of its framers. For those, like the present author, who are convinced that the Bavli is far less the work of traditionalist redactors than the idiom in which it is cast might lead one to believe, the feigned traditionalism of its surface rhetoric calls for explanation—when indeed feigned, of course. I shall not repeat the reasons given in preceding chapters for maintaining that the apparent traditionalism of the Bavli is at least part of the time indeed apparent. For the present I shall assume that it is; moreover, I shall assume that when it is, it is contrived and deliberate—that is to say, not merely a habitual manner of speaking, a mere mannerism. The question is then: what, when merely professed, might the Bavli's professed traditionalism have been *meant* to achieve? I have emphasized the word *meant* because what such a move might have been meant to achieve, and what it might have achieved in practice are, more times than not, two quite separate questions—as those who read Part 1 of this book are surely aware. Indeed, one of the more speculative and hesitant conclusions of the present study will be that the Bavli largely failed to achieve what it was arguably meant to achieve. Still, it is worth looking first at the most obvious and inevitable consequences of this sort of double-talk, the type of outcome that those responsible for it could hardly have failed to anticipate.

The most obvious effect of saying one thing and doing another is to mislead the obviously unwary. By the unwary I mean those who are most liable to take one for one's word, and by the obviously unwary, those who are almost certain to do so. In the case of the Bavli these would comprise two main readerships: trusting beginners and committed-in-advance traditionalists. Whatever its purpose, wherever the Bavli's doublespeak is indeed deliberate, those responsible had to realize that in so doing they were contributing extremely little to recruiting their less experienced readers to the antitraditionalist cause and doing even less to win over their more seasoned traditionalist adversaries. This is no way at all to do battle with the opposition. It might be a way to hide one's real intentions from one's opponents. But if the antitraditionalists among the framers of the Bavli were actively *campaigning* for their position, then one should assume that it would have been these two audiences that they would have sought most effectively to address. We may safely assume, therefore, that the Bavli's traditionalist rhetoric was not intended as a *polemic* device.[1]

Given the resoluteness with which the dispute between traditionalists and antitraditionalists is pursued in the tannaitic texts, it is curious that those responsible for the antitraditionalist voice of the Bavli would have thus given up the fight in advance. The reason for this could, in principle, be one of three. One possible explanation might be that the deliberate toning down and masking of the Bavli's antitraditionalist voice was due to traditionalist repression. On such a showing the Bavli's traditionalist language would be seen as a defense, a camouflage, a shield. Direct confrontation is avoided, perhaps because it is perceived as hopeless. There is little to support such an explanation. For one thing, there is more to the Bavli's antitraditionalist voice than its muffled presence in transgenerational negotiations. Had the composition of the Bavli been dominated by traditionalists to the extent of actually stifling the opposition, we would not expect it to include such corpora as the outspokenly antitraditionalist Jabne stories or the many disputes between the Houses with its firm amoraic resolution in favor of the antitraditionalist Hillelites. The Bavli also includes several legends of amoraic origin that seem as clearly and self-consciously as the Jabne stories to side with antitraditionalism. We shall have the chance to examine two of these stories in some detail below.

A second possible explanation of the Bavli's apparent reluctance to openly combat traditionalism in its first-order halakhic negotiations might be to view it as an expression of contempt, as the type of spurious, affected indifference that *is* a polemical move. Though somewhat more plausible than the first explanation—owing to the rabbis' tendency to ignore or snub those contenders who, like the Sadducees, are perceived not to belong within the boundaries of legitimate dispute—such an interpretation, I suggest, should also be repudiated. First, it seems quite ridiculous for one to express indifference to a position by doing one's best to sound as if one has accepted it! More seriously, traditionalists as such are not considered heretics and are nowhere treated with the kind of contempt reserved for the Sadducees and their like. The Bavli, as I have argued above, seems to me to be far too dedicated to *teaching* Torah-study to simply ignore an entire school of thought. If its framers wrote self-consciously from an antitraditionalist perspective, it is hard to imagine why they would have decided not to address the opposition openly. Unless, of course, and this is the third possible explanation, there was no real opposition to speak of.

The idea that the great dispute, made so much of in the tannaitic literature, had been largely resolved, or perceived to have been largely resolved, by the time the Bavli had reached the final and decisive stages of its composition, strikes me as the most plausible option. The amoraic literature—by which I mean the literature composed by *amoraim* about *amoraim*—does not contain legends that speak of bitter conflict between the two schools of thought as do the Jabne stories or the Houses' disputes. The amoraic literature contains several dramatic, legendary passages that vividly announce and advertise a clear antitraditionalist point of view. But, as we shall see, these accounts hardly acknowledge the existence of a traditionalist opposition. They proudly present a philosophy, characteristically couched in story form, at times in very bold terms, but, unlike their tannaitic counterparts, they do not seem at all to be *fighting* for it. These legends appear to expose and elaborate, not to dispute. It is not surprising perhaps that there exists no amoraic version of an ongoing metahalakhic controversy between two well-defined "Houses" or schools of thought. There is no amoraic parallel to the tannaitic House of Shammai, and no amoraic counterpart to the archtraditionalist tanna R. Eliezer b. Hyrqanus. The texts contain, of course, the occasional traditionalist statement and, as we have seen with regard to the legend of the young Hillel and *b'nei Beteira*, evidence of editors and compilers of traditionalist leaning. But by and large, the amoraic portions of the Bavli that do address these questions openly strike one quite clearly as being the product of a culture in which the great controversy is perhaps still remembered, but in which it is no longer being fought. All of which fares well with the conclusion of the preceding section in which we noted that by the end of the day *Berakhot* 19b reads more like an emphatic, if whispered, explanation of what *in fact* is at stake in transgenerational negotiations than a hostile exposition of what *ought* to be at stake in such units of discourse.

But if the antitraditionalist framers of the Bavli can be said to have plied their trade in a culture virtually free of traditionalist dissent, or at least to have conceived of it as such, then their apparent decision to hide their real views on *talmud Torah* behind a traditionalist facade remains all the more puzzling! To repeat, one inevitable consequence of such a decision is that the trusting beginner is almost certain to be kept in the dark. The type of doublespeak practiced by the Bavli all but ensures that inexperienced novices will be misled to think that they are indeed being introduced to, and partaking in, a traditionalist exercise. This is certainly an inevitable *outcome* of the Bavli's policy, but could it also have been its *objective*? Can the Bavli's double-talk be *explained* by its hoodwinking effect on an assumed readership of beginners?

Problem Two: Discerning a Role
for the Mishna

Before answering this question let us first consider another—one that has not yet been raised. Namely, what sort of impact might the Bavli's concealment of its antitraditionalism have on its assumed readership of *non*-beginners—that is to say, the Bavli's imagined seasoned reader who is assumed to somehow know in advance that its traditionalist rhetoric *is* indeed merely rhetoric. The answer, I suggest, depends on how this advanced reader is thought to perceive the aims of the facade. It should have no visible effect on seasoned students who perceive it as a device introduced by the *s'tam* for *the benefit of the novice* (for reasons to be described shortly). But for those who perceive the Bavli's rhetoric as also addressing them, the message they will inevitably receive is that although *talmud-Torah* is an essentially antitraditionalist undertaking, although senior members of the learning elite are supposed and expected to criticize all former authoritative documents, *one should never say so openly*—not merely for the benefit of the novices, but as a matter of principle! Those who are capable of hearing the Bavli's two voices, but who do not see them as addressing two separate readerships, will take the double-talk to convey a double standard, regardless of the intentions of its framers.

In keeping with the aims of the present study, the main question I wish to discuss is not that of the *possible* effects the Bavli's double-talk might have had on its *potential* readerships—at least not primarily—and it is certainly not that of the effects that it turned out *in fact* to have had on its *actual* readerships. Rather, I wish to ask what the antitraditionalist framers of the Bavli might have possibly thought to achieve by their employment of double-talk. As we have seen, beginners should be expected to be mislead by the surface rhetoric almost as a matter of necessity. It is inconceivable that the framers of the Bavli would not have taken that into consideration. So the question boils down to their possible intentions with regard to their more sophisticated readers. Broadly speaking, the two options are these:

(i) That the framers of the Bavli intended their seasoned readers to see through and past the hollow traditionalist idiom employed in their writings, to understand that its function is purely didactic, and to go on eventually to pursue their own halakhic enterprises in good, simple, and openly antitraditionalist fashion.

Or, conversely:

(ii) That the framers of the Bavli intended their seasoned readers to realize that the subtle masking of their antitraditionalism was meant to serve as an example and model for all future studies. That they believed that maintaining the sort of affected, almost entirely formal traditionalism practiced throughout the Bavli is essential for the perpetuation and proliferation of talmudic culture for reasons that transcend pedagogy,

Let me say at the outset, that I do not consider myself qualified to decide the issue even to my own satisfaction. In what follows I shall try to elaborate somewhat on the two alternatives and shall try to show that the former, the more modest proposal of the two, perhaps explains more. Though I should admit that I have so far failed to convince many of my colleagues and friends that it is even a viable option. The latter, certainly the more intriguing of the two, seems to many of those I have talked to, to be the obvious and natural choice.[2] They, in turn, have so far failed to convince me, however. But I do agree that (ii) presents attractive possibilities. Let us start with (ii) and work back to (i).

What would a culture have to gain by jealously preserving a vacuous and almost wholly formal traditionalist idiom even among those of its members who have long realized that it *is* merely formal? What can a community of learners, who perceive progress and the growth of knowledge as products of creative troubleshooting, be said to achieve by refusing to talk openly about it? What is to be gained by scrupulously preserving all former authoritative documents while granting to subsequent generations limitless freedom to reinterpret them as they see fit? What is the point of adhering willingly to the most advanced, novel, latter-day opinions, while doing one's utmost to harmonize them at almost all costs with those of yesterday?

One thing that can be achieved in this way is a form of canonization that is undogmatic and almost wholly unrestrictive. In his forthcoming *People of the Book: Canon, Meaning, and Authority*, Moshe Halbertal distinguishes between two types of canonical texts. On the one hand, there are texts that become "central for a profession or a culture." He calls them "central texts." Such texts "might affect directions and knowledge and shape the future of the field." They are the classics, the "Great Books" of a culture. They set standards, constitute paradigms, and serve as milestones for historians and educators. Central texts constitute the great steps forward, the pivotal events in the development of a field, landmarks with reference to which later progress is measured. Such are the works of Isaac Newton and James Clerk Maxwell for physics, William Shakespeare and George Bernard Shaw for literature, Johann Sebastian Bach and Gustav Mahler for music. But in text-centered communities and enterprises, such as talmudic or kabbalistic Judaism or certain systems of law, one also encounters a different kind of canonical work. Halbertal calls them "formative texts"—perhaps not the most successful choice of phrase because "central texts" may also be formative in a variety of ways. A "formative text," he writes, is a text such that progress in the field is made *via its interpretation*. "A text-centered culture that has formative texts proceeds in that mode; its achievements are interpretative."[3] Newton's *Principia* obviously figures prominently as a canonical work of Western science, but no one would claim that progress in physics is achieved by discovering new ways of reading Newton. If anything is formative of physics in this sense, it is the *phenomena* physicists investigate. This is true of any community of inquirers. The phenomena that function as the subject-matter of investigation, are, by definition, formative of the field, or the culture, in Halbertal's sense of "formative." In text-centered cultures, certain canonized texts will serve as the objects of constant reflection and interpretation. They constitute the "formative canon" of such cultures and are analogous to the persisting subject matters of scientific fields of inquiry.

There is no question that the fundamental distinction between the Written and the Oral Torah, as described above, fares well with Halbertal's analysis of the function of a formative canon within a text-centered culture. The all-inclusive canonization of the Written Torah, the bare, as-of-yet unread text of the Hebrew Scriptures, is axiomatic in talmudic culture. At the level of the bare "text-in-itself" so to speak, the twenty-four books of the Bible are sacred. They form an object of study that is itself untouchable, transcendental. As the rabbis put it, the Written Torah constitutes for talmudic Judaism a primordial Given imposed upon its interpreters "as a mountain pressed down upon their heads." As noted earlier, there exists no halakhic procedure by which to change a single letter of the sacred Text. But according to all but the most extreme traditionalists—so extreme, we now realize, as to be little more than mythical exaggerations—the Oral Torah is in fact an ongoing, open, and progressive project. In the realm of halakha, the less extreme traditionalist views the Oral Torah as an ever-growing, accumulating stock of authoritative rulings, while for the antitraditionalist, it is the frequently revised product of persistent troubleshooting.

Under the heading "The Concept of 'Torah' in 'Talmud Torah,' " Halbertal analyzes, among other things, the formative function (in his sense of "formative") that the Written Torah played for the talmudic Sages and that played by the Bavli and the Zohar for posttalmudic Judaism.[4] He does not apply these insights to the possible role that the tannaitic texts, especially the Mishna, might have been meant to play by and for the framers of the Bavli. Employing Halbertal's terminology we could restate the deliberation between (i) and (ii) by asking whether in their treatments of the tannaitic texts the framers of the Bavli sought to establish them as texts that were meant to be *central* or texts that were meant to be *formative* to talmudic culture. This is because a text that is central to a culture will be revised but seldom reinterpreted, while formative texts, while liable to be reinterpreted are expected always to be preserved.

One good reason for preferring the latter option—that the Bavli treats the tannaitic texts as a formative canon—is that the former, of itself, does not explain the Bavli's double-talk, while the latter, as we shall see immediately, does so nicely. For a text to be made central to a culture, in the sense in which Newton's *Principia* is central to physics, no aspect of it needs to be preserved and worked into the culture's current, day-by-day give and take. But when a text is considered formative for a realm of activity it will be preserved and constantly referred to by all involved. It is tempting, therefore, to explain the Bavli's seeming ambivalence toward the tannaitic literature by viewing it as part of a contrived effort on behalf of its framers to establish that literature as a formative canon. On such a showing what we have been calling the Bavli's traditionalist facade and traditionalist surface rhetoric would no longer be considered a facade or surface rhetoric at all but a genuine expression of traditionalism toward the bare, uninterpreted tannaitic text.

According to this a view, the aim of the Bavli was to preserve the tannaitic texts at all costs *in order for them to serve as the permanent objects of ongoing halakhic reflection and interpretation*. The framers of the Bavli will be viewed as partaking in the very enterprise they seek to establish: never contesting the tannaitic texts, only interpretations of them; constantly

seeking new, creative readings of them which are arguably less problematic. Thus construed, the Bavli's traditionalism is directed exclusively to
the form of the tannaitic document, to its wording, to the text itself—
never to its content. The Bavli's antitraditionalism, by contrast, asserts itself only in relation to *interpretations* of these texts—never questioning
the texts themselves. In this manner, the Bavli's so-called halakhic double-talk is rendered wholly analogous to the culture's differentiation between the words of the Bible and their meaning, between the Written
and the Oral Torah.

But we have so far been reasoning hypothetically. All we have said is
that *had* the framers of the Bavli intended to make the tannaitic texts
part of the culture's formative canon, *then* their apparent indecision with
regard to the traditionalist/antitraditionalist dispute in their treatments of
the tannaitic literature would be explained. But we have first to ask *why*
they would have sought in the first place to expand the corpus of their
formative texts in this fashion beyond the "Torah from Heaven."

One very good reason for them *not* to have done so is precisely that.
Unlike the tannaitic literature, they held the Written Torah to be "from
Heaven"—*min ha-shamaim.* By contrast, the tannaitic halakhic literature
is, on all accounts—*especially from the point of view of the Bavli's antitraditionalists!*—viewed by the rabbis as the work of human authors. The rabbis' distinction between the word and meaning of the Bible is predicated
and in a sense entailed by the unquestioned Divine origin of the former.
There would seem to be no analogous theological motive for granting the
tannaitic texts the same sort of status. If they are not the actual *words* of
the Almighty, and no one claimed that they were, what is there to justify
their transcendence in the eyes of their amoraic heirs? But then, they
might have been granted formative status for a quite different reason.

One must not forget that Torah-study, as understood and pursued by
the rabbis, was geared to and motivated by two closely related yet very
different objectives: on the one hand, the essentially midrashic, interpretative task of scriptural exegesis, and the essentially halakhic, judicial task
of establishing and developing the system of Jewish law, on the other.
Midrashei ha-halakha, the largely exegetical, largely tannaitic literature devoted to educing the Torah's halakhic import, serves to a great extent to
amalgamate the two great rabbinic undertakings. But it is impossible to
combine them fully. No amount of creative exegetical work is capable of
producing a systematic, comprehensive, consistent body of halakha, just
as no amount of creative legislation is capable of producing a systematic,
comprehensive, and consistent reading of Scripture.[5] The two undertakings are motivated by and develop in the course of constantly reacting to

two categorically different sorts of question. In terms of Halbertal's distinctions, the Written Torah may be said to provide rabbinic culture with a set of texts that is formative *only with regard to the interpretative, exegetical of the two projects*. For although it contains numerous halakhic and even metahalakhic passages, the Torah itself cannot be said to provide its learners with even the rudimentary elements of a systematic body of halakha. As much as one might want them to, the Hebrew Scriptures are incapable of functioning as formative texts for the task of halakhic system building. In other words, insofar as talmudic culture may be said to have been in need of a formative text for the sake of grounding its halakhic discourse, the Mishna would have been the perfect candidate.

The Mishna as a Formative Code

Needless to say, many (some of my colleagues say all) legal systems take form, function, and develop with reference to constitutive, formative canons. In many legal systems, acts of legislation are considered acts of interpretation of the system's formative documents or at least as elaborations thereof.[6] Codified systems are, of course, the paradigmatic example. I doubt, however, that this is true of every known legal system. To my limited understanding, constitutional systems and systems founded on common law are not text centered in the same sense. They do not necessarily premise a text of which each act of legislation is considered an interpretation. But even if it were true that *as a matter of fact* all legal systems are known to premise such formative texts, as some of my colleagues insist,[7] that would still not be sufficient reason to assume that the framers of the Bavli thought so too. Since only few would be willing to claim that there is something inherent to legal systems as such to *necessitate* their grounding in formative texts, no amount of theorizing about such systems is capable of deciding whether or not the Mishna is treated by the Bavli as a text formative of the halakhic world of discourse it reconstructs. On this the Bavli will have to speak as it were for itself.

In order to answer this question in the affirmative, to claim that the antitraditionalist framers of the Bavli do *not* treat the Mishna as a central canonical text but as a formative one, one needs to show that the Mishna is *not* treated by the Bavli as an elaborate first shot at the ideal, full-blooded halakhic system, that it does *not* make the Mishna the object of intense troubleshooting, and does *not* treat the Mishna as a point of departure for attempted improvement. To maintain that the Mishna is upheld by the Bavli as a formative halakhic text requires one to show that it is not revised by the Bavli but made to serve as what students of the law call a Continental-like authoritative code—Continental-like in the sense of it being all-encompassing as are thought to be the great French and German civil codes; authoritative in the sense of it being final and fully obligating.[8] Most Western European systems—the United Kingdom being the most conspicuous exception—have adopted one of two great codes; legal codes that are formative exactly in the sense that acts of new legislation, judiciary and institutional, are considered acts of application—interpretative, deductive, or analogical—of the code's relevant general ruling. The important point for the present discussion is that, as a rule, where such systems obtain, the code itself is not troubleshot or amended by the official legislative bodies but is only expounded upon by a variety of interpretative and explicative devices. Such codes are formative of and for the legal system in Halbertal's sense of the word. Halakhic reasoning today premises such authoritative codifications. Maimonides's twelfth-century *Mishne Torah* and R. Yosef Karo's sixteenth-century *Shulhan Arukh* function in current halakhic discourse very much as formative texts of this kind, although they were hard disputed in their time.[9] But there are good reasons to think that a major and constitutive voice of the Bavli treats the Mishna very differently.

Take for example the amoraic discussion of Mishna, *Berakhot* 6:3. The Mishna's general concern in this chapter is to define precisely which of the various benedictions is appropriate to the which kinds of foods. Before eating the fruit of the trees one says: "(Blessed art Thou, O Lord our God, King of the universe) Who creates the fruit of the tree," for wine, "(. . .) Who creates the fruit of the vine," for vegetables ("fruits of the earth") one says: "(. . .) Who creates the fruit of the ground," and so on. Mishna 6:3 states:

> Over anything whose growth is not from the earth, one says: "(. . .) By Whose words all things exist." Over vinegar, fruit which falls unripe from the tree, and locusts, one says "(. . .) By Whose words all things exist."

The *s'tam* begins by quoting another tannaitic source, a *beraita*, that enlarges upon the ruling of the Mishna:

> Our Rabbis have taught: Over anything whose growth is not from the earth, such as: the flesh of cattle, beasts, birds, and fish, one says "(. . .) By Whose word all things exist." Over milk, eggs, and cheeses, one (also) says "(. . .) By Whose word." Over bread which has become moldy,[10] and wine which has filmed over,[11] and cooked food which "has lost its form"

(i.e., gone bad), one (also) says "(. . .) By Whose word." Over salt, brine, truffles, and mushrooms, one (also) says "(. . .) By Whose word."

From which follows, continues the *s'tam*, that mushrooms and truffles are considered not to grow from the earth. However, a second tannaitic source clearly seems to imply the opposite.

But we are taught: He who vows not to eat the fruits of the earth is forbidden to eat [fruit or vegetables] but is allowed to eat mushrooms and truffles. However, if he declares: "All that grows from the earth is forbidden to me!" he is also prohibited from eating mushrooms and truffles.

In other words, this *beraita* implies quite clearly that mushrooms and truffles *are* considered to grow from the earth. At this point the *s'tam* cites the fourth-century *amora* Abaye who pointed out that although mushrooms and truffles indeed grow from the earth, they do not extract their nutriment from the earth. But of what assistance can that be, he asks, for the Mishna explicitly states that in order for anything to qualify for "(. . .) By Whose word" it has not to grow from the earth, and yet mushrooms and truffles apparently do qualify for this benediction despite the fact that the halakha casts them specifically among things grown from the earth![12] The solution proposed by the *s'tam* is to amend the Mishna: instead of the original text, the criterion by which to qualify for "(. . .) By Whose word" should be changed in accord with Abaye's botanical observation—or as the *s'tam* puts it: from now on "read instead: 'Anything which does not derive nutriment from the earth'!"[13]

This is not the dissenting view of a first-generation amora like Rav, nor is it a suggestion to reinterpret the Mishna. It is the view of an anonymous framer of the Bavli writing more than three hundred years after the Mishna was finalized and sealed, proposing, as the solution to a pressing halakhic problem, to *rewrite it* by appeal to the opinion of a fourth-generation amora.[14] The *s'tam* is not offering a new *reading* of the Mishna. He is not suggesting that when the Mishna refers to "growth from the earth," it really means something else. On the contrary, growth from the earth is a halakhic category that is employed in other matters.[15] Nor is it implied in any way that the *s'tam* is suggesting that the text of the Mishna might have been corrupted and that by replacing the phrase "anything whose growth is not from the earth" with the phrase "anything whose nutriments are not derived from the earth" we shall be *restoring* the original text. It is a simple amendment of the law—of the *word* of the law, not of one of its interpretations!—prompted by a legal problem, which does not involve reinterpreting the Mishna or any other canonical document and is not presented as such. It is also quite a common move in the

Bavli's treatment of the Mishna.[16] This is not the way a formative text is supposed to function or is expected to be treated. On the contrary, it seems typical of the way an antitraditionalist community would treat a central canonical text: taking it extremely seriously as a significant step towards the ideal, desired system, troubleshooting it with a vengeance for any conceivable type of flaw or shortcoming, and not hesitating to putting it right when necessary.

Not all sugyot of this kind are as well defined. In many cases in which more than one solution is proposed to a problem—cases in which the *s'tam* presents different views attributed to individual, named authorities—one frequently encounters the suggestion of a second-order, metahalakhic dispute in this respect between the various discussants. The Bavli's discussion of Mishna, *Sanhedrin* 7:1, provides a good and, again, quite typical example. Unlike most systems of law, the halakha requires that certain criteria of adequacy be met before a testimony is considered testifiable in court. These are the so-called "seven inquiries and examinations" designed to make sure that testimonies are refutable—namely, that witnesses do not merely report what they heard or saw, but that their testimonies also include clear and refutable statements of the precise time and place in which their sighting occurred.[17] Mishna, *Sanhedrin* 7:1, rules categorically that the same criteria of adequacy apply to all types of legal testimony. "Both civil and capital cases," it states, "demand inquiry and examination (of the witnesses), as it is written: 'You shall have one manner of law' (Lev. 24:22)." "Do civil suits really need inquiry and examination (of witnesses)," asks the Gemara,[18]

> Indeed, this (Mishna) opposes the following (*beraita*):[19] If a bond is dated the first of Nissan (even if it is the Nissan) of the *Shemita* (seventh sabbatical year),[20] and (other) witnesses come forward saying: "How can you have witnessed this bond, were you not with us on that day in such-and-such a place at that time?" (Still) The bond is valid, and its signatories remain competent (witnesses). (How can that be? The reason is that) we presume that they might merely have postponed writing it.[21] But if you suppose that inquiry and examination are requisite (even in civil suits), how come it is allowed to be witnessed with a later date in mind?[22]

The Mishna and *beraita* are clearly in conflict.[23] Inquiry and examination are required, by Jewish law, in order to ensure that testimonies are refutable. "An irrefutable testimony" (one that does not explicitly state its precise time for instance), states the Bavli, "is no testimony at all."[24] The "inquiry and examination" in question consists of seven questions placed to each of the witnesses in order for each of them to state the exact week, day, hour, month, date, year, and place in which whatever they saw or heard occurred. But if the date on a bond is allowed to diverge from that of its actual signing, its signatories cannot be contested by other witnesses. Three different solutions, attributed to three different Sages, are cited in response.

The solution to the problem of Mishna, *Sanhedrin* 7:1, proposed by the first-generation Palestinian amora R. Ḥanina implies that he treats the mishnaic system of law not as a formative halakhic text occasionally requiring reinterpretation but as a text liable to be modified in view of new problems and changing circumstances.

> R. Ḥanina said: By the law of the Torah, (witnesses in) both monetary and capital cases require inquiry and examination, as it is written: "You shall have one manner of law" (Lev. 24:22). Why then were civil cases exempted from this procedure? In order not to lock the door against them in need of a loan.[25]

R. Ḥanina's line of reasoning is commonplace and is employed time and again throughout the Bavli. Normally, however, it is done in order to explain why later authorities decided to go a stringent step further and impose additional restrictions that go *beyond* those of the Torah.[26] In cases such as these the original ruling of old is preserved basically intact and is simply supplemented. Moreover, the further restriction is explicitly presented as motivated by care for, and as a means of *bolstering*, the existing system.[27]

The case in hand, as explained by R. Ḥanina is different, however. Here a law of the Torah, presented by the Mishna as part and parcel of its halakhic system, is described, by R. Ḥanina, as having proved problematic in ways the Sages had not or perhaps could not have foreseen at the time the Mishna was written. The *beraita*, R. Ḥanina argues, represents a step forward in the development of the halakha away from that original legislation. The mishnaic system, representing that of the Torah in this instance, does not remain intact and is certainly not bolstered in the process. The earlier restriction, despite being "true of the Torah," is repudiated.[28] The move is not presented by R. Ḥanina as a new *reading* of the halakhic system it aspires to put right or as an attempt to retrieve the original intentions of his predecessors. It is a modification pure and simple and is duly presented as such. For R. Ḥanina, the Mishna does not seem to function as a formative canonical text.[29]

The other two solutions to the problem are quite different. The fourth-generation Babylonian amora Rava suggests that from the start the Mishna's ruling that "both civil and capital cases demand inquiry and examination" was not intended to apply to *all* cases of property law. The Mishna, he submits, relates only to criminal law, including such 'property' aspects of it as fines. The *beraita*, on the other hand, addresses civil suits proper such as contracts and bonds. The fifth-generation Babylonian amora R. Papa disagrees with Rava and suggests that the Mishna and *beraita* both apply to civil cases proper. The reason that they rule differently is not because they address different areas of the law, as Rava proposed, but because they address different types of case. The less stringent *beraita* refers, according to R. Papa, to civil cases in which no foul play is suspected, while the Mishna is about cases suspected of fraud. Unlike R. Ḥanina, Rava and R. Papa prefer to explain the discrepancy between the Mishna and *beraita* not by viewing them as representatives of different stages in the critical and dynamic process of halakhic development, but by interpreting them as representatives of different areas of a static, undeveloping halakhic reality. (The great medieval codifications remain interestingly divided in their treatment of this amoraic debate, but that is beside the point.)[30]

This is not to say, of course, that by attempting to harmonize the *beraita* with the Mishna in this way, Rava and R. Papa necessarily cast themselves with the definitive opposition to R. Ḥanina. There is nothing inherently inconsistent for an amora to treat the tannaitic literature he has received, like R. Ḥanina, as a preliminary system liable to be modified if and when found wanting and to argue locally that two specific, seemingly contradicting tannaitic rulings may not have addressed the same issue. The same, however, does not apply to R. Ḥanina. Specific and local as it may be, his suggestion to view the *beraita* as evidence of the annulment and amendment of a former decree of the Mishna is, in principle, irreconcilable with the view of the Mishna as a formative text in Halbertal's sense of the term.

Generally speaking, to view the tannaitic literature as a canon formative for the halakhic enterprise of the Bavli is to maintain that as a rule, with the rare exception of genuine lacunae, the Bavli's treatment of the general principles laid down by the Mishna will be *applicatory*. At the most basic, uncritical level, such principles will serve amoraic halakhists as premises for simple deductive inferences to lower-level generalizations and particular single cases. In the event of apparent contradictions within the system of the Mishna or of new questions apparently not covered by it, general rulings of the Mishna thought relevant to the question at hand will be reinterpreted—their initially thought scope of application stretched or contracted as the case requires. Such reinterpretation can be radical. The rule in question may acquire a totally new meaning in the process, or, as we have seen, may even be supplemented. But if the system is truly considered formative, a general principle will seldom, if ever, be withdrawn, openly modified, or replaced. As Ferdinand Fairfax Stone writes, a major feature of a system of law that is regarded as a code is that it be "considered (. . .) and in all ways dealt with as a single fabric." Such systems have many advantages, but their main shortcomings, writes Stone, are those of the human agents who construct them.

> Man has consistently tried to build a frame for the universe and to squeeze it into that frame however much it may bulge over the sides or pinch in the corners. Always there are the bulges, the parts left over (. . .) So too when man essays to imprison the law into a single, homogeneous fabric, or when he tries to encompass the whole of man's legal personality in a single set of articles, either he spells out the particulars in such detail, providing for every conceived eventuality, that man is lost in the forest, or else he retreats into generalities so vague that man shrugs them off as too indefinite to constitute notice.[31]

In order for a code to avoid both undesirable extremes and to remain, as far as possible, capable of performing its formative role for future generations of jurists and legislators, its rulings will have to be general but not to the point of vacuity. More importantly, the concepts and definitions with which they are formulated will have to be flexible enough to accommodate the unavoidable bulging and pinching of changing circumstances and human sensibilities.

We have encountered amoraic considerations of certain general rulings that were found inadequate to the point of actual revision. The revisions, however, had been local. The general criterion for reciting the benediction "(. . .) By Whose word all things exist" was revised, according to the *s'tam*, in the light of new botanical data; that of the laws of evidence in civil suits, according to R. Ḥanina, in the light of changing realities in the money market. But the Bavli contains a critical discussion of a general ruling that does not lead to a revision, but to a clear and forceful declaration that the general principles contained in the Mishna should, *as a rule, never be inferred from*. To cast it in terms of the present section, such a declaration adds up to no less than a most significant call for refraining in general from canonizing the Mishna as a formative code. The occasion is provided by the well-known ruling of Mishna, *Kidushin* 1:7:

> All affirmative precepts (commandments to do, as opposed to prohibitions) limited to (a specific) time,[32] men are liable and women are exempt. But all affirmative precepts not limited to time are binding upon both men and women.

The amoraic discussion of this passage of the Mishna begins innocently enough by citing a *beraita* in which the notions of affirmative precepts that are and are not limited to specific times are exemplified.[33]

> Our Rabbis taught: Which are affirmative precepts limited to (a specific) time? (Precepts such as dwelling in a) *Suka* and (taking a) *Lulav* (during the Festival of Booths, blowing the) *Shofar* (on *Rosh Ha-Shana*, wearing) *Tzitzit* (fringes on one's four-cornered garments) and (putting on) *Tefillin* (during the hours of the day).[34] And what are affirmative precepts not limited to time? (Those such as fixing a) *Mezuza* (to one's door posts, the requirement to make a) Parapet (to one's roof), returning lost property, and "the dismissal of the nest" (the requirement to drive away the mother-bird before removing the eggs).

Is the rule stated by the Mishna supposed to be a general halakhic principle, asks an exasperated *s'tam*?

> But then (eating) *Matzo* (on the first night of Passover), rejoicing (during the pilgrimage festivals, participating in the) *Hak'hel* (the Great Assembly that is called during the first feast of tabernacles of the shemita cycle) are all affirmative precepts that are limited to time, and yet they are all incumbent upon women! Furthermore, *talmud-Torah*, procreation, and redemption of the (firstborn) son are affirmative precepts that are not limited to time, and yet women are exempt from them all!

This is not a simple incongruity between two contemporary tannaitic sources. All the counter examples are such that no *tanna* would deny. No attempt is made here to solve the problem by distinguishing, for example, one type of time-dependent precept from another, or by suggesting that the apparent deviations from the rule laid down by the Mishna owe their origins to some later development. In fact no attempt is made to solve the problem at all. The two generalizations stated by the Mishna, asserts R. Yohanan, are, qua generalizations, pure and simply false! Astonishingly, rather than attempt to somehow salvage the two faulty generalizations, *their fault is generalized* to all manner of such general principles! General principles stated by the Mishna, declares R. Yohanan, *all* general principles stated by the Mishna, not merely these two, should never be taken to apply to what they claim to apply—even when they explicitly state their own exceptions!

> Says R. Yohanan: we cannot learn (infer) from (tannaitic) general principles, even where exceptions are stated. For we have learned:[35] "An *Eruv* and a partnership[36] may be made with all comestibles—excepting water and salt." And are there no other exceptions (i.e., other foodstuffs that also fail to qualify for an *Eruv* or partnership)? Lo, there are mushrooms and

truffles! Hence, we cannot learn from general principles, even where exceptions are (explicitly) stated.

The *s'tam* goes on to inquire from where could the *tannaim* have learned the two supposedly general rules in the first place. The details of this inquiry do not concern us here, for it does not in any way challenge R. Yohanan's bold conclusion, which seems not only to have been wholly adopted, but actually applied to other cases. Not surprisingly the issue is raised again in the Bavli's discussion of the above-cited Mishna, *Eruvin* 3:1.[37] Here R. Yohanan's principle (about learning from general principles!) is shown to apply to yet another, previously unmentioned case and is subsequently hailed by the *s'tam* as a general principle in its own right—not merely as an ad hoc explanation of the two problematic *mishnayot*.

R. Yohanan's stunning conclusion and its apparent acceptance by later authorities,[38] including, so it seems, the *s'tam* himself, is of utmost significance to our present discussion. Needless to say, those who accepted it would have clearly opposed the idea that the tannaitic halakhic system could serve as a formative code for their own halakhic enterprises. It is one thing to claim along with R. Hanina that certain general mishnaic principles, such as the one concerning testimonies in civil suits, were initially not only enacted in all seriousness but were even fully justified in their time. However, it is another thing entirely to claim that the general principles stated by the Mishna are suspect of inaccuracy on their authors' own admission! If R. Yohanan is right—and as we have seen, there are several *amoraim* who evidently thought he was—it is doubtful that the Mishna could have even served as a central text, let alone a formative one!

The Yerushalmi does not attribute such a position to R. Yohanan, but on three occasions relates a similar declaration to the fourth-generation amora R. Yona, who likewise does not mince his words in claiming that "(tannaitic) generalizations do not (truly) amount to (or aspire to be) general principles."[39] On the other hand, the Yerushalmi has on record R. Zeira citing the first-generation amora Shmuel who formulates a position that is even more extreme: "We cannot learn (infer) from the *halakhot* (of the Mishna, of our predecessors?), or the *aggadot*, or from the *tosafot* (Tosefta?)—but only from the talmud!"[40] In other words, each case must to be worked out for itself in accord with the most recent, up-to-date opinion. Important as it may be, implies Shmuel, the corpus of a latter-day halakhist's inherited literature must not be allowed to dictate or direct his day-by-day halakhic deliberations. Whatever function this literature was meant to serve according R. Yohanan, Shmuel, R. Hanina, and their many followers, it was certainly not that of a formative halakhic code.

Rational Rabbis, or The Mishna as Textbook

In other words, for a major, if not the most prominent voice of the Bavli, the Bavli's double-talk remains unexplained. If the tannaitic litera- ture was considered at most to be a central halakhic text, in the technical, above-mentioned sense of the word, to be preserved as a landmark, as the object of troubleshooting, but not as a text formative of the Bavli's halakhic enterprise, what is gained by pretending differently? To this I would like to join a second question. If the Mishna was not meant to be taken as a code, why is its code-like structure preserved? Let me say im- mediately, this question is not posed to the framers of the Mishna, so much as to the framers of the Bavli. The Bavli deals with areas of tannaitic writing that extend far beyond the Mishna. But while it treats all other tannaitic material sporadically, citing the short passages it sees fit to dis- cuss in accord with its own agenda, system, and program, it subordinates itself to those of the Mishna. The Bavli allows the Mishna to preserve its code-like structure and to dictate to it its framework, order, and agenda. Why?

In order to answer both questions let us return for a moment to the closing paragraphs of Part 1 and allow the line of thought I have sought to develop in this study to come full circle. I argued there for an approach to rational endeavor—constructive skepticism—according to which to act rationally is to prudently consider the goal-directed systems on which one works as potentially imperfect regardless of their past successes; that it is imperative, therefore, that they be tested as relentlessly and thor- oughly as possible, and that those involved be willing at all times to revise or replace them as the case requires. Modern science, I claimed, is, on such a showing, a model of rational inquiry. According to this Popperian model, scientists seek truth but are aware that they lack the means to recognize it. They see science, therefore, as an enterprise geared to ame- liorating the scientific end product by persistent troubleshooting—an en- terprise whose short-term objectives are forever to discover the shortcom- ings of current theories with a view to replacing them with supposedly less problematic alternatives. The task of establishing and maintaining an *ongoing* tradition of rational scientific inquiry, I argued there, is never an easy one. It requires the delicate and balanced transmission, from each generation to the next, of two different yet complimentary messages or lessons. For the next generation of inquirers ever to be in a position to effectively continue the task of troubleshooting the theories of today, they must be well versed in state-of-the-art versions of those very theories. And in order for this knowledge not to be received dogmatically but to indeed be made the object of keen troubleshooting, it is equally pertinent for a second tradition—a critical tradition—to be nurtured and promulgated. The balance is not an easy one. The two traditions of learning that nourish the logic of rational discourse are required to combine in such a way as to toe the delicate line between dogma and mere fancy. Scientific theories need to be taught as tentative though not frivolous; to be taken in all seriousness yet not as infallible; as the very best we have though not as the best possible.

These two central and complementary concerns of the scientific learning process—indeed of the learning process of any genuinely rational endeavor—find their expression in both the scientific curriculum and the scientific library. Concerning the former, undergraduates are normally set the task of studying, plainly and quite uncritically, the theories, techniques, tools, and methods that will only later be made the objects of reflection and revision. Only at the graduate or postgraduate levels do students normally begin the process of critically reflecting upon the knowledge they had formerly acquired. To a great extent the seriousness and effectiveness of this second stage of their studies will depend on the seriousness and singlemindedness with which the first phase was pursued. In most institutions of higher education the scientific curriculum reflects this ordered, antithetical two-phase structure.

And the same can be said for the scientific library. Here the dividing line manifests itself in the difference between the two basic sorts of publication that typically adorn the scientific bookshelf: the scientific textbook and the scientific logbook or research report. Together they combine to document the two halves of a complete scientific cycle. While the former, the textbook, constitutes the main instrument for passing scientific knowledge on from one generation to the next, the latter, the logbook or research paper, documents the processes of trial and revision *of* the knowledge thus received, as part of the debate that will in turn motivate the framing of the textbooks of tomorrow. In an ideal community of learners, one in which the rules of the game are known and understood, there will be little need for the textbooks to teach them. An ideal textbook, designed to initiate the new generation of an ideal community of learners, will be mostly devoted to its main purpose—namely, that of handing down a body of knowledge as effectively and efficiently as possible. But in such a community, the main purpose of learning the achievements of one's predecessors is for the sake of subjecting them as soon and as effectively as possible to the keenest possible trials. Textbooks will therefore be purposefully framed to hand down knowledge that is organized and systematized *with a view to making its future testing the easiest and most effectual.*

There will always be a difference between texts that are designed to serve as vehicles of the truth and those designed to hand down intelligent guesses at the truth for the sake of further testing. Only the former will seek to *convince* their readers! Textbooks of the latter kind will ideally not aspire to foster agreement, rally support, create consensus. One expects such textbooks to sin deliberately on the side of simplicity and systemization, to strive for clean, symmetrical, and concise formulations even at the price of rounding some of the corners and ignoring some of the bulges. Such texts organize and systematize their materials in neat, hierarchical, and nested structures. And all this is done even at the risk of inaccuracy. The ideal textbook will typically concentrate on general principles to the partial neglect of their exceptions.[41] Geared to effectively exposing the system's premises, fundamental ideas, and basic principles, this sort of exposition will tend to oversimplify, idealize, and oversystematize. Because they aim to opening the system they describe to question, to setting it up efficiently as the object of systematic troubleshooting, textbooks such as these will normally not take themselves with utmost seriousness—other than as teaching aids of course. Their generalizations cry out to be questioned rather than to be endorsed; they will be almost self-challenging in their artificial simplicity, cleanliness, and symmetry.

Legal codes also strive to achieve simplicity, systematization, and a logical, hierarchical structure and to frame the law by means of principles sufficiently general to be fruitful and sufficiently detailed not to be vacuous. But unlike the sort of textbook we have described, legal codes do take themselves with utmost seriousness. They are not written in order to attract criticism. On the contrary, they present themselves and are treated by others as the last, definitive, and comprehensive word on the subject. A legal code cannot afford not to be accurate, not to appear as catering in advance for every conceivable eventuality. A legal code found to harbor the kinds of problems found to plague the three mishnaic rulings considered above would be considered defective. The sort of textbook we have been describing would not.

I have maintained throughout this study that the Bavli be considered a work centrally motivated by didactic concerns. I suggested that over and above their wish to summarize the *fruits* of their learning, the framers of the Bavli sought to document the *process* by which they learned with the intention of teaching their future readerships how to study Torah as they did. One reason I offered for favoring this conclusion was the way in which the Bavli resembles a protocol rather than a summary—in the light of our present discussion, perhaps a logbook is a better analogy. I was also puzzled at the time about why the framers of the Mishna apparently preferred to present the issues it deals with in the form of a code rather than in the form of a protocol, and why the Bavli decided to preserve this structure of the Mishna unlike any other tannaitic compilation it cites. But in view of our previous discussion one might answer this question by proposing that the framers of the Bavli contrived to present their future readerships not only with countless true-to-life reconstructions of study sessions but with paradigmatic examples of the two types of text that constitute the full learning cycle. Straddling the Talmud's great generational divide, the Mishna is presented as what they took to be a paradigmatic example of the way the conclusions of one generation of learners should be handed down to the next, and the Gemara as a paradigm of the manner in which knowledge thus received should be treated thereon: the Mishna as the model textbook, the Gemara as the model logbook.

The conjecture is a bold one, but it explains much. It explains why the Mishna resembles a code but is not treated as one. Textbooks of the sort described exhibit code-like features but are not meant to be reinterpreted when found wanting. The textbooks of an ideal community of learners are *born* wanting in every applicatory respect. They are not meant to be used as manuals or handbooks. Engineers do not use physics and chemistry textbooks for their work. A textbook of this kind *must not be used* as a vade mecum or a guide. Textbooks should be concerned with the gist of the field, with its basic structures and ceteris paribus conditionals, far less with its details. They are teaching aids and are intended for little more.

Viewed thus, the Mishna's code-like structure ceases to be puzzling, and R. Yoḥanan's warning against using the Mishna as a handbook for practical halakha is rendered no longer surprising. Even such gross and obviously false mishnaic generalizations, such as the one concerning the exemption of women from time-dependent affirmative precepts[42] that apparently prompted R. Yoḥanan's remark, no longer seem so strange. Good textbooks are simply not supposed to duplicate the systems they aspire to teach, especially if they are as vast and complex as Jewish law, but to capture and convey the essentials of their spirit and substance. But if that is case, the Mishna should *not* be viewed as the halakhic system that is actually troubleshot by the Gemara but only as a textbook exposition of that system. On such a showing, the tannaitic system troubleshot by the Bavli achieves its full elaboration in a whole series of supplementary tannaitic sources—most of them truly oral. Viewed thus the Mishna is not the system that is central to Bavli's halakhic undertaking but the one designed to teach it.

The suggestion to view the Mishna's role for the antitraditionalist voice of the Bavli primarily as a textbook would also explain its name, which apparently derives from the Hebrew word for learning by repetition and which some take to have actually meant "textbook."[43] More importantly, viewing the Mishna as a text designed to teach the halakhic system that is subsequently troubleshot and revised in the course of the extended process documented by the Bavli would explain two rather peculiar features of the amoraic phase of the cycle: why, on the one hand, the Bavli never finished the job so to speak, why it only treats thirty-seven of the Mishna's sixty-three tractates, and why, on the other hand, the amoraic Sages, after completing their revision of the tannaitic system, did not compose a Mishna of their own (as Maimonides would do for their writings six centuries later). The answer to both questions is that if the Bavli was primarily intended to teach future generations how to teach and how to learn, how the Oral Torah is at any one moment best transmitted and how it should be received, the sort of text that teachers should aim at producing and the type of process to which their teachings should be then subjected, if this is what the Bavli aimed at accomplishing, there would have been no need for its framers to start a second cycle by formulating a new textbook. Nor would it have been necessary to treat the entire system in order to explain the nature of the undertaking. Few would care to deny that during the period of three hundred years or so during which they functioned antitraditionalist amoraic centers of study continuously discussed, debated, troubleshot, and revised the entire system of tannaitic halakha. The point is that in order for the framers of the Bavli to *teach* future generations their view of the nature of their undertaking, there would have been no need at all for them to document the entire process. All of this fares well with the Bavli in its finished form. It also renders the Bavli's antitraditionalist voice dominant and formative in and of the entire enterprise. But it still does not answer our question.

The idea to view the Bavli as an attempt to teach *talmud-Torah* by combining a model textbook and model logbook explains much but does not answer our initial question: why the double-talk? It is still not clear why a text, intentionally composed in order to teach a decidedly antitraditionalist vision of Torah-study, would invest so much effort to conceal its main message. Indeed, insofar as the Bavli's antitraditionalism looms larger, the question is rendered all the more glaring. Why the double-talk? The answer to this question lies, I believe, in what was previously said about the typical scientific curriculum. Even in as excessively critical and undogmatic a culture as the world of antitraditionalist talmudic discourse, it is quite unproductive, if not impossible to reverse the order of the learning process. One cannot develop a healthy skepticism without first having something to be skeptical *about*. Effective troubleshooting is a creative art that requires training, preparation, and encouragement. One cannot learn it in theory. Troubleshooting can only be learned and experienced on real goal-directed systems, preferably on those to which it is meant to apply and on those closest to heart. To this end the relevant systems themselves have to be studied first.

The difference between a real logbook and a text contrived to act as a *model* logbook is that the latter is inevitably a textbook of sorts. And I believe that the Bavli was to a significant degree framed self-consciously as a textbook of sorts, with a view to teaching *talmud-Torah* by extended example. However, rather than cater separately for the beginners, as do the textbooks of most institutions of learning—including the actual academies that functioned in talmudic Palestine and Babylonia![44]—the "virtual" academy constructed by the Bavli, the course of "studies by correspondence" as it were that it aspires to provide, *consists of one extensive text that is addressed simultaneously to students of all levels*. This, I submit, may very well have been the reason for the Bavli's persistent double-talk. It is also the best explanation I have to offer at this stage.

From a purely pedagogical perspective, it is of utmost importance for beginners, setting out to train themselves as future halakhists, to acquaint themselves thoroughly with the tannaitic halakhic system before they can be introduced to the full force of its critical review and revision in the hands of its amoraic commentators and discussants. The Mishna needs no disguise, although it occasionally requires explanation. The Gemara does need disguising and is, therefore, superficially presented as if it were no more than an extensive, searching, but largely uncritical exposition of the system taught by the Mishna. The technique, as we have seen, especially when compared to the Yerushalmi's far less nuanced narrative, is ingenious. Great revisions are achieved in the guise of exposing the tannaitic system's allegedly real intentions. Transgenerational incongruities—potentially the most conspicuous telltale evidence of amoraic revision and reform but in fact the most thoroughly muffled—are elegantly harmonized out of existence in ways through which only the most advanced student could be expected to see. I believe that the more seasoned students were not intended to be misled by the Bavli's surface rhetoric any more than graduate students in science are expected to be incapacitated by the textbooks of their undergraduate days.

The more I study the Bavli the more I am convinced that this was the plan and strategy of a great many of its framers and subsequently of much of the text as we have it. It was brilliantly conceived and is ingeniously executed. Unfortunately, it was also to a great extent unsuccessful. The Bavli succeeded in establishing a remarkably resilient community of Torah-learners that flourishes to this very day without showing any signs of weakening. But by and large the community's attitude to halakha is, and has been, largely traditionalist. As noted above, one of my critics wisely pointed out that even those of the Bavli's readers who succeed in seeing through to the antitraditionalist heart of *Berakhot* 19b will still have learned from the *s'tam* that one is not supposed to make one's true attitude known. But even that hardly ever happens. The Bavli has been almost equally ineffective in establishing a tradition of covert antitraditionalism. Habitually almost, the vast majority of its students, even the most clever and advanced, simply take its facade at face value. Throughout the ages there have been many rabbinic exceptions, some of them very impressive,[45] but they have had little effect on the way the vast majority of religiously motivated Jews have studied the talmudic literature. The great eleventh-century commentator Rashi, to take one example, never wrote a treatise on the Bavli, but, as we have seen above,[46] scattered throughout his illuminating and unique commentary to the text one does encounter remarks that suggest that he did not always take the Bavli's traditionalism for its word.

By contrast, and closer to the present time, the writings of the late-nineteenth-century R. Naftali Tzvi Yehuda Berlin, known as the NaTZYB (pronounced *Natziv*) of Volozhin, offer an intriguing example of the extent to which, with regard to halakha, even the boldest thinker remains basically traditionalist. In the preface to his commentary on the Pentateuch he explains his undertaking of such a work.

> Just as it is not possible for the wise student of nature ever to boast knowledge of all of nature's secrets (. . .), and just as there is no guarantee that what he has accomplished in his investigations will not be invalidated by colleagues in this generation or the next who elect to study the same things differently, so it is not possible for the student of the Torah to claim that he has attended to each and every point that claims attention, and even that which he does explain—there is never proof that he has ascertained the truth of the Torah!

Berlin's bold statement would seem at first blush to anticipate major themes of the present work—an analogy between science and Torah-study as two truth-seeking endeavors that in principle lack the means ever to confirm their findings. Both enterprises, argues the legendary head of the great Volozhin Yeshiva, are forever beset by *two* persisting doubts: *first*, that there is probably more than meets the eye, and *second*, that whatever one might think *has* been accomplished could well turn out to be mistaken. The NaTZYB's preface was first published during the 1860s, long before the upheavals in physics that would lead to the Popperian turn in philosophy of science. His description of the advancement of science in revolutionary rather than accumulative terms is, therefore, extremely interesting. His description of Torah-study in the same terms would be equally, if not more surprising, if it could be said to include the development of halakha. But apparently Berlin's revolutionary model was reserved exclusively to the nonhalakhic aspects of Torah-study. When he does talk specifically about the advancement of halakhic knowledge, the second of the two doubts associated with science and biblical exegesis vanishes. In his commentaries to Exod. 34:1, 27 and Lev. 18:5, his preliminary remarks to Deuteronomy, and most importantly his lengthy introductory essay to the *She'iltot de-Rav Aḥai Gaon*, entitled "The Way of Torah," he describes the development of halakha as a creative, innovative enterprise that is governed by strict, God-given rules of inference—such as the thirteen *midot*—by which new conclusions are derived from old. One generation of halakhists is capable of halakhic innovation only within, and by building upon, the existing system, much as progress is commonly described in mathematics: creative, innovative, but by no means revolutionary. In other words, despite his instructive, suggestively antitraditionalist comparison between science and biblical exegesis, the NaTZYB's vision of halakhic development as the accumulation of valid halakhic conclusions remains overly traditionalist.

Rabbi Berlin's understanding of the process of halakhic development, especially as it is recorded by the Bavli,[47] is commonplace. It also differs greatly from the conclusions of the present study. Insofar as the framers of the Bavli sought to convey and promote an antitraditionalist approach to halakhic discourse, they can hardly be said to have succeeded. But then, however well intended and however well executed, one has to admit that their plan was almost bound to fail. If there is any truth to the suggestions made and the guesses hazarded in the present chapter, the Bavli, at least substantial portions of it, acquired its final form with a view to relegating the inevitable two phases of the learning process to two separate levels of reading of the same text, rather than, as is normally the case, to one level of reading of two separate texts. Such a solution is probably fine for beginners, who will easily be fooled to think that Torah-study is traditionalist, but it is far too sophisticated for the more advanced who will hardly know what to expect—unless, of course, they are guided by teachers to whom all of this is clear to begin with. However, since today's teachers are yesterday's students, lacking a clear, systematic, and unambiguous statement of antitraditionalist intent, I suspect that confusion swiftly bred more confusion to the point of largely obscuring the Bavli's real, but too well-hidden agenda.

If the tannaitic portions of the talmudic corpus[48] attest to an open, vital, and very much alive dispute between dedicated traditionalists and antitraditionalists, one that manifests itself, as we have seen, in halakhic, metahalakhic, and legendary corpora, the strictly amoraic texts, as I remarked earlier, no longer bear witness to a controversy of this kind. As I put it above, the amoraic writings strike one as being the product of a culture in which the great debate is perhaps still remembered, but one in which it is no longer being fought. It follows from this that the Bavli's traditionalist idiom was in all probability *not* introduced as a concession to the opposition. I doubt that, by the time the composition of the Bavli was entering its final stages, there was an opposition to talk of. It was introduced, I have argued, as a didactic device designed to facilitate the initial stages of a learning process geared in the long run to producing a cadre of keen antitraditionalist troubleshooters. Paradoxically, the plan backfired, and the Bavli's traditionalist idiom appears to have contributed instead to the wholly unintended revival and reinstitution of a major traditionalist school of thought that would later come to dominate post-talmudic culture.

"Turn Thee Around" (Bavli, *Menahot* 29b), or Back to the Future

For the reasons we have discussed it is perhaps less puzzling now why the framers of the Bavli elected not to make a clear statement of their antitraditionalist point of departure within the actual framework of halakhic debate. The tannaitic literature, as we have seen, does contain such statements in both Mishna, *Sanhedrin* and *Eduyot*—as opposed to the parallel Tosefta—and somewhat less explicitly in the discursive practices attributed to the Hillelites in their debates with the Shammaites. Still the amoraic literature does contain legendary material that seems centrally and intentionally geared to proclaiming a definite antitraditionalist ideology. To repeat, unlike the tannaitic Jabne stories, these amoraic legends seem not to be written with a view to dramatizing, personifying, or even scoring points in a metahalakhic *controversy*. Again, they appear to have been designed to serve pedagogical and didactic rather than polemical purposes. And yet some of them are striking. Hence, instead of summarizing this book by recapitulating its arguments and claims yet again, I shall bring it to conclusion by presenting and briefly discussing two such legends. They are both quite well known and will be familiar to many of my readers. But in light of what we have said so far I think they acquire a significance and a focus rather different from what they are normally granted. The first is about the prospects, pitfalls, and overall goal of day-by-day halakhic negotiation. The second offers a unique and rare glimpse of the way the *amoraim* conceived of their project in a dramatically dynamic, historical context.

The first legend, related in Bavli, *Bava Metzia* 84a, tells the story of the commencement and tragic ending of the extraordinary relationship between the two great first-generation Palestinian *amoraim*: R. Yoḥanan and Resh (R. Simeon b.) Lakish.

> One day, as R. Yoḥanan was swimming in the Jordan, he was spotted by Resh Lakish who jumped in and joined him in the water. Your powers befit (the study of) Torah, said R. Yoḥanan.[49] Your beauty, replied Resh Lakish, befits women! Should you repent and study, he answered, I shall give to you my sister who is even more beautiful than I. Resh Lakish agreed. But when he tried to leap back across the river to retrieve his clothes, he found he was unable to do so.[50] R. Yoḥanan taught him Scripture and Mishna and [Resh Lakish] rose to greatness. One day, at the bet midrash they were debating the following question: (we know that) the sword, knife, dagger, spear, shear, and scythe are liable to 'receive' ritual impurity from the moment the process of their production is completed[51]—but when exactly is that? R. Yoḥanan says: from the moment they forged in the fire; Resh Lakish said: only after they are polished in the water.

The question is a classic and the two answers typical, if, apparently, new. R. Yoḥanan opined that metal instruments of this kind become what they are the minute they acquire their final form. In Resh Lakish's view an instrument becomes what it is only from the moment it can be put to its intended use. A red hot piece of metal may look like a sword in all respects, but it cannot yet be used as one. It is significant perhaps, especially from an antitraditionalist perspective, that the debate between the two *amoraim* ignores the fact that the Mishna would seem to have decided the issue differently.[52] As we shall see, a far stronger and explicit claim is made in the story against having halakhic positions prove their mettle merely by indicating tannaitic precedent. In all events, the discrepancy between the two answers and the Mishna, if there ever was one, is not raised in the story or by its narrator.

At this point, we are told, R. Yoḥanan, perhaps unwittingly, managed to gravely insult Resh Lakish in the heat of the argument. He said to him: "a robber would know his weapons wouldn't he?!"[53] Resh Lakish was shattered. "What good has following you done to me?" he grieved, it is as if nothing had changed, "here (among the rabbis) they call me Master and there (among the thieves) they call me Master—I thought, at least for you, there was a difference!" "I did you a favor in bringing you under the wings of the Almighty," R. Yoḥanan retorted crossly, apparently unaware of how deeply he had offended his colleague. He had lost his judgment, we are told, and became more and more insensitive to his brother-in-law's plight. Resh Lakish, on the other hand, fell gravely ill.

> Pray for him, pleaded his sister, that my sons not be orphaned. It is written, "Leave thy fatherless children, I will preserve them alive" (Jer. 49:11) he preached her. Do it for me then, lest I be widowed, she begged. "And let thy widows trust in me" (*ibid.*), he went on quoting (heartlessly). And thus Resh Lakish passed away, and R. Yoḥanan (who had now come to realize what he had done) knew no end to his grief. The rabbis said who should be the one to revive his sanity? Let R. Elazar b. Pedat approach him, whose halakhic mind is the sharpest.[54] He went and sat before him, and for every opinion that R. Yoḥanan would utter, he would provide him with confirming tannaitic sources. He said to him: are you supposed to replace Resh Lakish?! For every thing I opined Resh Lakish would put to me four and twenty questions to which I would have to seek four and twenty answers! In this way the halakha was ameliorated.[55] And you, you say 'here are tannaitic sayings that agree with you'! Do I not know that my views are feasible?! And he walked away, and tearing at his clothes he wept: where art thou son of Lakish, where art thou son of Lakish?! And he screamed so until he lost his mind, and the rabbis prayed that he be pitied, and so he (too) passed away.

This powerful, and, if I may say so, very Popperian, tale contains almost all an antitraditionalist could wish for. The doors of the academy are open to all who sincerely wish to join it regardless of their background or former employment. Halakhic discourse is itself genuinely open: it does not merely tolerate different positions, but is truly pluralistic. Even a great and knowledgeable Sage as R. Yoḥanan is well aware that even he is incapable of effectively putting his own views to the test. For this one needs a friend, a study companion whose views he opposes. In the realm of Torah-study, confirmation is useless. Finding that one's views happen to fare well with those of former generations can teach nothing from which the halakhic system can be said to profit. Keen, prudent, relentless troubleshooting is the only way to improve the system. And this can be adequately achieved only by group effort, and only by groups sufficiently heterogeneous and sufficiently divided in their opinions to allow real debate. It is not truth that one can ever hope to achieve only progress, improvement, amelioration. The halakhic system can and should be constantly enhanced by trial and error. The story does not speak merely of going beyond the existing system and building upon it. There is nothing of the traditionalist, accumulative model of halakhic development in the way the tormented Sage pooh-poohs corroboration from tannaitic quarters in favor of new and unexpected challenges to his well-versed conjectures. His young and inexperienced study companion does not even offer corroborating *arguments* only corroborating evidence. His yes-manship is distinctly of the traditionalist kind. And it is flatly rejected.

But the story tells us more. We also learn just how delicate the dividing line is between keen and constructive criticism and arguing ad hominem; how easy it is, even for the closest of friends, for a heated, well-intended, critical exchange to deteriorate into a bitter feud; how sensitive one must be to the feelings of one's colleagues, especially in a culture that lays such value on criticism. The story is a true celebration of antitraditionalism, but at the same time it issues a serious warning against the two main perils that such a culture will always face. A truly antitraditionalist discourse is fragile, the story forcefully implies. On the one hand it is easily misunderstood. On the other hand, it is extremely difficult to teach and maintain. It can easily self-detonate when tempers flair and feelings are hurt. And it can equally easily deteriorate into smug acceptance of the ways of old. In view of the hesitant conclusions of the present section, this is perhaps the most telling aspect of the story: the fact that when a new study partner was sought for R. Yoḥanan, it turns out that, to the Master's anger and dismay, even the brightest and most well-versed graduate of the academy had not begun to grasp the very basics of the school's antitraditionalist philosophy!

Read thus, this quite amazing story emerges as a highly significant, constructive criticism of antitraditionalism itself. There is nothing smug, boastful, or moralizing about it. It is not a celebration, but an extremely sober and wary examination of constructive skepticism. It is an antitraditionalist attending reflectively to the fundamentals of his own position as he would have any other system attended to: with a view to exposing its perils and problems rather than to singing its praise. And yet its criticism is not destructive. Nothing in the story would indicate that it is aimed at discrediting R. Yoḥanan's constructive skepticism in favor of the opposition. It is keenly critical of antitraditionalism but not hostile toward it. And in its criticism the story, it seems to me, anticipates two of the main problems that talmudic culture will face in the wake of the Bavli: a perpetual difficulty, on the one hand, to accommodate genuine halakhic pluralism within stable communities, with differences of opinion frequently leading to fragmentation. And, on the other hand, the type of difficulty already encountered in the Bavli of effectively perpetuating the very culture of halakhic antitraditionalism. As some stories do, this one forcefully raises the problems but has no solution to offer.

The second and final legend I shall discuss is, in a sense, far more daring, if somewhat less explicit. Appropriately enough, it is attributed in Bavli, *Menaḥot* 29b, to the second-generation Babylonian amora R. Yehuda relating it in the name of his great teacher Rav.

> Said R. Yehuda in the name of Rav: when Moshe ascended on the High (to receive the Torah for the second time)[56] he found the Almighty (still) busy affixing coronets to the letters.[57] "Who is delaying you Lord of the Universe?" inquired (the puzzled) Moshe.

Moshe is asked politely to be patient. These unfinished subtleties of the Text, he is told, are intended for someone else. Notice also how the Almighty's answer is made to play havoc with historical time:

> He said: There *is* a man, who *will arise* at the end of many generations, Akiva son of Joseph *is* his name, who *will* expound upon each coronet heaps and heaps of laws. Lord of the Universe, Moshe requested, show him to me. Turn thee around, the Almighty replied, whereupon Moshe (was transported into the future, and) found himself sitting at the end of the eighth row (of R. Akiva's academy).

We shall return to the curious phrase "turn thee around" shortly.[58] Its meaning in and for the story is clear enough, however: Moshe is invited to enter the time machine as it were, travel into the future, and peruse the great Akiva at work. Had the story been told by a traditionalist, the main point of having a great Sage of the likes of R. Akiva have the good fortune of Moshe, the lawgiver himself, miraculously appear in his classroom, would have undoubtedly been for Moshe to demonstrate how much halakha had been lost en route and to assist Akiva and his students in filling the gaps and mending the flaws that would have inevitably accumulated since the Torah was first handed down. But not here. This particular legend is anything but traditionalist in this respect. Not only does Moshe have nothing to contribute to the discussion at Akiva's academy, but he sat there, we are told, failing to understand a single word that was said. So great was his shock and confusion that he weakened at the knees.[59] Until, at one point, the Sage, offering his opinion on a certain issue, was asked by a student "Whence, Master, do you know that?" Whereby Akiva answered: "That (particular law) is a law given to Moshe on Sinai." Hearing this, though apparently still not understanding a word,[60] Moshe, we are told, regained his wits and was comforted.

But why was Moshe comforted? If we are meant to believe that the initial cause for his uneasiness had merely been his personal failure to comprehend what was going on, then being informed that what was being said was in fact something he was supposed to have heard at Sinai the first time around would have only made things worse! If the reason for his recovery, as the story seems clearly to indicate, was his being assured that the strange things he had heard, and could still not grasp, had indeed originated at Sinai, then we must conclude that his initial breakdown was not brought about by the thought that he might be losing his cognitive abilities, but by the thought that the future Sage might be teaching something different than the Torah he was about to receive. Upon learning that Akiva perceived a direct link between the Torah he knew and the issues under discussion, Moshe recovered, realizing, to his immense relief, that the Torah had not been replaced, and that his failure to comprehend the proceedings attested to no more than to his apparently inferior understanding of these matters. This also explains his reaction upon returning to his former place and time. "Lord of the Universe," he said, "Thou hast such a man as Akiva and Thou giveth the Torah by me"?! The Almighty's reply is extraordinarily harsh by talmudic standards. "*Shtok!*" (be silent, or even stronger: shut up!), replied the Lord, "for such is my decree!"[61] The story could well have ended here, but its author takes the extraordinary exchange a seemingly unrelated step further:

> Lord of the Universe, he went on to say, you have shown me his learning (his Torah) show me his reward. Turn thee around, He said. And he turned to his behind and saw [Akiva's] flesh being weighed as if by a butcher.[62] He said to Him: is this Torah and this its reward?! Be silent, He replied, for such is my decree!

This extra twist to the story would seem to be quite uncalled for. The only obvious relevance it has to bear on what comes before it is to imply that the Almighty's decision as to who is the person most worthy to receive the Torah is as incomprehensible to humans as His moral judgments. This would be true, though still not very helpful, only if the cause for God's exasperation could indeed be said to have simply been Moshe's exaggerated low self-esteem. But this does not at all seem to be the point of the story. (Indeed, Moshe is praised by the Bible for his great modesty.) I believe that the sudden shift from the difficulties he experienced when attempting to comprehend a future phase of halakhic development (and the story *is* expressly about halakhic development) and those he experienced in attempting to comprehend a divine future decision provides a key to the origin, meaning, and significance of this unique tale.[63]

Moshe is on record more than once for requesting of the Almighty privileged, inside information regarding the future. And each time his requests are politely, if firmly denied. At the burning bush he is appointed the future task of confronting his brethren with the news of their impending salvation. Not satisfied with knowledge of the Almighty's past identification—"I am the God of thy father, the God of Avraham, the God of Yizhaq, and the God of Ya'aqov" (Exod. 3:6)—Moshe demands to know in advance the identity he should expect *at the time of his future mission*: "Behold, when I come to the children of Yisra'el, and shall say to them, the God of your fathers has sent me to you; and they shall say to me, What *is* his name? What shall I say to them?" (Ibid. 3:13). The Almighty's enigmatic answer is *"Eheye asher eheye"*—I shall be what I shall be![64]

The second time that Moshe is said to have similarly inquired into the future occurs immediately after the Torah informs us of the extraordinarily intimate relationship that had developed between him and the Almighty after the affair of the golden calf: "And the Lord spoke to Moshe face to face, as a man speaks to his friend" (Exod. 33: 11). Apparently shaken by not having anticipated either the intensity of the people's desire for a concrete, tangible Deity or the ferocity of the Almighty's reaction to it, and encouraged by his obviously successful damage control, Moshe, as Rashi nicely puts, decided to take advantage of the situation.

> And Moshe said to the Lord, See thou sayst to me, Bring up this people: and thou hast not let me know whom (what) thou wilt send with me.[65] Yet thou hast said, I know thee by name, and thou hast also found favour in my sight. Now therefore, I pray thee, if I have found favour in thy sight, show me *now* thy way, that I may know thee (. . .)

The Almighty's answer (to a great extent lost on the available English translations) is a polite refusal: I cannot grant you such knowledge of my future plans and ways. You will have to make do with knowing, as all humans inevitably do, only after the event: *Panai yelekhu va-hanihoti lakh*—"My face (front) shall go (forward) and I shalt thus give you direction." But Moshe is not deterred, and he persists: "I pray thee show me thy glory." This time, however, he is given a detailed answer, which according the interpretation suggested here would loosely translate as follows: Even if I should present to you the entire picture and should inform you even of my decisions to whom I have decided to be gracious, to whom merciful, you will be incapable of comprehending my actions prospectively—for no living person is capable of seeing me "from the front." However, there is a place by me here upon the rock, and I shalt put you in a cleft of that rock, and my glory shall pass before you, and I shall shelter your eyes with my palm, and only when it passes shall I remove it, and you shall see me from behind but my front shall not be seen (referring to Exod. 33: 19–23).[66] In other words, Moshe is firmly informed of humankind's inability, in principle, of adequately comprehending anything prospectively. We can only learn by hindsight. There is no sense at all, he is told, of even providing you with the data. No living person could make heads or tails of it.

The story related by Bavli, *Menahot* 29b, I propose, originated from a midrashic reading of Exod. 33: 13–23 along the lines suggested above. However, while the original dialogue between Moshe and the Almighty is about the nature of God's moral management of the world and future plans for His people, R. Yehuda in the name of Rav relocates it with stunning force in the realm of halakhic development. In the hands of Rav and R. Yehuda, halakhic discourse is rendered an essentially historical entity the future trajectory of which is humanly unpredictable. But the story says much more. Even if a great halakhist, as great even as Moshe himself, were to be transported miraculously to an academy functioning several generations into the future, there is little chance he would have understood what they were discussing. Not only are humans incapable of knowing the future, in certain areas—the development of halakha being one—they would be incapable of comprehending future developments even if they were miraculously made known to them! Even if foresight were made humanly possible, prospective *understanding* would not.

Such a notion of halakhic development as that strongly implied by this story goes against all brands of traditionalism. If halakha is considered a body of knowledge that accumulates by the recursive application of a fixed and accepted set of rules to the conclusions of former applications, then Moshe should not have had any difficulty in understanding what went on. Only in areas in which development is liable to involve the *revision* of formerly accepted conclusions, areas in which knowledge does not necessarily accumulate, but is frequently modified or replaced, only in areas of discourse in which revolutions are possible, would one expect the founding fathers to be truly perplexed by future developments—and the further into the future they travel, the greater the perplexity.

Our understanding of the theological, metaphysical, and especially the legal meaning of the Written Torah, implies R. Yehuda in the name of Rav, will necessarily shift and turn in time. As historical circumstances change, as new and unforeseen problems come to light, responsible Torah-study will necessarily yield new and unpredictable insights. In this respect, and in sheer contradiction to the traditionalist point of view, Akiva, privileged by the historical hindsight of his time and place, is infinitely better positioned than was even Moshe to carry the day at his academy. Akiva was looking back into the past and learning from the mistakes and the inappropriateness of his predecessors' teachings, planned for the present and near future. Moshe, unable to appreciate the difference, begged to be given the opportunity of traveling to the future in order to look back and plan for his own present. "Turn thee around" he is ordered with the view that he will learn his lesson the hard way. Instead of "looking" forward to the future from his present vantage point in time, Moshe is given the opportunity of "turning around" and gazing back at his present from R. Akiva's vantage point. What he fails to understand is that it is nevertheless his inevitably historically situated self that is transported in time. It is a self whose experience, outlook, vocabulary, background knowledge, and common sense are inevitably the product of *its* past and rooted in *its* present. Moshe is rendered by R. Yehuda and Rav a talmudic Rip van Winkle as it were, who by miraculously relocating in Akiva's academy had not gained any new knowledge. What causes the Almighty to lose His temper is Moshe's apparent inability to comprehend all this. Incapable of learning the lesson he is being taught, Moshe is ordered to be quiet and to accept the situation as an arbitrary divine decree.

For a halakhic ruling to be considered one "given to Moshe at Sinai," the story implies, it does not necessarily have to be one that had been passed down whole and intact from the days of Moshe to the present time. Suffice it to be a ruling such that the process of troubleshooting and revision can be traced back to the time of Moshe. Akiva, in his time, is capable of recognizing some rulings as such—although from the necessarily prospective point of view of Moshe himself they will be incomprehensible. And by the same token, implies R. Yehuda, in the name of the founding father of the amoraic era Rav, writing over a hundred years after Akiva, had Akiva miraculously happened upon *our* academy today, he would probably not have understood much and not have agreed to much either.

NOTES

Introduction

1. See, for example, Mishna, *Bekhorot* 4:4; Tosefta, *Aholot* 4:2, *Berakhot* 1:1–6, *Nida* 1:3; Bavli, *Sanhedrin* 33a, *Nazir* 52a, *Nida* 22b, *Pesaḥim* 94b.

2. As a rule, the talmudic literature refrains altogether from pronouncing final judgments on all issues other than those pertaining directly to the performative aspects of the ritual and the law—i.e., halakha. We shall return in some detail to this point in the pages that follow.

3. Jacob Neusner, *The Making of the Mind of Judaism: The Formative Age*, Brown Judaic Studies, vol. 133 (Atlanta: Scholars Press, 1987), p. ix.

4. Cf. Amos Funkenstein, *Perceptions of Jewish History* (Berkeley: University of Claifornia Press, 1993), esp. pp. 216–17. Though one would be pressed to name any Jew who contributed to the scientific revolution, there were some Jewish *responses* to early modern science first mainly in Italy and Prague and later also in England. Jewish reactions and lack of reaction to these developments have just recently become the focus of serious study in the wake of David Ruderman's pioneering work. See, for example, David B. Ruderman, *Jewish Thought and Scientific Discovery in Early Modern Europe* (New Haven: Yale University Press, 1995), and Noah J. Efron, "R. David b. Solomon Gans and Natural Philosophy in Jewish Prague" (Ph.D. diss. Tel Aviv University 1995).

5. Neusner, *The Making of the Mind of Judaism*, p. xi. See also, Jacob Neusner, *The Formation of the Jewish Intellect*, Brown Judaic Studies, vol. 151 (Atlanta: Scholars Press, 1988).

6. While in *The Making of the Mind of Judaism* Neusner concentrates on tracing what he takes to be the fundamental differences between the logic governing talmudic and philosophico-scientific discourse, some of his more recent work takes the opposite course. I am referring in particular to two books: *Judaism as Philosophy: The Method and Message of the Mishnah* (Columbia: University of South Carolina Press, 1991), and his most recent and, in my opinion, most perceptive work in this area: *Jerusalem and Athens: The Congruity of Talmudic and Classical Philosophy* (Leiden: E. J. Brill, 1997), the title of which speaks for itself. In the preface to *Jerusalem and Athens*, Neusner describes the joint project of the two books thus: "The Talmud—the Mishnah, a philosophical law code, and the Gemara, a commentary upon the Mishnah—works by translating the principle mode of Western intellectual inquiry into the analysis of the rules of rationality governing concrete, this-worldly realities. That is philosophy, including science. Science, in particular natural history, supplies the method of making connections and drawing conclusions to the Mishnah, the law code that forms the foundation-document of the Talmud. Philosophy, specifically dialectical analysis, defines the logic of the Gemara and guides the writers of the Gemara's compositions and the compilers of its composites in their analysis and amplification of some of the topical presentations, or tractates, of the Mishnah" (p. x). The claim, central to *The Making of the Mind of Judaism*, that a "logic of fixed association" singularly characteristic of talmudic modes of presentation renders it so different from those of science and philosophy as to account for "why no science in Judaism," is nowhere reexamined in these two later works. I asked Neusner in private conversation whether, with the completion of *Jerusalem and Athens*, which he was generously willing to share

with me prior to its publication, he now considered his earlier claim superseded. I pointed out that, as suggested above, the very different realms to which the two modes of discourse are applied, coupled with the enormous weight attached by talmudic Judaism to Torah study in comparison to science or philosophy, would fully explain why, despite their congruity, no science in Judaism. However, his answer to date remains negative.

7. Amos Funkenstein and Adin Steinsaltz, *The Sociology of Ignorance* (in Hebrew) (Tel Aviv: Ministry of Defense Press, 1987), esp. chap. 3–5.

8. David Kraemer, *The Mind of the Talmud: An Intellectual History of the Bavli* (New York: Oxford University Press, 1990) pp. 7–8.

9. Although a central aim of this study is the retrieval of one of the two main talmudic voices I shall examine, it is not judgmental of the other. It does not seek to contextualize and historicize away the voice I find less palatable. In this respect, despite my personal leanings, I have indeed done my utmost to remain impartial and descriptive. There are two related senses, however, in which this book does adopt a critical stance toward the material with which it deals. These concern, on the one hand, the ways in which the Bavli undertook to convey to its future readerships the full extent of its antitraditionalism, and the way it was read by later generations, on the other. The first point is made explicitly in the third section of Part 2, the second is merely implied in passing. In this respect, to paraphrase Daniel Boyarin, the present study aspires to retrieve from between the pages of the talmudic literature, a "usable past" of which, for various reasons, we have to an extent lost sight. To this end I found Boyarin's notion of cultural critique both inspiring and useful. See Daniel Boyarin, *Carnal Israel: Reading Sex in Talmudic Culture* (Berkeley: University of California Press, 1993), esp. pp. 19–21.

Part 1: Science as an Exemplar of Rational Inquiry

1. The term "empirical adequacy," first introduced in Bas van Fraassen, *The Scientific Image* (Oxford: Clarendon Press, 1980), is normally associated with the antirealist view that to accept a scientific theory "involves no more belief (. . .) than that what it says about observable phenomena is correct" (p. 57). According to this view, all a theory postulates regarding unobservables (e.g., forces, photons, fields, egos, ids, etc.) are not held in advance as aspiring to describe real features of the world. I use the term here less limitingly and less technically to denote no more than the commonsense claim that scientific theories are expected by realist and antirealist alike to withstand the tests of observable phenomena.

2. See, for example, Francis Bacon, *Advancement of Learning*, 1, p. i. Thomas Sprat, celebrating the new experimental method, speaks in his *History of the Royal Society* (1667) of the "two reformations" that each prized the original copies of God's two books, nature and the Bible, bypassing the corrupting influence of scholars and priests. Cited in John H. Brooke, *Science and Religion: Some Historical Perspectives* (Cambridge Eng.: Cambridge University Press, 1991), p. 111.

3. Much of the discussion that follows in the present section can be found in some of my earlier work. See in particular Menachem Fisch, "The Problematic History of Explaining Science" (in Hebrew), *Alpayim*, 1 (1989), pp. 194–208, and esp. "Towards a Rational Theory of Progress," *Synthese*, 99 (May 1994).

4. Pragmatists normally view truth not as a semantic relation between what a statement asserts and the state of affairs it describes, but rather as resting upon

widespread consensus or convention among specialists as to the nature of such a relationship. For many pragmatists, the scientifically true reduces in one way or another to whatever is thought to be true by the scientific community. For a cogent, if extreme, summary and exposition of the pragmatist's theory of language and truth, see Richard Rorty, *Contingency, Irony, and Solidarity* (Cambridge, Eng.: Cambridge University Press, 1989), esp. chap. 1, tellingly entitled: "The Contingency of Language."

5. "Instrumentalists" is the termed coined by the American philosopher John Dewey to describe those who view scientific theories merely as conceptual tools whose sole purpose is to systematically organize low-level empirical generalizations. The belief that science can truly explain the phenomena is viewed by instrumentalists as a conceit. Instrumentalism confines scientific realism to generalized low-level raw sense data. Unobservable and hence theoretical entities employed in scientific reasoning, like space, time, forces, fields, and even atoms, are not regarded as representing anything real; they are at best viewed as useful conceptual schemata designed to facilitate the efficient organization of observable and experiential knowledge. This view, that was first seriously realized in the scientific work of the French mathematician and physicist Joseph Louis Lagrange and his school, and articulated philosophically first by Auguste Comte and Ernst Mach, and later by Pierre Duhem, achieved some prominence during the second quarter of the twentieth century in the writings of the school of philosophy called 'logical positivism'.

6. A statement is said to be analytically true (or simply analytical) if its truth resides wholly in the definition of its terms. "Bachelors are unmarried" and "2 + 2 = 4" are normally cited as typical examples.

7. Michael Friedman, *Foundations of Space-Time Theories* (Princeton: Princeton University Press, 1983).

8. The view of science dubbed hypothetico-deductivist earned its nickname by insisting that the scientific worth of a conjecture has little to do with the way at which it is arrived, and all to do with the way it is borne out by experiment. Rather than condemn scientists for hypothesizing (or as Francis Bacon put it: for "anticipating nature"), many nineteenth-century inductivists realized that imaginative conjecture is indispensable for science. Once proposed, hypotheses are then carefully studied for the observable predictions they entail. They gain in scientific status and credibility according to their predictive successes.

9. For a cogent presentation of this argument see, Hilary Putnam, *Reason, Truth, and History* (Cambridge, Eng.: Cambridge University Press, 1981), pp. 119–24.

10. Imre Lakatos, "History of Science and its Rational Reconstructions," in Yehuda Elkana, ed., *The Interaction Between Science and Philosophy* (Atlanta Highlands, N.J.: Humanities Press, 1974), pp. 195–241 (quotation p. 196). For an eloquent application of Lakatos's strategy see Eli Zahar, *Einstein's Revolution: A Study in Heuristic* (La Salle, Ill.: Open Court, 1989).

11. Lakatos's methodological papers were collected posthumously in J. Worrall and G. Currie, eds., *The Methodology of Scientific Research Programmes*, vol. 1 of *Philosophical Papers* (Cambridge, Eng.: Cambridge University Press, 1978).

12. Lakatos, "Rational Reconstructions," pp. 196–97.

13. See for example Pierre Duhem, *The Evolution of Mechanics*, trans. Michael Cole (Dordrecht: Sijthoff and Noordhoff, 1980), chaps. 6–8.

14. Lakatos, "Rational Reconstructions," p. 214 and passim.

15. Notably, Joseph Agassi, *Science and Society: Studies in the Sociology of Science*, Boston Studies in the Philosophy of Science, vol. 65 (Dordrecht: D. Reidel, 1981), chaps. 16 and 17; Daniel Garber, "Learning from the Past: Reflections on the Role of History in the Philosophy of Science," *Synthese*, 67 (1986), pp. 91–114; Ron Curtis, "Are Methodologies Theories of Scientific Rationality?" *British Journal for the Philosophy of Science*, 37 (1983), pp. 135–61; and Ronald N. Giere, *Explaining Science: A Cognitive Approach* (Chicago: University of Chicago Press, 1988), pp. 38–43.

16. This is not to deny, of course, that scientists, like any creative thinkers, frequently grope in the dark and indeed occasionally perform canny feats of sleepwalking. These, however, cannot be considered, in my opinion, part and parcel, let alone prime examples, of the *rational* component of their work. For an intriguing analysis and appreciation of this arational aspect of scientific research see Arthur Koestler, *The Sleepwalkers* (London: Hutchinson, 1959).

17. I owe this powerful turn of phrase to Ruth Anna Putnam.

18. Michael Williams, *Unnatural Doubts: Epistemological Realism and the Basis of Skepticism* (Princeton: Princeton University Press, 1996), esp. chap. 1.2. Although this is one of the most thorough discussions of skepticism and its modern responses, the Popperian option is hardly touched upon.

19. Thomas Nagel, *The View from Nowhere* (Oxford: Oxford University Press, 1986), chap. 5. See also Williams, *Unnatural Doubts*, pp.15 f.

20. Stanley Cavell, *The Claim of Reason: Wittgenstein, Skepticism, Morality, and Tragedy* (Oxford: Oxford University Press, 1979), pp. 217 ff.

21. The literature for and against confirmation theory is enormous. For an excellent summary and critical analysis of the main issues afloat and a useful bibliography, see L. Jonathan Cohen, *An Introduction to the Philosophy of Induction and Probability* (Oxford: Clarendon Press, 1989).

22. The position I attribute here to Popper reflects the theory of science presented in his 1934 classic *Logik der Forschung* (translated into English in 1959 as *The Logic of Scientific Discovery* [London: Hutchinson]. In doing so I knowingly ignore the theory of verisimilitude developed in his later work. The theory of verisimilitude was first proposed by Popper in his "Truth, Rationality and the Growth of Scientific Knowledge," *Conjectures and Refutations: The Growth of Scientific Knowledge* (New York: Harper & Row, 1963), chap. 10. The notion, elaborated in detail by Popper, that verisimilitude—i.e., nearness to truth—can be formally expressed despite the fact that truth itself is unknown has been called to task by several writers. See for example William Newton-Smith, *The Rationality of Science* (London: Routledge & Kegan Paul, 1981), pp. 52–59.

23. Realism, as Popper uses the term, is the thesis, contrary to idealism, that the world which science aspires to describe exists. It is not the view, often attributed to realists, that what science proclaims to be the case at any given moment is in fact the truth. Popper's realism in no way mollifies the fundamental assertion that all scientific theory is conjecture. See, for example, Karl R. Popper, *Objective Knowledge: An Evolutionary Approach* (Oxford: Clarendon Press, 1972), pp. 32–33.

24. Lakatos advocates scientific research programmes in which hypotheses are generated by rigid "hard cores" insulated by "protective belts" of "auxiliary hypotheses," actively and deliberately defended, at times even "pig-headedly" by "negative heuristics" designed intentionally to shield them from disconfirming findings. While Thomas Kuhn depicts exemplar mature normal science as governed by a single paradigm, Lakatos envisions science at its best when different "research programmes" compete for the same ground. For Lakatos's methodol-

ogy, see references in notes 10 and 11 above. For Kuhn's theory of normal science, see his *The Structure of Scientific Revolutions*, 2d ed. (Chicago: University of Chicago Press, 1970), chaps. 2–6. For a comparison of their views to that of Popper, see Menachem Fisch, "Trouble-Shooting Creativity," *History and Philosophy of the Life Science*, 16 (1994), pp. 141–53.

25. "My thesis is that realism is neither demonstrable nor refutable. Realism like anything outside logic and finite arithmetic is not demonstrable; but while empirical scientific theories are refutable, realism is not even refutable. . . . But it is arguable, and the weight of the arguments is overwhelmingly in its favour." Popper, *Objective Knowledge*, p. 38.

26. Not every system is necessarily goal-directed. The solar system, for example, though a system, is no longer thought of as a means to an end—with the possible exception of natural-theological apologetics. As far as man-made and man-applied systems are concerned, their goal-directedness is wholly determined by the intent of their designers/users. A discarded appliance or theory used by no one ceases to be a goal-directed system. On the other hand, it is not unreasonable, in my opinion, to consider an *unintentionally functioning* system as goal directed. Organisms, for example, are now widely regarded as goal-directed systems, i.e., as complex structures geared to maximize gene distribution, although neither their design nor function is ascribed by modern biology to anything akin to intention. Goal-directedness is undoubtedly in the eyes of the beholder, though I seriously doubt that it is thereby wholly rendered a merely secondary quality. But I am well aware that this is a controversial issue. Many philosophers of biology would claim that the whole point of evolution theory is to free the life sciences of teleology and resist any talk of organistic goal-directedness. I disagree, but since the only reason for discussing goal-directed systems in the context of the present study is in order to elucidate the notion of rational action, which is here taken to be wholly intentional, I shall press the point no further.

27. I am referring to my particular car and my German, as opposed to the general notion of an automobile and the German language as such. For an account of this model with special reference to its historiographical implications see Menachem Fisch, *William Whewell Philosopher of Science* (Oxford: Clarendon Press, 1991), pp. 9–12, and "Towards a Rational Theory of Progress."

28. It is important in my opinion to distinguish sharply between problems and questions. Questions are parts of speech that endeavor to state queries (epistemic or cognitive perplexities) and subsequently call for answers. Problems, on the other hand, are usefully described as objective features of the situation. Questions are sometimes attempts to articulate problems—and in a sense could be assigned truth values insofar as they correspond or fail to correspond to real problems— but problems are problems regardless of whether or not they are articulated. R. G. Collingwood stressed the need for empathetic reenactment on behalf of historians in order for them to understand an agent's action as an answer to a question— that is to say, as a solution to what *he or she*, the agent, took the problem to be. Popper, on the other hand, sought to establish a historiography of science based on an "epistemology without a knowing subject," in which theories are regarded objectively as solutions to problems, which are in turn taken to be inherent features of situations regardless of the questions any particular agent might have had in mind. For Collingwood's approach, see in particular his *An Autobiography* (Oxford: Oxford University Press, 1939), esp. chap. 5. For Popper's "epistemology without a knowing subject," see Popper, *Objective Knowledge*, chaps. 3 and 4. For

an eloquent discussion of the difference between questions and problems, see J. N. Hattiangadi, "The Structure of Problems," parts 1 and 2, *Philosophy of the Social Sciences*, 8 (1978), pp. 345–65, and 9 (1979), pp. 49–76, esp. section 1.

29. John Kekes, "Rationality and Problem-Solving," *Philosophy of the Social Sciences*, 7 (1977), pp. 351–66.

30. Even Popper occasionally sounds as if theories or hypotheses may be considered immanently rational: "the problem of the arguability, *and thus of the rationality, of scientific hypotheses*—is, of course, linked with the problem of their testability." Karl R. Popper, *Realism and the Aim of Science* (part 1 of *Postscript to the Logic of Scientific Discovery*) (1983; reprint, London: Hutchinson, 1988), p. 161 (my italics).

31. There are several writers, both within and without the Popperian camp, who seem willing to accept as rational any effective mobilization of means to a desired end. Ronald Giere, the most prominent spokesman in recent years for a naturalized philosophy of science, writes of a "sense of 'rational', which simply means using a known effective means to a desired goal. This is hypothetical, or instrumental rationality, and it does come in degrees. When cognitive scientists investigate people's judgmental strategies, they are at most evaluating the instrumental rationality of their subjects. What strategies are subjects using in pursuing their goals? Are they doing as well in achieving their goals as they could in the circumstances? And if not, why not?" Giere, *Explaining Science*, p. 7. See also the opening paragraphs of Joseph Agassi and Ian C. Jarvie, "The Problem of the Rationality of Magic," *British Journal Of Sociology*, 18 (1967), pp. 55–74: "An action is rational by and large, if it is goal-directed." Are we to understand from this, as previously noted, that breathing in one's sleep, one's unaware digestion of food, or even the instinctive behavior of animals and plants are to be regarded rational simply because they are effective means to obviously desired goals? Or is the requirement that the goal be "desired" meant to rule out cases in which we are unaware of what we are doing? But then is the mere awareness of our desires enough to rate any successful eventuation of them rational as such? Or, as Agassi and Jarvie submit, is the fact that "there is an explanation readily available which shows these acts to be goal-directed" enough to rank the habitual planting of seeds or carving tree trunks into canoes rational acts? (p. 55f.). Again, are we to consider the successful digestion of our food a rational act simply because we consciously desire to nourish ourselves, or because there exists an explanation that it does? For a detailed criticism of Giere's approach, see Noah J. Efron and Menachem Fisch, "Science Naturalized, Science Denatured: Reflections on Ronald Giere's *Explaining Science*," *History and Philosophy of the Life Sciences*, 13 (1991), pp. 187–221.

32. Michael Polanyi, *Personal Knowledge: Towards a Post-Critical Philosophy* (1958; reprint, London: Routledge & Kegan Paul, 1983), p. 56.

33. I agree, therefore, with Kekes's proposal to regard problem seeking and problem solving as the external standard of rationality. Kekes, "Rationality and Problem-Solving," p. 268. I disagree, however, with his view of problems proper. I believe Kekes is mistaken in claiming that "there are no problems in nature, problems occur when facts disappoint one's expectations" (p. 269). If problems are taken as I suggest as the inherent shortcomings of systems, then they are themselves facts. To solve a problem rationally, as I shall argue below, certainly requires theory. But it is not the case that "what counts as a problem depends on

[a] theory," as Kekes insists (p. 269). Theories, I agree, are normally developed as solutions to problems, but that does not render theories themselves rational as Kekes proposes (p. 266) (since, as I have argued, there is no such thing as a rational theory per se). Indeed, theories may be developed, assented to, or rejected rationally. But what makes for the rationality of such moves is not merely the fact that the resulting theory happens to solve a problem, but that the treatment of the problem was rational—i.e., conscious, deliberate, and involving an appeal to reason.

34. This was for Lakatos the ultimate criterion for choosing between theories of rationality. "The hallmark of a relatively weak internal history (in terms of which most actual history is either inexplicable or anomalous)," he writes, "is that it leaves too much to be explained by external history. When a better rationality theory is produced, internal history may expand and reclaim ground from external history." Lakatos, "Rational Reconstructions," sec. 2B. Lakatos's ultimate metacriterion of rationality theories has been forcefully called to task in Daniel Garber, "Learning from the Past," pp. 91–114.

35. Since I do not regard rationality as a category of anything other than specific actions, the terms "rational agent," "rational reformer," "rational rabbi," etc., should be taken throughout this study as shorthand for: "an agent, reformer, rabbi, etc., engaged in rational activity."

36. With regard to scientific theories, this, again, is partyline Popper. See, for example, *Realism and the Aim of Science*, pp. 243–44). Subsequently, the severity of the tests an hypothesis is known successfully to have passed—that is, the initial improbability of its passing them—coined by Popper its "degree of corroboration," will form the grounds for its (tentative) adoption. Despite his reservations regarding "Popperian anti-inductivism," L. Jonathan Cohen has made this admittedly Popperian strategy the basis for a series of detailed studies of the logic and methodology of eliminative modes of reasoning in both natural science, linguistics, jurisprudence, and analytical philosophy. For acknowledgment of his Popperian leanings, see Cohen, *Induction and Probability*, pp. 183–85. Contrary to Cohen, Joseph Agassi adopts Popper's anti-inductivism wholeheartedly but rejects the requirement for severity. In Agassi's view the severity of tests is prescribed in technology by law, and directed in science by personal preference. His view is perhaps not as far from Cohen's as it is from Popper's, since for the former the list of variables deemed relevant for the testing of an hypothesis is in the last analysis a matter of expert convention. For Agassi's position, see for example: *Technology: Philosophical and Social Aspects* (Dordrecht: D. Reidel, 1985). For Cohen's system, see L. Jonathan Cohen, *The Implications of Induction* (London: Methuen, 1970); idem, *The Probable and the Provable*, (Oxford: Clarendon Press, 1977); idem, *The Dialogue of Reason: An Analysis of Analytical Philosophy* (Oxford: Clarendon Press, 1986); and idem, *An Introduction to the Philosophy of Induction and Probability*. For critical appraisals of Cohen's position, see L. Jonathan Cohen and Mary B. Hesse, eds., *Applications of Inductive Logic* (Oxford: Clarendon Press, 1980), and Menachem Fisch, "Learning from Experience," In Ellery Eells and Thomas Maruszewski, eds., *Probability and Rationality: Studies on L. Jonathan Cohen's Philosophy of Science*, Poznan Studies in the Philosophy of the Sciences and the Humanities, vol. 21 (Amsterdam: Rodopi, 1991), pp. 65–79.

37. For an illuminating discussion of the crucial difference between evaluating the credence of witchcraft beliefs and judging the rationality of those who held

them in sixteenth-century Europe, see Quentin Skinner, "A Reply to My Critics," in James Tully, *Meaning and Context: Quentin Skinner and his Critics* (Oxford: Polity Press, 1988), pp.242–46.

38. The relationship, however, is not commutative. If S is problematic, it follows necessarily that T_D is problematic. But the fact that T_D is found problematic does not necessarily imply that S is incapable of eventuating G.

39. It should not be concluded from the above that rational reform is limited to troubleshooting, that is to say to situations in which S is tested to ensure that it effectively eventuates G. Rational 'Research and Development' pertains to exactly the same schema. In these cases an otherwise unproblematic system S will typically be tested with respect to some new goal G'; as, for example, when a drug, known to be effective in its prescribed domain, is tested for ailments different from those for which it was originally developed, or when an otherwise successful scientific theory is applied to phenomena different from those it was originally meant to accommodate. Again, in the first instance T_D, which will normally not predict the desired results, will be all the researcher will have to go on. If the outcome of the tests is positive, and the drug or theory proves to be equally adequate in the new cases, T_D will be rendered problematic for failing to predict and explain the new results and will require modification. If, however, S fails the new tests, both S and T_D may need to be rethought in order to achieve their new desired objective. In other words, the initial problem in S may itself be conjectural.

40. Popper makes this point forcefully: "[W]hat distinguishes the attitude of rationality is simply openness to criticism (. . .) But the criticisms (. . .) offered in my approach are in *no* sense ultimate; *they too are open to criticism*; they are conjectural. One can continue to examine them infinitely; they are infinitely open to reexamination and reconsideration. Yet no *infinite regress* is generated: for there is no question of proving or justifying or establishing anything; (. . .) It is only the demand for proof justification that generates an infinite regress. (. . .) This is the heart of the difference between justification and criticism." Popper, *Realism and the Aim of Science*, pp. 27–29.

41. An agent's initial choice of a goal is, of course, an action in itself—though one quite different from that of attaining it—the rationality of which may be determined only with reference to the goals of some higher-level system. For example, it is one thing to assess the manner in which researchers approach the development of a specific appliance, but quite another to assess the directors of the firm's decision to produce that type of appliance rather than another.

42. Martin Hollis, "The Social Destruction of Reality," in Martin Hollis and Stephen Lukes, *Rationality and Relativism* (Cambridge, Mass.: MIT Press, 1986), pp. 67–86, pp. 83–84 (quotation; my italics).

43. Skinner, "A Reply to my Critics," pp. 248–55, p. 249 (quotation).

44. Ibid., p. 255.

45. Ibid., pp. 257–58.

46. Unnecessary even from Skinner's own perspective in view of his explicit adherence to Collingwood's erotetic historiography. To Collingwood, writes Skinner, "indeed, I am directly indebted for what remains my fundamental assumption as an intellectual historian: *that the history of thought should be viewed not as a series of attempts to answer a canonical set of questions, but as a sequence of episodes in which the questions as well as the answers have frequently changed.*" Ibid., p. 234, my italics.

47. There is, therefore, an important distinction to be made between knowing that a proposition is true and believing it. Had we had at our disposal a foolproof procedure for ensuring truth, knowing and believing would be less distinguishable. But we don't. I have argued along Popperian lines that in science the recognition of truth may be simply dispensed with. In other areas, such as technological standards committees and courts of law, where rulings as to the truth of certain assertions must be made, certain publicly upheld criteria are adopted to ensure credibility "beyond reasonable doubt." Hence, one could easily conceive of a jurist who agrees that in view of the evidence, a point has been proven beyond reasonable doubt, *and yet wholly disbelieves it.* This gives rise to the intriguing question whether a juror or judge is required to rule on the basis of what he or she *knows* or on the basis of what they believe! L. Jonathan Cohen discusses this question in detail in *An Essay on Belief and Acceptance* (Oxford: Clarendon Press, 1992). For those acquainted with his book, my debt to him will be apparent.

48. Popper, *Conjectures and Refutations*, chap. 4. "Towards a Rational Theory of Tradition."

Part 2: The Jewish Covenant of Learning

1. There are obviously many committed Jews who would have little difficulty in accepting my view of scientific rationality, but who would strongly oppose the idea that anything resembling it was ever entertained by a talmudic sage with regard to halakhic development. The opposite is rare. Still, I can testify to at least one eminent man of science who in public debate wholeheartedly accepted the latter while vehemently opposing the former.

2. Boyarin makes this point in several places, but elaborates upon it most fully in the introduction to *Carnal Israel*. The paragraph I paraphrase is in Daniel Boyarin, "Homotopia: the *Fem*inized Jewish Man and the Lives of Women in Late Antiquity," *d i f f e r e n c e s: A Journal of Feminist Cultural Studies*, 7 (1995), pp.41–81, p. 44 (quotation).

3. I state my own latter-day motivations for adopting the theory proposed fully aware that Popper's original motivations were, to the best of my knowledge, very different. The set of oppositions in defiance of which Popper originally formulated his philosophy were (in chronological order, not in order of gravity) the academic school of logical positivism that prompted the formulation his theory of science during the 1920s and 1930s and the ruthless dogmatism of both the Nazi and Communist regimes that provoked his social and political writing during the 1940s and 1950s. For the latter, see esp. Karl R. Popper, *The Open Society and Its Enemies, 2 vols. (London: Routledge and Sons, 1945).*

4. Exod. 34:27–28. See also, Exod. 19:5; Lev. 26:15; Num. 10:33; Deut. 4:13, 23, 5:2, 3, 9:10, 11, 15, 17:2, 28:69, 31:9, 26, and 33:9. Unless indicated otherwise biblical quotations are drawn from Harold Fisch, ed., *The Holy Scriptures* Jerusalem: Koren Publishers Jerusalem, 1969). Quotations from rabbinic sources are translated by the author.

5. See, for example, Harold Fisch, *Jerusalem and Albion* (London: Routledge & Kegan Paul, 1964), pp. 96 ff.

6. See Mishna, *Peah* 1:1, and Bavli, *Shabbat* 127a-b, where Torah-study is deemed superior to all deeds "the fruit of which a person may enjoy in this world and still their capital is invested in the next" (which in turn, are superior to those whose capital is not similarly preserved). Maimonides, following Bavli, *Kidushin*

40b, rules that: "No commandment equals that of Torah study, rather Torah study is comparable to all other commandments put together, since study leads to deeds, and therefore has precedence over deeds everywhere" (*Mishne Torah, hilkhot talmud Torah*, iii, 3), to which he adds: "and deeds do not lead to study." Ibid., i, 4.

7. A third-century tannaitic commentary to Numbers and Deuteronomy.

8. *Sifri*, Deut. 6, in *Sifre to Deuteronomy*, ed. Saul Harowitz and Louis Finkelstein (New York: Jewish Theological Seminary of America, 1969, p. 14. Herafter cited as Finkelstein ed. Throughout the midrashic literature the word "*levanon*" is taken by the rabbis consistently to designate the Temple rather than a geographic location, as suggestive of the cedars imported from Lebanon by King Solomon for building the First Temple. See for example: *Mekhilta* to Exod. 17:14; *Sifri* to Num. 27:12; Deut. 3:25 and 34:1; *Gen. Raba* to Gen. 2:8 and 11; Bavli, *Yoma* 39b, *Ketubot* 62a, *Gitin* 56b; Yerushalmi, *Yoma* iv, 41d and vi, 43c; *Ketubot* v, 30b.

9. The *Sifri's* attitude here reflects the book's basic exegetical strategy toward the opening chapters of Deuteronomy. The first verse of Deuteronomy states that "These are *the* words which Moshe spoke to all of Israel (. . .)." How can one book of the Torah be described as *the* words he spoke, asks the *Sifri*, certainly he spoke all five!? The *Sifri*, therefore, abandons the literal meaning of "the," and scanning the Bible for other cases in which but a portion of an author's words are described as *the* words he spoke (e.g., Jer. 30:4; 2 Sam. 23:1; Amos 1:1), concludes that in all such cases the "the" designates rebuke (*Sifri*, Deut. 1, Finkelstein ed., pp. 1–3). Moshe's entire opening speech is accordingly interpreted as conveying a series of reprimands—verse 7 included.

10. The biblical subordination of the Sinai covenant to political objectives, is stated perhaps most explicitly in Exod. 19:3–6:

> And Moshe went up to God, and the Lord called to him out of the mountain, saying, Thus shalt thou say to the house of Ya'aqov, and tell the children of Yisra'el; You have seen what I did to Mizrayim, and how I bore you on eagle's wings, and brought you to myself. Now therefore, *if* you will obey my voice indeed, and keep my covenant, *then* you shall be my own treasure from among all peoples: for all the earth is mine: and you shall be to me a kingdom of priests, and a holy nation.

To the best of my knowledge, nowhere in the talmudic literature do the rabbis ascribe a literal political meaning to the last phrase.

11. This is, at present, the widely accepted interpretation of the Talmud's account of Rabban Yoḥanan b. Zakai's decision to abandon and surrender the besieged Jerusalem on condition that Jabne, its sages, and the dynasty of Rabban Gamliel be saved (e.g. Bavli, *Gitin* 56b). Gedaliahu Alon, *The Jews in their Land in the Talmudic Age*, vol. 1 (Jerusalem: Magnes Press, 1980), pp. 107–18, argues in considerable detail that Rabban Yoḥanan's regulations and reforms all represent an attempt to salvage as much as possible of the Temple cult in order to minimize the trauma of the fall of Jerusalem. But not all agree with Alon's assessment. His contenders argue that the move from the Temple to the synagogue inaugurated by Rabban Yoḥanan's reforms was perhaps not altogether reluctant (see following paragraph), and that his new regulations represented a contrived and deliberate attempt to reform Judaism. Shmuel Safrai, "*Behinot Ḥadashot le-Ba'ayat Ma'amado u-Ma'asav shel [Rabban Yoḥanan] le-Aḥar ha-Ḥurban*," in Menahem Dorman, et al., eds., *In Memory of Gedaliahu Alon: Essays in Jewish History and Phi-*

lology (Tel Aviv: Hakibbutz Hameuchad, 1970), pp. 203–26, argues, contrary to Alon (and quite convincingly in my opinion), that while many of the Temple rituals could have been preserved after the destruction, Rabban Yohanan appears to have decided not to do so. According to Safrai, there existed historical and halakhic precedents that would have allowed him not to discontinue many of the animal sacrifices for instance. From the fact that Rabban Yohanan apparently chose not to do so, Safrai concludes that the move out of the Temple and into the synagogue was regulated by him with a deliberate view to partly discontinue the Temple cult.

12. For a detailed account of R. Yehuda's patriarchy and the relevant bibliography, see Aharon Oppenheimer, *Galilee in the Mishnaic Period* (in Hebrew) (Jerusalem: Zalman Shazar Center, 1991), pp. 60–78.

13. See for example Bavli, *Avoda Zara* 10b, and Yerushalmi, *Megila* iii, 74a.

14. The ninth of *Ab* is the Jewish day of mourning in commemoration of the destruction of the two Temples. For R. Yehuda's suggestion, see Bavli, *Megila* 5a-b; Yerushalmi, *Ta'anit* iv, 69c; *Megila* i, 70c; *Yevamot* vi, 7d. His opinion was rejected, however, as premature.

15. A fuller discussion of this obviously controversial issue lies beyond the scope of the present study. Notwithstanding the suggestive legendary material of the sort I have mentioned—from which so much can be learned of a culture's self image—it is equally instructive to scan the rabbis' halakhic writings for the kind of Temple-related material one would expect them to have contained, yet omit to mention. Thus, while they elaborate in minute detail the laws and regulations related to the sacrifices themselves, there is hardly mention and certainly no attempt to establish a systematic body of halakha devoted to the proper running and maintaining of the Temple. Had a fully-fledged functioning Temple been regarded, even theoretically, as a genuinely desired religious reality, one would have expected the Mishna, for example, to include an entire Order laying down the law regarding such issues with tractates devoted to the training and employment of personnel, the treasury, supplies and suppliers, the maintenance of the implements and the buildings, etc. The talmudic sages seem to have made no serious attempt to preserve the Temple within the system of halakha they established, as the type of functioning institution it would have to be.

16. For a detailed analysis of the role played by the canonization of the Hebrew Scriptures in the replacement of the priest by the sage as the principal religious authority, see Moshe Halbertal, *People of the Book: Canon, Meaning, and Authority* (Cambridge, Mass.: Harvard University Press, forthcoming), Part 1.

17. The epoch of Babylonian Jewish learning is marked by the return to Babylonia in 219 A.D. of R. Yehuda *Ha-Nasi's* important student Abba Arikha, afterwards known simply as Rav. Leaving the existing academy in Nehardea to his friend Shmuel, Rav founded a new academy in Sura. Nehardea was destroyed by Odenathus in 259 A.D., and its place was taken by the neighboring town of Pumbedita where R. Yehuda b. Yehezqel, a pupil of both Rav and Shmuel, founded a new school. The Pumbedita academy was transplanted a century later by Rabba to his home town of Mehuza. The original academies of Nehardea and Pumbedita were both restored shortly after their initial closure.

18. David Weiss Halivni, *Peshat and Derash: Plain and applied Meaning in Rabbinic Exegesis* (New York: Oxford University Press, 1991), pp. 123–25 and 134–35.

19. On the possible epistemological significance of the way the talmudic literature describes the Sadducees' dissent from the teachings of the Pharisees, see

Harold Fisch, "Teach Us to Count Our Days—a Note on Sefirat Ha-Omer," in Jacob Neusner, et al., eds., *From Ancient Israel to Modern Judaism: Intellect in Quest of Understanding—Essays in Honor of Marvin Fox*, Brown Judaic Studies, vol. 159 (Atlanta: Scholars Press, 1989), vol. 1, chap. 12, § 2, and Menachem Fisch, *To Know Wisdom: Science, Rationality and Torah Study* (in Hebrew) (Tel Aviv: Hakibbutz Hameuchad, 1994), pp. 45–59.

1. The Great Tannaitic Dispute: The Jabne Legends and Their Context

1. For details of the structure and the content of the Mishna and two Talmuds see n. 2 of the following section.

2. Such halakhic codifications as that of Maimonides in the twelfth century and of R. Yosef Karo in the sixteenth century encountered fierce opposition in their day. In both cases critics accused their authors of violating the essential open-endedness of the Bavli's approach to halakhic deliberation. For detailed surveys of both debates and of the relevant secondary literature, see Menahem Elon's monumental *Ha-Mishpat Ha-ivri: Toldotav, Mekorotav, Ekronotav*, (Jerusalem: Magnes Press, 1973), vol. 2, part. iii, chaps. 34 and 37, and Halbertal, *People of the Book*, part 2.

3. For an illuminating discussion of the difference between learning how to do something mimetically, by being shown, and learning how to do it from written instructions in the context of contemporary halakhic Jewry, see Haim Soloveitchik, "Rupture and Reconstruction: The Transformation of Contemporary Orthodoxy," *Tradition*, 28, no. 4 (1994), pp. 64–130, esp. p. 72.

4. *Genesis Rabbah* 31:12 and 34:7, ed. Theodor Albeck, *Midrash Bereshit Rabbah me-et Yehudah Teodor ve-Hanokh Albek* (Jerusalem: Vahrman, 1965), pp. 286 and 316. All citations refer to this edition.

5. Ibid., 35:1, p. 328. To the best of my knowledge, nowhere else in the talmudic literature does one find a clear and unambiguous declaration in favor of monastic abstinence comparable to that of R. Nehemiah. In this sense the homily before us is unique. But in the way it abstains from ruling on even the most fundamental extrahalakhic question, it is commonplace.

6. Bavli, *Avoda Zara* 2b.

7. *Genesis Rabbah* 1:4, p. 6.

8. For a different, anti-Platonic rendering of this midrash, see Ephraim E. Urbach, *The Sages, Their Concepts and Beliefs* (Jerusalem: Magnes Press, 1969), pp. 176–77.

9. *Genesis Rabbah, op. cit.*, 1:1, p. 2.

10. See for example Bavli, *Gittin* 6b; *Bava Metzia* 86a.

11. In several instances, however, the rabbis' midrashic renderings of Scripture would appear in fact to infringe upon the Bible's purely syntactical stratum. Two examples will suffice to make the point. Bavli, *Berakhot* 32a, relates the following: "R. Elazar said: Moses addressed the Heavens insolently, for it is said: '[And the people cried out to Moshe; but] Moshe prayed to the Lord [and the fire was quenched]' (Num. 11:2). Do not read 'to (*el*) the Lord' but rather 'on (or against, *al*) the Lord'—for the school of R. Eliezer b. Yaakov (frequently) read (interchange) *alephs* instead of (for) *ayins* and *ayins* instead of (for) *alephs*." This particular school of exegesis is apparently reported to have taken the liberty of tampering with the text itself. In similar fashion, though for entirely different reasons, Bavli, *Berakhot* 57a, and *Pesahim* 49b urge us to read Deut. 33:4: "Moshe com-

manded us a Torah, the inheritance (*morasha*) of the congregation of Ya'aqov" as "engaged to be wed (*me'orassa*) (to) the congregation of Ya'aqov"—a reading that requires adding the letter *aleph* to *morasha*. Several such examples exist. Note, however, that the recurring format of such *midrashim* is "*al tikri:* . . . *ela:* . . . " (do not *read:* . . . but rather: . . .). They propose novel *rereadings* of the text, but do not suggest that it be *rewritten!* They interpret Scripture; they do not amend it. For a thorough discussion of the talmudic concept of a "written text" see Jose Faur, *Golden Doves with Silver Dots: Semiotics and Textuality in Rabbinic Tradition* (Bloomington: Indiana University Press, 1986), esp. chap. 4.

12. See n. 10 above. For similar taxonomies of rabbinic attitudes to the onto-logical and epistemological foundations of *talmud-Torah* especially in the post-talmudic literature, see Halbertal, *People of the Book*, part 2; Avi Sagi, " 'Both Are the Words of the Living God': A Typological Analysis of Halakhic Pluralism," *Hebrew Union College Annual* (summer 1995); and Yohanan Silman, "*Torat Yisrael le-or Hidushea—Beirur Phenomologi,*" *Divrei Ha-Akademia Ha-Amerikait Le-Mada'ai Ha-Yahadut*, 57 (1992), pp. 49–67. All three studies limit their discussion to the realm of halakha. Rabbinic conventionalism, however, asserts itself even in the realm of historical disagreement, where one would normally least expect to encounter such a view. For while some could still argue that the law—halakha—should remain, at least in part, open and capable of adapting itself to unperceivable social, political, and cultural changes, and that, therefore, perhaps the purely halakhic component of the Oral Torah may have been intentionally not fully fixed in advance, most would still insist that the same could not be said of the factual status of past occurrences. A state of affairs either obtained or failed to obtain (regardless of whether or not it can be humanly *known* to have obtained). And to deny God perfect knowledge of the past, most would argue, would be seriously to encroach upon His omniscience. And yet the following is recorded in Bavli, *Gitin* 6b: R. Evyatar and R. Yonatan, it is told, were in disagreement as to the precise nature of the concubine of Bethlehem's misconduct described in Judges (19:2) simply as: "*va-tizne alav pilagsho*" (And [the Levite's] concubine played the harlot against him). Both agree that her Levite master needlessly overreacted to a minor neg-ligence on her part—the former submitted that he found a fly in his plate, the latter, that he discovered a hair she had neglected to remove. R. Evyatar is then told to have happened upon Elijah and to have asked to know "What The Holy Blessed be He was doing." "He is busy [studying] (the episode of) the concubine in Gibeah" answered Elijah. "And what does He say (about it)?" inquired the sage. He said: "Evyatar my son says so and so, Yonatan my son says so and so." "God forbid!," exclaimed the sage, "that there should be doubt before the Heav-enly One!?" To this Elijah answered: "Both are the words of the living God"! (For the last phrase, compare Bavli, *Eruvin* 13b.) I wish to thank Daniel Boyarin for drawing my attention in private conversation to the significance of this remark-able legend.

13. For a cogent latter-day presentation of the two traditionalist schools of thought, see, for example, *Ha'amek Davar* (commentary on the Pentateuch by Rabbi Naftali Tzvi Yehuda Berlin) to Exod. 34:1.

14. Whether or not one can at all speak, as does the realist, of there existing a single "one true meaning" of Scriptures, that is to say, of there at all being a canonical, humanly independent reading of the Written Torah—in the mind of God, for instance—is, of course, a different question entirely. On this question, as noted above and discussed below, the talmudic texts intriguingly speak in more

than one voice. My goal at this point is to analyze the traditionalist view in contradistinction to its most conspicuous rivals, namely, those that do premise truth values to proposed interpretations of the Written Torah while, at the same time, denying that they are, in principle, humanly accessible. The very existence, in talmudic thought, of such an option is a point of considerable importance, because it establishes a significant dividing line between the two great intellectual traditions of late antiquity. While much of the rabbis' writings proclaims a deep sense of epistemic optimism, they do so, unlike the Greek philosophers, without presupposing that truth (if it at all can be said to exist) is humanly recognizable. Western philosophy recognized and began seriously to entertain the possibility of a theory of knowledge which is at once both skeptical *and* constructive only during the present century. Popper, as I have argued, was the first Western philosopher to propose and partly develop a progressive theory of intellectual accomplishment while insistently denying scientists the capacity to ever verify, or even confirm their conjectures. Those of the rabbis who adopted the third of the realist options listed above appear, as we shall see, to have seriously contemplated such an option.

15. The traditional distinction between *aggada* and *halakha* tends to blur the more philosophically significant distinction between the rabbis' exegetical (i.e., midrashic) and purely halakhic concerns. While the only practical difference between aggadic and halakhic issues is that only the latter are decided, the second distinction is grounded in different types of problems. Because the Jabne debate, discussed in the present section, was primarily nonexegetical, there is no need at this point to differentiate between conventionalist and realist oppositions to traditionalism. For a similar though apparently unintended conflation of conventionalism and antitraditionalist realism with regard to halakhic development, see Sagi, "Both are the Words of the Living God," in which both schools of thought are jointly subsumed under the heading "The Authority Model."

16. As dramatic depictions often do, the Jabne stories of antitraditionalist triumph tend to ridicule and caricature the opposition. They are not intended as studies of the defeated traditionalists but as studies of their defeat. In them, the archtraditionalist—personified in the figure of R. Eliezer b. Hyrqanus—is reputed to declare that in his many years as an active halakhist he never uttered a thing he had not previously heard from his own Masters. Despite its currency in posttalmudic culture, such extreme traditionalism is reserved by the framers of the talmudic texts almost exclusively to these legendary materials. As we shall see shortly, when halakhic rather than legendary texts give voice to traditionalist leanings, they tend invariably to be of the less extreme variety. Traditionalism as it is represented in the Jabne tales is to a significant extent atypical of most 'real' manifestations of traditionalist bias found elsewhere in the rabbis' writings.

17. The many debates recorded in the talmudic texts between the "Houses" of Shammai and Hillel are a case in point. In no other case of pre-amoraic disagreement does one find the debates themselves recorded in such detail and variety. Throughout the talmudic corpus over thirty such debates are on record, and since the rabbis clearly and wholly aligned themselves with the Hillelites (see Bavli, *Eruvin* 13b, and Yerushalmi, *Suka* ii, 53—I shall discuss the nature of this alignment below), it is tempting to speculate as to the purpose that the reconstruction of these debates was meant to serve. Interestingly, although the arguments leveled by each House against the other are for the most part equally devastating, only the Hillelites are reported ever to have relinquished their initial

position in the light of Shammaite criticism. See, for example, Mishna, *Yevamot* 15:2; *Eduyot* 1:12, 1:13, 1:14; *Kelim* 9:2; Yerushalmi, *Gitin* iii, 45b. The Shammaites, by contrast, are presented as having stubbornly held to their original views irrespectively of the problems they are shown to harbor. (On one occasion, R. Jacob b. Aha suggests [Yerushalmi, *Yevamot* i, 3b] that, in order to prevent social fragmentation, the Shammaites did in fact systematically concede to the Hillelites whenever the latter's rulings were more stringent than their own on matters relating to marital status. His suggestion, however, is flatly rejected.) In the light of our present discussion, it is tempting to propose that the two Houses, or schools of thought, function in these reconstructed debates as (latter-day contrived) representatives of diametrically opposed positions regarding the aim and value of halakhic disputation. While the Shammaites are frequently described by the talmud as steadfast traditionalists firmly committed to their received views, the Hillelites are consistently portrayed in these debates as *developers* of halakha through the rational employment of logical and analogical argument. These two conflicting modes of halakhic decision making are forcefully contrasted in both the Bavli and the Yerushalmi's accounts of the young Hillel's encounter with the Shammaite family of High Priests, *b'nei Beteira* (Bavli, *Pesaḥim* 66a, Yerushalmi, *Pesaḥim* vi, 33a)—a fascinating story I shall discuss in some detail below. The Hillelites appear to enter these disputes with a clear sense that their halakhic commitments are, in principle, refutable, and, therefore, as prudently seeking and welcoming criticism. Most significantly, in 15 percent of the debates they prove willing to revise their views when disproved. The Shammaites, firmly committed in advance, appear systematically to treat counterarguments as objections to be either warded off or simply ignored, but never accepted. The Bavli states its preference in favor of the Hillelites by attributing the essential open-endedness of their learning to their intellectual modesty: "Why did the House of Hillel merit having the halakha established according to them? Because they were genial and modest (humble), and taught their sayings (together) with those of the House of Shammai. Furthermore, they put the words of the House of Shammai before their own" (Bavli, *Eruvin* 13b). The Yerushalmi is even more explicitly antitraditionalist on this point: "Why was it granted to the House of Hillel that the Halakha be decided according to them? R. Yuda b. Pazi said, because they put the words of the House of Shammai before their own, and in addition would, after hearing the words of the House of Shammai, (frequently) relinquish their own position" (Yerushalmi, *Suka* ii, 53). For a delightful description of the shrewd nonreflective polemical manner of the Shammaites, see Bavli, *Yevamot* 16a, Rashi's commentary there to the term *bekhor Satan*, and n. 81 below. For a detailed evaluation of the primary and secondary literature concerning the two Houses, see Israel Ben-Shalom, *The School of Shammai and the Zealot's Struggle against Rome* (in Hebrew) (Jerusalem: Yad Izhak Ben-Zvi and Ben Gurion University Press, 1993), chaps. 5–8. On the possible significance of the debates themselves, see Moshe Weiss, "*Ha-Otentiyut shel ha-Shakla ve-Tarya be-Maḥlokot Bet Shammai u-Bet Hillel,*" *Sidra* 4 (1988), pp. 53–66; idem, "*Ha-G'zeira Shava ve-ha-Kal va-Ḥomer ba-Shakla ve-Tarya shel Bet Shammai u-Bet Hillel,*" *Sidra* 6(1990), pp. 41–61; Haim Shapira and Menachem Fisch, "*Pulmusei ha-Battim*" (forthcoming).

18. Tosefta, *Sanhedrin* 7:1 and *Ḥagiga* 2:4, cited Bavli, *Sanhedrin* 88b. See also Bavli, *Sotta* 47b and Yerushalmi, *Sanhedrin* i, 19a. Yerushalmi, *Ḥagiga* ii, 77d brings the point home the most forcefully of them all: "Thus, as the disciples of Shammai and Hillel who had not attended upon their Masters sufficiently in-

creased in numbers, and the number of (halakhic) disputes in Israel increased, they became two (separate) sects—these would rule a thing unclean while these would rule it clean, and it will not be settled until the coming of the Son of David (i.e., the Messiah)!"

19. This, again, is reported as a point of contention between the schools of Hillel and Shammai. "The House of Shammai declare that one should not teach Torah but to the wise, humble, well-born and well-off, while the House of Hillel declare that a person should teach to anyone, for many criminals were among the people of Israel who were drawn close to *talmud-Torah* and eventually became righteous, benevolent and worthy." *Abot de-Rabbi Nathan*, A iii, ed. Soloman Schechter (New York: Feldheim, 1945), pp. 14–15. All citations to this edition.

20. For a survey and discussion of talmudic pronouncements that appear to affirm or to oppose the idea of "the decline of the generations" see Menachem Kellner, *Maimonides on the "Decline of the Generations" and the Nature of Rabbinic Authority* (New York: SUNY Press, 1996). Kellner, however, limits his analysis to the question of whether or not and in what respects are later generations perceived as intellectually or morally inferior to their predecessors. Yet it is possible for a steadfast traditionalist to contest the idea that later generations are in any way inferior to earlier ones, while still having to admit that the Torah itself, the transmitted body of their teachings, is forever and necessarily dwindling. For him human *knowledge* of the Torah is forever declining regardless of whether or not the generations themselves exhibit decline.

21. For a comprehensive exposition and evaluation of Eliezer as "a bearer of ancient tradition and early *halakha*" see Yehudah D. Gilat, *R. Eliezer ben Hyrqanus, A Scholar Outcast,* (Ramat Gan: Bar Ilan University Press, 1984). Eliezer's traditionalism is generally far more pronounced in later post-Mishnaic depictions than in the authentic tannaitic literature (see below n. 25), as is, if to a lesser extent, the equally adamant antitraditionalism attributed to his adversaries. The historical credence of the Talmud's portrayal of the Jabne debate is, therefore, doubtful to say the least. However, the aim of this study, as I have explained at the outset, is to analyze the Talmud's historiosophy and mythologies regarding its origins for what they might have been intended to convey, and not in order to judge their historical accuracy. To repeat, the talmudic texts, especially the Bavli, are treated in the present study as intentionally framed wholes rather than as potential repositories of historical data.

22. Mishna, *Abot* 2:8.

23. Bavli, *Sanhedrin* 68a; *Abot de-Rabbi Nathan*, A xxv, Schechter ed., p. 81.

24. Bavli, *Suka* 27b-28a; *Yoma* 66b.

25. The amoraic version of the story cited in the form of a *beraita* by the Bavli differs considerably both in detail and tenor from the tannaitic source in Mishna, *Yadayim* 4:3. In the version cited by the Mishna, Yossi b. Dormaskit is not even mentioned, Eliezer's tears seem to have been tears of joy, and most importantly, the clause "*al taḥushu le-minyankhem*"—do not heed your voting—translates there, equally well as "Be not concerned about your voting." The Mishna's account of Eliezer's response to the Jabne decision easily reads as an attempt to praise and encourage, rather than to reprimand his colleagues.

26. Bavli, *Sanhedrin* 68a, and slightly modified in *Abot de-Rabbi Nathan*, A xxv, Schechter ed., p. 81.

27. Bavli, *Bava Kamma* 74b.

28. Bavli, *Berakhot* 27b, *Rosh Ha-Shana* 25a, and *Bekhorot* 36a.

29. Bavli, *Rosh Ha-Shana* 25a.

30. I interpret R. Yehoshua's response as humorous rather than evasive in view of the fact that his public denial of his former ruling was offered after Raban Gamliel had settled the matter. As noted above although he evidently disagreed with the President on certain matters, R. Yehoshua never challenged his authority once Gamliel had issued a ruling. Interestingly, with regard to the evening prayer, the halakha was eventually decided according to R. Yehoshua rather than Raban Gamliel, although later generations "took it upon themselves" to treat it as obligatory. See Maimonides, *Mishne Torah, hilkhot tephila* 1:6.

31. Bavli, *Berakhot* 27b. See also Yerushalmi, *Berakhot* iv, 7d.

32. Raban Gamliel, who eventually repented and apologized to R. Yehoshua, was later reinstated and served, according to the Bavli, alongside R. Elazar b. Azaria, as rotating President, preaching three *Shabbatot* to every one of Elazar's (Bavli, *Berakhot* 28a). According to the Yerushalmi, Gamliel was fully reinstated as President and Elazar b. Azaria was appointed Head of the *Bet din* (Yerushalmi, *Berakhot* iv, 7d). There is no historical evidence, however, that Raban Gamliel was ever removed from office, or that R. Elazar ever served as President.

33. Yerushalmi, op. cit.

34. For R. Yehoshua's enthusiastic support of his young disciple's philosophy of *talmud-Torah*, see Bavli, *Hagiga* 3b, and below in this chapter.

35. *Abot de-Rabbi Nathan*, A iii, Schechter ed., pp. 14–15. See also n. 19 above.

36. Bavli, *Berakhot* 28a.

37. Mishna, *Eduyot 3:12*, for example, documents three cases in which the R. Elazar's halakhic opinion was presented at Jabne and voted down by his colleagues. On the special significance of this particular tractate for the present discussion see below p. 71ff. R. Elazar's apparent revocation of the President's special power of arbitration seems not to have been an innovation on R. Elazar's part, but a return to former custom. See the following paragraph.

38. Tosefta, *Sanhedrin* 7:1; *Hagiga* 2:9; Bavli, *Sanhedrin* 88b; Yerushalmi *Sanhedrin* i, 19a.

39. Although there appear to be no substantial differences between the four sources, the version cited by the Bavli is clearer and better structured, except for the reversal of the last two sentences.

40. *Lishkat Ha-Gazit*, the official seat of the *Sanhedrin* within the Temple complex.

41. Tosefta, *Hagiga* 2:9.

42. As we shall see shortly, the Mishna's account of the hierarchical relationships between the three Temple-based *batei dinim* is significantly different.

43. This in itself is a rather curious statement, for it is hard to imagine how the halakhic cleavage referred to by the Tosefta and Bavli could be attributed solely to the inadequacy of their students when Shammai and Hillel themselves are known to have disagreed so fundamentally on so much. It is as if we are meant to believe that if only their students had been more attentive, halakhic disagreement would have remained confined to the truly new. This is extremely unconvincing. Add to this the fact, previously noted, that the disputes between the two Houses recorded in the talmudic literature strongly tend to portray the Hillelites as decided antitraditionalists to begin with. If this were the case, it is doubtful that, as members of the Great *Bet din*, they would have ever agreed to give unquestioned precedence to received views on any matter—old or new. One seems forced to conclude, therefore, that the traditionalist ideal and its eventual

diminution described by the Tosefta either owes its origin to latter-day tannaitic authors of traditionalist bent (writing in conscious disagreement with those responsible, for example, for the version in the Mishna, and for reconstructing the debates between the two Houses), or should not be taken as a serious portrayal of either a mythical past or an ideal future. The first option—according to which the fundamental tension between traditionalism and antitraditionalism remained an open and debated question even for the latter-day redactors of the tannaitic and amoraic literatures—is surely the more appealing one, if only for it being the more charitable of the two. We shall have chance to pursue this point further when comparing below Tosefta and Mishna *Eduyot*, and the Bavli's and the Yerushalmi's accounts of the young Hillel's encounter with the Shammaite *b'nei Beteira*. In conceding to the first option, however, I do not wish to imply that the traditionalist tendencies evident in both the content and editorial policies of certain of the talmudic texts remained on equal footing with those of their opponents. What I wish to show in the remainder of this part of the book is that the myth of R. Elazar's antitraditionalist campaign at Jabne took root to the extent that, though evidently audible in the talmudic literature, the traditionalist voice is rendered marginal—especially at the editorial level.

44. It would seem unfair then that an elder may be found guilty by a lower ranking court that is powerless to contradict a former ruling. This is the reason perhaps for Maimonides's unusual interpretation of the Mishna (*Mishne Torah, hilkhot mamrim*, 3:8). He reads the phrase "If they heard they told them" as if it applied to the elder rather than to the court. If *he*, the elder, takes heed of the court's ruling, and retracts his own ruling to the contrary, then the case is closed with no further ado. But if he refuses to do so, the case is referred to a higher instance. In other words, according the Maimonides, a rebellious elder can only be found guilty by a Sanhedrin.

45. Ben-Shalom, *The School of Shammai and the Zealot's Struggle against Rome*, pp. 273–76, argues forcefully that "in Jabne of Rabban Yoḥanan b. Zakai and in Jabne of Rabban Gamliel there is no trace at all of any activity of the two Houses" (p. 273). According to Ben-Shalom, the House of Shammai virtually vanished from history as a result of the utter failure of the Great Revolt, while the House of Hillel, firmly represented by Rabban Yoḥanan b. Zakai, swiftly dominated Jabne, and therefore, with the establishment of the Jabne center, ceased to exist as a separate "House." See also Zecharia Frankel, *Darkhei Ha-Mishna* (Tel Aviv: Sinai, 1959), pp. 55–56; J. N. Epstein, *Introduction to Tannaitic Literature: Mishna, Tosefta, and Halakhic Midrashim* (in Hebrew) (Jerusalem: Magnes Press, 1957), pp. 60–61; Isaac H. Weiss, *Dor Dor Ve-Dorshav* (Vilna: Rom, 1911), vol. 1 pp. 173–4; Shmuel Safrai, "*Ha-Hakhra'a ke-Bet Hillel be-Yavne*," in *Proceedings of the Seventh World Congress for Jewish Studies, Halakha and Midrash* (Jerusalem: Magnes Press, 1981), pp. 21–44, esp. p. 38.

46. Cf. Haim Shapira, "The Deposition of Rabban Gamliel: Between Myth and History" (in Hebrew), forthcoming.

47. On intertextuality as a theory of talmudic midrash, see Daniel Boyarin, *Intertextuality and the Reading of Midrash* (Bloomington: Indiana University Press, 1990). On the application of Boyarin's theory to an analysis of some of the stories about the Jabne sages see Azaria Baitner, "Conflicts and Dialectics in the Tales of Yavneh's Scholars as a Model of a Literary Phenomenon with Ideological and Historiosophical Meaning" (in Hebrew) (Ph.D. diss., Bar Ilan University, 1994).

48. The verb *nishneyt* derives from *shana* (plural: *shanu*) that is normally em-

ployed to denote the study or teaching of an existing text or saying. The passive present tense *nishneyt*, however, is used invariably by the Bavli to connote the teaching process involved in the *establishment* of a Mishnaic text. See, for example, Bavli, *Shabbat* 123b; *Pesahim* 19a; *Hagiga* 24a; *Bava Kama* 94b; *Bava Metzia* 33b; *Bekhorot* 30b; *Nida* 25b. The one exception is Bavli, *Shabbat* 83b. The Soncino translates: "Eduyyoth was *formulated* on that day."

49. Bavli, *Berakhot* 28a.

50. The one exception is (2:4): "Three things were stated by R. Yishma'el in the presence of (before) the sages at *Kerem be-Yavne*." On the possible origin of the term *Kerem be-Yavne* see next note.

51. Significantly perhaps, the term *Kerem be-Yavne*—lit. the vineyard at Jabne—is explicitly associated by the Yerushalmi with the day of R. Elazar b. Azaria's election to office, with reference to the great number of benches that were needed to be added: "this (the story of the added benches) is a midrash related by R. Elazar b. Azaria to the sages at *Kerem be-Yavne*: for was there a vineyard there? Rather, (it refers to) the learned sages who (now) formed row after row like a vineyard" (Yerushalmi, *Berakhot* iv, 7d).

52. *Divrei Sofrim*, a term denoting laws of early rabbinic origin not explicitly stated in the Torah text.

53. Tosefta, *Eduyot* 1:1.

54. For a detailed elaboration of this point see, J. N. Epstein, *Introduction to Tannaitic Literature*, pp. 422–29.

55. Mishna, *Eduyot* 2:5 and 8, Tosefta, *Eduyot* 1:7 and 8.

56. Significantly, all three are disagreements between Hillel and Shammai themselves and not between the two "Houses" of their disciples, some of whom are reputed (by the traditionalist tannaitic describer of the court system) to have misrepresented the original views of their Masters

57. One *kav* is approximately 2.4 liters. According to the halakha, one is not allowed to fill a ritual bath with drawn water. The ritual bath of purification, the *mikve* (Lev. 11:36), has to contain at least forty se'ahs (approx. 12.148 litres) of originally running water. The problem at hand is to decide the minimum amount of drawn water sufficient to disqualify the *mikve*.

58. Mishna, *Eduyot* 1:4. Unlike the former two rulings, where the Jabne sages rule differently from Hillel and Shammai seemingly on their own accord, it is still possible to interpret their ruling here as an act of arbitration between three traditions—those of Hillel and Shammai and that of their teachers Shema'aya and Avtalyon. While the Tosefta seems clearly to support such a reading, the Mishna, we shall see, evidently does not.

59. This is the accepted version (Heb., *lebatlan*). In some manuscript sources one finds *lebatala*, i.e., superfluously, for apparently no reason.

60. According to Maimonides, one *Bet din* will be regarded greater in wisdom than another if and when "the head of the assembly of the one is deemed wiser than that of the other" (see his commentary on the present *mishna*). Elsewhere, he explains how one *Bet din* of seventy-one can be considered greater in number than another. In this case, he opines, the basis for the calculation should be "the number of sages of that generation" who conform to the *Bet din*'s ruling (*Mishne Torah, hilkhot mamrim* 2:2). Among the post-talmudic commentators on this passage there are some who attribute to it an even more radically antitraditionalist statement. Thus, for example, the early-seventeenth-century commentator Rabbi Yom Tov Lipmann Heller interprets the Mishna to be saying that the requirement

of one *Bet din* to be greater in wisdom and number in order to overthrow the
ruling of another only obtains in the absence of support from a former minority
opinion. If such an opinion is on record, he asserts, the Mishna allows any *Bet
din* to endorse it (*Tosafot Yom Tov, Eduyot* 1:5).

61. The nature of R. Yehuda's explanation for mentioning minority opinions
is somewhat ambiguous. Most later commentators (e.g., Maimonides's and the
Tosafot Shantz's commentaries to Mishna, *Eduyot*) understand it as adding to rather
than as disputing that of the anonymous *tanna kama*, claiming that the reason
for recording dismissed minority opinions is *not only* in order to provide grounds
for future *Batei dinim* to overturn former decisions, but *also* in order to bar latter-day traditionalists from relying on formerly rejected positions. It is possible,
however, to read R. Yehuda as claiming, contrary to the *tanna kama*, that the *only*
reason for mentioning minority opinion as such is to ensure, in traditionalist fashion, that the traditions of the majority reign supreme. (Such, for example, is the
twelfth-century R. Abraham b. David, the RABAD's reading of R. Yehuda's comment. The RABAD, however, tends questionably to interpret the entire passage
along traditionalist lines in harmony with the Tosefta.) But even if R. Yehuda is
taken to have disputed the Mishna, his view must then be regarded a minority
opinion and be duly rejected!

62. Tosefta, *Eduyot* 1:3.

63. That is to say, for the benefit of future generations seeking to overrule former majority rulings.

64. Compare Bavli, *Nida* 6:a-b.

65. Tosefta, *Eduyot* 1:1.

66. Tosefta, *Eduyot* 1:6.

67. Mishna, *Eduyot* 1:12–14.

68. See above, n. 17.

69. At one point, however, the Tosefta seems to hint that the Jabne reformers
did encounter some resistance from extreme traditionalist quarters. See below n.
80.

70. See also Bavli, *Berakhot* 19a, and *Pesaḥim* 74b.

71. On the way in which the Mishna generally serves to "create a world of
discourse quite separate from the concrete realities of a given time, place, or society," see Jacob Neusner, *Judaism: The Evidence of the Mishnah* (Chicago: University
of Chicago Press, 1981), p. 245 and passim.

72. But even these are often juxtaposed in ways that thoroughly distort the
historical sequence of events. The way in which the story about Akavia b. Mehalal'el, a sage active about a century prior to the establishment of Jabne, is listed
without comment along with others clearly related to second-generation Jabne
is typical.

73. It goes, of course, without saying that the "chronology" implied in this
and other passages has virtually nothing to do with the actual dating of the texts
under consideration. Even the most ahistorically minded latter-day reader, contemplating the entire talmudic corpus whole and intact, is well aware that the
Mishna and the Tosefta were composed and collated hundreds of years before
the Bavli. Nor am I suggesting that because the *beraita* alludes to the origination
of *Eduyot*, one may assume that it was in fact composed earlier. For all we know,
the Bavli could easily be citing a source that was written long after the event as a
reflective, retrospective commentary on an existing well-established *Eduyot*. As a
purely literary device, however, the *beraita*'s allusion to the composition of *Eduyot*

is significant and should not be ignored. I take the imagined chronology constructed by the *beraita* solely as a literary means for establishing its relevant literary-philosophical context, as an invitation that I readily accept to place the two texts side by side, to ignore the historical sequence of events, and to view *Eduyot* philosophically as the (imagined ideological) outcome of Elazar's reforms. On the much labored question of the historical significance of the talmudic narratives see, for example, Moshe D. Herr, *"T'phisat ha-Historia etzel Hazal,"* in *Sixth World Congress of Jewish Studies* (Jerusalem: Magnes Press, 1976), vol. 3, pp. 129–42, and Yona Fraenkel, *"She'elot Hermeneutiyot be-Kheker Sipurai ha-Aggada,"* *Tarbitz* 47 (1978), pp. 139–72.

74. See also Bavli, *Berakhot* 19a, and Yerushalmi, *Moed Katan* iii, 81c. For a comprehensive survey of the various interpretations of this fascinating story see I. Englard, "The 'Oven of Akhnai': Various Interpretations of an Aggada" (in Hebrew), *Annual of the Institute for Research in Jewish Law*, 1 (1974), pp. 45–57. See also Susan Handelman, *The Slayers of Moses: The Emergence of Rabbinic Interpretation in Modern Literary Theory* (Albany: SUNY Press, 1982), pp. 40 ff.

75. The fact that only fully fledged utensils, *kelim*, are liable to "receive" ritual impurity—*tum'at kelim*—gives rise to many intriguing questions concerning the precise definition of "utensilhood" with respect to the seven principal categories of *kelim* (items of clothing and sacking, leather-goods, and utensils made of bone, metal and wood and pottery). Thus it becomes of crucial halakhic importance to determine the exact moment in which various objects become utensils in the course of their construction and, conversely, that moment in which they cease to be utensils in the course of their breakage or dismantlement.

76. Bavli, *Berakhot* 19a, and *Bava Metzia* 59b—the former has *"halakhot,"* the latter "words."

77. Lit. he taught, or stated—a term frequently used by the Bavli to introduce a brief *beraita* to explain or expound upon a tannaitic source. What follows is therefore to be considered an additional and independent source.

78. Mishna, *Eduyot* 7:7, Tosefta, *Eduyot* 2:1.

79. For two recent such readings see Daniel Boyarin's superb analysis of the story in *Intertextuality and the Reading of Midrash*, pp. 33–77, and the editors' introduction to J. P. Rosenblatt and J. C. Sitterson, eds., *Coherence and Complexity in Biblical Narrative*, (Bloomington: Indiana University Press, 1991), pp. 1–5.

80. Following a brief description of the dispute, Tosefta, *Eduyot* 2:1 states: "And it was named *tanuro shel akhnai,"* and adds: "(and) it gave rise to numerous disagreements in Israel." This is the only tannaitic source I am aware of that so much as indicates that this particular disagreement became the focus of wider contention. In fact, it is the only tannaitic text that even hints that perhaps not all was settled and agreed upon at *kerem be-Yavne*. As we have seen, Tosefta, *Eduyot*, portrays Jabne quite differently from the Mishna, as engaged primarily in arbitrating between well-established halakhic traditions by majority vote—a minimal breach of the strict traditionalist norm in order to unite the people. And yet it reports that the sages' dismissal of R. Eliezer's opinion regarding the oven was not universally accepted and that as a result *rabu makhlokot be-Yisrael*—the very same term used in Tosefta, *Sanhedrin*, and elsewhere to describe the crisis due to the rift that allegedly developed earlier between the two Houses. This could only mean that R. Eliezer and his followers apparently refused to accept the authority of the Jabne majority to dismiss his teachings—certainly those related to the oven. On this point at least (despite other major differences), Tosefta, *Eduyot*, and

Bavli, *Bava Metzia*, appear to converge. As we shall see immediately, the Bavli, unlike the Tosefta, clearly follows Mishna, *Eduyot*, in describing the Jabne reforms in decidedly antitraditionalist terms. Moreover, unlike the Tosefta, the Bavli goes on to tell how Jabne dealt with the situation.

81. As noted at the outset, even the most avid traditionalist—and the R. Eliezer of the Jabne stories is portrayed as the most avid traditionalist imaginable—is liable to engage in lively halakhic polemics. What a traditionalist will never do is to repudiate a received position by force of counterargument. But, as we have seen with regard to Bet Shammai, for example, traditionalists, even as zealous as the Eliezer of the legends, will normally have no problem arguing *against* the positions of their rivals. I say all this so that the phrase used here: *"heshiv R. Eliezer kol t'shuvot she-ba'olam"* (lit. "R. Eliezer answered [offered, submitted] all rejoinders in the world") will not be taken as an indication that R. Eliezer might have been equally willing, like an antitraditionalist, to heed to *t'shuvot* leveled against *his* legacies. The term *t'shuvot*, used by the *beraita*, is an almost technical term reserved in both the tannaitic and amoraic literatures to countermoves in halakhic debate. Boyarin renders *t'shuvot* 'refutations', which is clearly better than the Soncino's 'arguments'. I prefer 'counterargument', which seems to be the most widespread usage of the word in texts of both tannaitic and amoraic origin. (See, for example, Tosefta, *Eduyot* 2:10, *Aholot* 2:7, 2:8 and 3:3, *Sanhedrin* 7:6–7; Bavli, *Shabbat* 47a, *Pesaḥim* 71b, *Ketubot* 87b. Far less frequently the term is used to denote explanations (e.g., Bavli, *Shabbat* 77a), 'proofs by attesting to traditions' (Bavli, *Shabbat* 153b, *Yoma* 12b, 23b, Yerushalmi, *Yoma* vii 44d), or simply 'answers' (Bavli, *Megila* 25b) but always, without exception, as countermoves in halakhic debates.) The most fascinating occurrence of the term, and certainly the most relevant to our present concerns, is found in the following Jabne story cited Bavli, *Yevamot* 16a. Apparently it was rumored that, in his time, the now ailing and blind R. Dosa b. Hyrqanus had ruled in favor of Bet Shammai on a matter that was known to have been formerly decided in accord with the view of the Hillelites. Seeking to settle the matter once and for all, R. Yehoshua b. Ḥanania decided to pay the old man a visit, accompanied by R. Elazar b. Azaria and R. Akiva. After elaborate introductions and lengthy preliminaries they finally got round to the main purpose of their visit:

> What is the halakha, they asked him, in the case of a daughter's rival? This, he answered them, is a question in dispute between Beth Shammai and Beth Hillel. In accordance with whose ruling is the halakha? The halakha, he replied, is in accordance with the ruling of Beth Hillel. But, indeed, they said to him, it was stated in your name that the halakha is in accordance with the ruling of Beth Shammai! He said to them: Did you hear (them refer it explicitly to) "Dosa" or (merely to) "the son of Hyrqanus"?—By the life of our Master, they replied. We heard [only the latter]. I have, he said to them, a younger brother who is a so-and-so, and his name is Jonathan and he is one of the disciples of Shammai. Take care that he does not overwhelm you on questions of halakha, because he has three hundred *t'shuvot* to prove that the daughter's rival is permitted.

But I, he continued, call heaven and earth to witness that he is wrong! The very fact that the story assumes that the two traditions were not supposed to remain deadlocked, and that a firm ruling had been made in favor of Bet Hillel, speaks of at least a weak antitraditionalist bias. Rashi intriguingly renders the curious phrase *bekhor Satan* (which, contrary to the Soncino's somewhat anemic "daredevil," comes closer to "son-of-a-gun") as follows: "He was shrewd, and stuck to his traditions, and was willing to take action in order not to renounce his traditions *in favor of majority opinion.*" This is precisely the way I wish to read the *t'shuvot*, the counterarguments and rejoinders presented on that crucial and fateful "day" at Jabne by the other great alleged Shammaite, R. Eliezer b. Hyrqanus: as the polemic moves of a dedicated traditionalist, in a heroic attempt to uphold his traditions in the face of a majority vote against them.

82. On this point also the present *beraita* appears to differ from Tosefta, *Eduyot* (see above n. 80), which implies that the point of contention between R. Eliezer and his colleagues was precisely their authority to arbitrate between traditions. In order to maintain a traditionalist portrayal of Jabne, the Tosefta is hence paradoxically forced to present R. Eliezer as an even more resolute traditionalist than does the Bavli!

83. A talmudic euphemism for anathematization, excommunication, banishment.

84. From a traditionalist perspective, this is not at all self-evident. Thus, for example, in order for a prophet to prove his trustworthiness, he is not required to perform supernatural feats at all, but only to accurately predict natural phenomena (Maimonides, *Mishne Torah, hilkhot yesodai ha-Torah* 10:1–3).

85. Nonetheless, many traditional commentators on the talmud go to great lengths to salvage an interpretation of the story along traditionalist lines. See, for example, R. Nisim Gaon to Bavli, *Berakhot* 19a, who suggests that the heavenly voice's intervention on behalf of R. Eliezer should be understood either as confirming R. Eliezer's halakhic opinion in all cases *other than* the oven of *Akhnai*, or as having issued forth merely in order to test the Jabne assembly. Subsequently, he explains R. Yehoshua's exclamation "it is not in heaven" as an assertion that the Torah itself informs us that once it was handed down to Moses perfectly whole and complete at Sinai, there will be no further heavenly additions or amendments!

86. See n. 80 above.

87. The only exception being, as suggested above (n. 17), those of the debates recorded in the tannaitic literature between the Houses of Hillel and Shammai in which the former are said to have abandoned their initial position as a result of criticism. But even if the argumentative strategies employed by the two Houses can be said to premise the two conflicting philosophies in question, the issue itself is never discussed in this context directly. Indeed the well-known sweeping ruling in favor of the Hillelites related in Bavli, *Eruvin* 13b is of alleged amoraic origin. Interestingly, the only tannaitic source in which it is mentioned that the halakha always follows the Hillelites, is Tosefta, *Eduyot* 2:2. However, owing to the Tosefta's marked traditionalist presentation of both *Eduyot* and the dispute between the Houses (lamented in *Sanhedrin*), the Tosefta's ruling in favor of the Hillelites cannot be read as a declaration in favor of antitraditionalism.

88. For an insightful discussion of the philosophical import of the Jabne stories along the lines suggested by the work of M. M. Bakhtin see Baitner, "Conflicts and Dialectics," pp. 269 ff. See also Ofra Meir, "The Acting Characters in the Stories of the Talmud and the Midrash" (in Hebrew) (Ph.D. diss., Hebrew University Jerusalem, 1977).

89. Boyarin, *Intertextuality and the Reading of Midrash*, p. 34.

90. In "the Torah as it is written," the verse seems to assert only that the fulfillment of the Torah's commandments is manageably within human capacity:

For this commandment which I command thee this day, it is not hidden from thee, neither is it far off. *It is not in heaven*, that thou shouldst say, Who shall go up for us to heaven, and bring it to us, that we may hear it, and do it? Nor is it beyond the sea, that thou shouldst say, Who shall go over the sea for us, and bring it to us, that we may hear it, and do it? But the word is very near to thee, in thy mouth, and in thy heart, that thou mayst do it. (Deut. 30:11–14)

91. Boyarin, *Intertextuality and the Reading of Midrash*, pp. 34–35.

92. See above n. 15 and text.

93. Boyarin, *Intertextuality and the Reading of Midrash*, p. 35.

94. Again, the plain verse as it is written, seems to assert exactly the opposite: "Thou shalt not follow a multitude to do evil; neither shalt thou speak in a cause to *incline after a multitude* to pervert justice" (Exod. 23:2)! I thus disagree with Boyarin's assertion (p. 36) that the verse "says explicitly (i.e. in its own context) that the law is in accordance with the view of a human majority."

95. Boyarin, *Intertextuality and the Reading of Midrash*, pp. 35–36.

96. On the way tannaitic exegetes frequently decide between equally viable readings of a verse on the basis of moral considerations, see Moshe Halbertal's pathbreaking *Interpretive Revolutions in the Making: Values as Interpretive Considerations in Midrashei Halakha* (in Hebrew) (Jerusalem: Magnes Press, forthcoming).

97. Thus Bavli, *Mena*ẖ*ot* 65b-66b, cites six different, equally plausible midrashic proofs that "the morrow after the Sabbath" means the morrow after the first day of Passover, without being obliged to decide which is the "correct" one. While rulings regarding the practical performance of the Torah's commandments are reached and issued by the rabbis wherever possible, their exegetical grounding, reasoned justification, theological significance, and symbolical meaning, though vigorously debated, are never decided.

98. See David Stern, "Midrash and Indeterminacy," *Critical Inquiry*, 15 (1988), pp. 132–61.

99. For an insightful contrasting of Hobbes's and the rabbis' approaches to biblical interpretation, see Harold Fisch, "Authority and Interpretation: Leviathan and the 'Covenantal Community'," *Comparative Criticism*, 15 (1993), pp. 103–23.

100. For a somewhat different statement of this point see Stern, "Midrash and Indeterminacy," pp. 140–41.

101. Paul Feyerabend, *Against Method* (London: Humanities Press, 1975).

102. Some philosophers, however, do tend to speak of science in this way. A good example is Richard Rorty's *Contingency, Irony and Solidarity*. Following Donald Davidson, Rorty rejects the idea that language may at all have any representational value—the languages of science included. One scientific vocabulary is replaced by another, according to Rorty, in ways ungoverned by argument or by criteria of any sort.

Europe did not *decide* to accept the idiom of Romantic poetry, or of socialist politics, or of Galilean mechanics. Rather, Europe gradually lost the habit of using certain words and gradually acquired the habit of using others. . . . we did not decide on the basis of some telescopic observations, or on the basis of anything else, that the earth was not the center of the universe, that macroscopic behaviour could be explained on the basis of microstuctural motion, and that prediction and control should be the principal aim of scientific theorizing. Rather, after a hundred years of inconclusive muddle, the Europeans found themselves speaking in a way which took these interlocked theses for granted. Cultural change of this magnitude does not result from applying criteria any more than individuals become theists or atheists, or shift from one spouse or circle of friends to another. (p. 6)

The problem is that Rorty extends this view of scientific development from the ways new vocabularies gain public currency to the ways in which they are originally devised. He portrays Galileo and Newton not as seriously troubleshooting earlier theories, but as playfully redescribing the world and devising new metaphors merely for the sake of novelty. Part 1 of the present study was written partly with a view to combating Rorty who writes as if the Popperian alternative did not exist.

103. A position curiously attributed to the rabbis by David Kraemer, *The Mind of the Talmud*, pp. 6–7.

104. "[O]n my account of intellectual progress," writes Rorty, in an attempt to explain his own discursive strategy, "rebutting objections to one's redescriptions of some things will be largely a matter of redescribing other things, trying to outflank the objections by enlarging the scope of one's favorite metaphors. So my strategy will be to try to make the vocabulary in which these objections are phrased look bad thereby changing the subject, rather than granting the objector his choice of weapons and terrain by meeting his criticism head-on." Rorty, *Contingency, Irony, and Solidarity*, p. 44.

105. See above text to n. 25. David Stern is the only other writer on midrash who takes the three stories as a related whole. His reading of *Hagiga* 3a-b, however, is rather different than my own. See Stern, "Midrash and Indeterminacy," esp. pp. 156 ff.

106. The entire story is related with minor differences in Tosefta, *Sota* 7:9–12. See also *Mekhilta* to Exod. 13:2; Yerushalmi, *Hagiga* i, 75d; *Bamidbar Rabbah* to Num. 7:48.

107. The attribution to R. Yehoshua of the Hebrew term *hidush*, innovation, strongly implies that study sessions at the *Bet midrash* are typically not confined to the mere preservation and transmission of former teaching, but are thought to yield real new knowledge. This decidedly antitraditionalist bias is rendered explicit further down.

108. The Yerushalmi's account, according to which Gamliel was fully reinstated as president with R. Elazar serving as head of the *Bet din*, appears to be the more historically credible of the two. As noted above, however, some historians have come to doubt whether Raban Gamliel was ever in fact removed from office even temporarily. Composed in Palestine almost a hundred years closer to the events and drawing on local lore, the Yerushalmi is normally regarded a more reliable historical source than the Bavli on matters related to the Palestinian sages. However, the objectives of the present study, to repeat yet again, are not historical, and at this juncture our main concern is to establish the theological and philosophical significance of the Bavli's account of the Jabne reforms, its historical credibility notwithstanding.

109. Stern, "Midrash and Indeterminacy," p. 156.

110. For a detailed analysis of the homilies and their parallels elsewhere in the talmudic literature, see Shraga Abramson, "Four Topics in Midrash Halakha" (in Hebrew), *Sinai*, 74 (1973), pp. 1–7.

111. As one would expect, the suggestive allusion to that which is given from the one shepherd was not lost on the traditionalists. One such reading, recorded in *Qohelet Rabbah* 12:1, asserts playfully: "What (is it meant by) 'spurs'

(*dorbanot*)? (It means) like a girl's playing ball (*ke-khadur shel banot*). For as this ball is passed from hand to hand without falling to the ground, so Moshe received the Torah from Sinai and passed it to Yehoshua, and Yehoshua passed it to the Elders."

112. The standard translation better captures the Hebrew allusion to plants and planting on which, as we shall see, Elazar elaborates: "The words of the wise are like goads; like nails *well-planted* are the words of Masters of assemblies; they were given by one shepherd." Perry's rendition of the verse, on the other hand, misses R. Elazar's reading by interpreting "Masters of collections" as "collectors of sayings" rather than of assemblies of learners, but better emphasizes the allusion to them (the sayings) being given by the one shepherd: "The words of the sages are like goads. But like well-fastened nails are collectors of sayings: They all come from the one shepherd." Theodore Anthony Perry, *Dialogues with Kohelet: The Book of Ecclesiastes—Translation and Commentary*, (University Park: Pennsylvania State University Press, 1993), p. 171.

I have elsewhere speculated on the legendary R. Elazar's general attitude to *Qohelet*—a book deeply concerned with the limits of human wisdom and the possible "worth" of human endeavor. Apart from choosing *Qohelet* as his text for presenting his philosophy of halakhic development, he is credited by Mishna, *Yadayim* 3:5, for being responsible for the canonization of the book that is reputed to have occurred on none other but "that very day" of his appointment at Jabne. For a preliminary reading of *Qohelet* along these lines see Fisch, *To Know Wisdom*, part 3, and "Ecclesiastes (Qohelet) in Context - A Study of Wisdom as Constructive Scepticism," in Ian C. Jarvie and Nathanel Laor, eds., *Critical Rationalism, the Social Sciences, and the Humanities: Essays for Joseph Agassi*, vol. 2, (Dordrecht, Holland: D. Reidel, 1995), pp. 167–87.

113. Stern, "Midrash and Indeterminacy," p. 138–39.

114. The slightly different wording of the original version of the homily in Tosefta, *Sota* 7, seems to localize and contextualize the various contrastive rulings even more clearly than the Bavli. "These pronounce the unclean unclean, and these pronounce the clean clean; with regard to the unclean *in its place* and with regard to the clean *in its place*." The term in its place is ambiguous, but it does lend itself to the view that such halakhic rulings are appropriate as a rule only to the particular circumstances in which they were originally issued.

115. Rashi, R. Shlomo b. Yitzhak the great eleventh-century commentator on the Bible and the Bavli, frequently describes the halakhic enterprise of the talmud in revisionist terms decidedly uncharacteristic of much of the post-talmudic literature. On the phrase "make your ear like the hopper" his commentary reads: "For since the hearts of all disputants are prudently directed to the heavens, therefore acquire an attentive ear, study (their words) and acquaint yourself with each of their (conflicting) opinions, and when you are capable of discerning which of them is best suited (to the situation in hand), declare the halakha to be as he says." Elsewhere (Bavli, *Ketubot* 57b), Rashi explains the nature of talmudic controversies that would seem to concern the meaning of opinions voiced by former sages. His outspokenly antitraditionalist interpretation of this type of talmudic polemics is worth quoting at length. In the event of two *amoraim* supposedly disagreeing about the opinion of another, we should read them as offering their own opinions on the matter at hand rather than as commenting on that of their predecessor. "For when two dispute the views of another, one of them claiming that he said this and the other claiming that he said that, (at least) one of them is lying

(i.e. misrepresenting the truth). But if two *amoraim* disagree on a matter of law, on what should be permitted and what should be forbidden, no lying is involved. Each of them is (merely) presenting his own opinion. One of them finds reason to permit, while the other finds reason to forbid, one argues one way, and the other argues differently. And (in such cases) we say 'both are the words of the living God.' *For at times one particular line of reasoning applies, while at (other) times a different line of reasoning applies. Indeed, any reason could be wholly reversed at the slightest change of circumstance"*!

116. See above pp. 63–64.

117. Most important among the other Jabne stories that bear directly on the traditionalist/antitraditionalist dichotomy are those that detail the Jabne sages' visit(s) to the estranged and ailing R. Eliezer b. Hyrqanus close to his death. The story is told and retold in several versions throughout the canon. Four of them are related in the Bavli: *Sanhedrin* 101a (twice); *Berakhot* 28b; *Sanhedrin* 68a. All of them are intertextually related to the three *beraitot* studied in the present chapter in various degrees and respects. I have dealt with them briefly in this context in Fisch, *To Know Wisdom*, p. 94, n. 25. For a detailed and comparative analysis of these and other Jabne stories along somewhat different lines, see Baitner, "Conflicts and Dialectics."

118. In the present work I have concentrated exclusively on the Bavli's versions of these stories. Here, indeed, one finds no disagreement at the editorial level. This is not necessarily true of all versions of the three stories. The Yerushalmi parallels seem not always to have been composed or adapted with the same clear awareness of the epistemological issues afloat. The most interesting departures from the Bavli's approach are found in the Yerushalmi's version of the deposition of Raban Gamliel (*Berakhot* iv 7d) and the excommunication of R. Eliezer (*Moed Katan* iii 81c). For an illuminating comparison of the former with reference to the relevant secondary literature, see Shapira, "The Deposition of Rabban Gamliel."

119. See above, esp. n. 17. Like the other texts dealt with in the present chapter, all the versions of all the thirty-three debates appear either in tannaitic texts proper, or as *beraitot*, citations of tannaitic material in one of the two *talmudim*.

120. My main reasons for not pursuing the issue in the present context are two: First, while the two versions of *Sanhedrin* and *Eduyot* address central aspects of traditionalism directly—e.g., the extent to which a Great *Bet din* is committed to the received view and the nature of the discretion it possesses in the face of conflicting testimonies—the respective traditionalism and antitraditionalism of the Shammaites and Hillelites is far less explicit. In the debates between the Houses the two metahalakhic positions manifest themselves only indirectly (though, to my mind, convincingly) in their quite different polemic strategies, the discussion of which would have taken us too far afield. Secondly, as noted in n. 87 above, the implied metahalakhic disagreement that allegedly informs these controversies is not resolved at the tannaitic level, but only in a later passage attributed by the Bavli to the first-generation *amora* Shmuel, and by the Yerushalmi, to the third-generation *amora* R. Y(eh)uda (b. Simeon) b. Pazi. For both these reasons, the concerns of the present study are better served by the two corpora discussed.

121. Bavli, *Pesaḥim* 66a; Yerushalmi, *Pesaḥim* vi 33a; Tosefta, *Pesaḥim* 4:11. Both the *Sifra* on Leviticus and *Abot de-Rabbi Nathan* mention the story without actually telling it. In both cases Hillel is alleged to have taught *b'nei Beteira* seven

rules of halakhic reasoning (whereby rulings may be inferred without need for direct transmission—see n. 125 below). As Finkelstein remarks, these versions appear to be Hillelite adaptations of an originally Shammaite story. See, for example, Louis Finkelstein, ed., *Sifra on Leviticus* (New York: Jewish Theological Seminary of America, 1983), vol. 4, p. 11.

122. During the year more than two hundred sacrifices are offered on the Sabbath: the two daily burnt offerings (*temidin*) along with the two additional sacrifices of every Sabbath, besides the special sacrifices offered on the Sabbaths that fall in the middle of the feasts of Passover and Tabernacles.

123. The Hebrew wording of their question, "from where do you know this?" (*minayin lekha?*), better captures its traditionalist presupposition than the translation—i.e., their desire to know the origin of his assertion, rather than his reasons for so asserting, as is apparent from the continuation of the story.

124. Hillel is here employing the rule of biblical interpretation known as *gezera shava*, by which an analogy is established between two quite different matters on the basis of verbal congruities found in the biblical texts in which they are mentioned.

125. Here a second rule of inference is employed, the *kal va-homer*—literally: light and, or versus, severe. Hillel, as mentioned above, is alleged on this occasion to have taught the *b'nei Beteira* seven such rules of reasoning although only two are actually mentioned here, and a third in the other two versions of the story. The "seven things taught by Hillel in the presence of *b'nei Peteira*" are enumerated in Tosefta, *Sanhedrin* 9:11, and *Sifra* on Leviticus. Hillel's seven *midot* form the nucleus for the thirteen *midot she-ha-Torah nidreshet bahem* attributed to R. Yishma'el. For the full list, including those of Hillel, and examples of their application see *beraita de-Rabbi Yishma'el* appended to the *Sifra on Leviticus*, ed. L. Finkelstein, vol. 2, pp. 3–11.

126. The position attributed here to Hillel falls short of the full-blown antitraditionalism expressed by some of the other texts we have examined. Subsequently, the traditionalism of his adversaries is less flexible than that of Tosefta *Sanhedrin* according to which the Great Bet Din is always authorized to rule on its own accord whenever a question arises for which no existing tradition is known to apply. The point of contention between Hillel and *b'nei Beteira* is wholly limited to the question of halakhic *warrant*: is one required only to rely on direct mouth-to-ear reception, or is one at liberty to rationally work things out for oneself. At no point in the story are Hillel's arguments put forth with a view to *challenging* the received view. The issue here is not at all that of halakhic modification, or the need to rethink former edicts. Neither is the issue that of halakhic authority. Hillel is not arguing for the supremacy of rational argument over oral transmission, only for its legitimacy in the process of halakhic decision-making.

127. For although it has been established that one is allowed to slaughter the Pascal Lamb on the Sabbath, it remains unclear whether one is therefore allowed to carry one's knife (because transporting a utensil between private and public domains is normally prohibited on the Sabbath).

128. Such appears to be Rav's understanding of the passage cited by the Bavli further down (66b): "Says R. Yehuda in the name of Rav: He who acts arrogantly, if he is wise, his wisdom abandons him; if a prophet, his prophetic powers abandon him."

129. For a similar though not identical interpretation of this passage, see David Halivni, *Sources and Traditions: A Source Critical Commentary on the Talmud—Tractates*

Erubin and Pesaḥim (in Hebrew) (New York: Jewish Theological Seminary of America, 1982), pp. 465–69.

130. According to the rule that the more frequent normally supersedes the less frequent, the frequent continual offerings should supersede the less frequent Sabbath, which is not the case with regard to the *Pessaḥ* offering, which occurs less frequently than the Sabbath.

131. Owing to its grounding in the biblical text, reasoning on the basis of *gezera shava* is considered a stronger proof than reasoning by *kal va-ḥomer* and normally overrules it in case the two lines of reasoning are found to conflict.

132. The idea that one may make halakhic inferences from one issue to a quite different one, merely on the basis of verbal congruities in the biblical text, affords the exegete an extremely powerful tool. In its developed form there is no need, as in Hillel's case, for the similar words to even remotely imply the type of halakhic analogy derived on their basis. Given the vast number of repeatedly recurring words and phrases throughout the Bible, the opportunities for employing *gezerot shavot* are endless. The antitraditionalist interprets all this along the lines of R. Elazar's homily discussed above—as if all such possible inferences were legitimized by God in advance and that since the Written Text offers infinite opportunities to develop the halakha this way or that, it remains, as it should, for "the Masters of assemblies" to debate matters seriously, and to rule (tentatively) as they see fit.

For the traditionalist, by contrast, the license to employ *gezera shava* is perceived as a real danger to the entire system. Should a person be at liberty to construct *gezerot shavot* on his own accord, writes Nahmanides, it would be possible for him to contradict the entire body of halakha. "For it is impossible for as large a book as the Torah to consist entirely of new words." Therefore, he concludes quoting the above mentioned passage, the talmud informs us that "one is not permitted to devise a *gezera shava* on one's own accord." *Gezera shava*, is thus not regarded by the traditionalist as a rule of inference at all, but as yet another item of certified knowledge received at Sinai (*Hasagot ha-Ramban le-Sefer ha-Mitzvot, Shoresh* II, see also Rashi to Bavli, *Pesaḥim* 66a). This appears to be the opinion of the vast majority of post-talmudic commentators. The talmudic literature itself, however, gives voice to the antitraditionalist approach even to *gezera shava*. Bavli, *Temura* 16a, for example, takes note of the fact related in the Mishna that seventeen hundred halakhot inferred by *kalin va-ḥamurin, gezerot shavot* and other scribal inferences were apparently forgotten during the period of mourning following the death of Moses, all of which, according to R. Abahu, were restored by Otniel ben Kenaz, *by the force of his argumentative brilliance ("be-khoaḥ pilpulo")*—implying, in other words, that he indeed did devise *gezerot shavot* on his own accord. Likewise, it is arguably the case that the *gezera shava* offered by Hillel himself, was also of his own construction. I shall return to this last point shortly.

133. While the continual offerings comprise exactly two lambs daily, no such restriction applies to the *Pessaḥ*. In other words the two types of offering are not entirely analogous.

134. While the former are "offered up" whole (burnt completely) on the altar, the latter is mostly roasted and eaten by common folk outside the Temple compound.

135. The phrase "converged upon him" or "crowded around him" (*ḥavru alav*) is employed in the talmudic literature without exception to denote a gathering motivated by apprehension. In using it the toseftist implies that those present in

the *Azara* gathered around Hillel demanding an explanation. (In his commentary to the Tosefta, the eighteenth-century Vilna Gaon adds the words "and they asked him, how do you know this?") Hillel's answer suggests that he was indeed confronted with such a query. Still, there is nothing to indicate that his *answers* were challenged as in the other two versions of the story. He appears at most to have surprised his audience, but not to have antagonized them. For similar occurrences of the phrase see: Bavli, *Beitza* 20a; *Yevamot* 84a; *Bava Batra* 11a; *Hulin* 6b; Yerushalmi, *Beitza* ii, 61c and *Hagiga* ii, 78a.

136. See *Sifra* to Lev. 1:5 and its parallels in Tosefta, *Zebahim* 1:8, and Bavli, *Zebahim* 13a, and *Sifri* to Num. 10:8 and its three identical parallels in the Yerushalmi: *Yoma* i 38d, *Megila* i 72b, *Horayot* iii 47d. The dating of the *Sifri* and *Sifra*, not to say the Tosefta, is still a matter of scholarly dispute. Still, the time of their final redaction need not coincide with that of specific passages they contain. Regardless of one's position regarding the dating of these documents, most would agree that the versions of these stories contained in the Tosefta and Yerushalmi postdate those of the *Sifri* and *Sifra*. On the dispute regarding the dating of these texts see, for example, Jay M. Harris, *How Do We Know This?: Midrash and the Fragmentation of Modern Judaism* (Albany: SUNY Press, 1995), p. 4, esp. n.12.

137. See also Tosefta, *Mikva'ot* 1:9; *Makhshirin* 2:14; *Shabbat* 13:5; *Hagiga* 3:33; *Aholot* 15:12; Yerushalmi, *Shabbat* xvi, 15c; and Bavli, *Kidushin* 66b; *Shabbat* 17a and 116a.

138. Num. 3:3. This verse is not about sprinkling, but by reasoning that receiving requires an appropriately adorned legitimate priest, Akiva has established a formal analogy between the two rituals.

139. This is the Soncino's rendition of this unique coinage of R. Tarfon spoken in a slightly different context in Bavli, *Yoma* 76a. The original Hebrew is harsher still, connoting something more like: "leave me alone with your long-winded heaping of words"!

140. The prohibition inferred by Akiva regarding the trumpets applies only to those blowing "over" the festival sacrifices. On all other occasion, any priest is permitted to blow.

141. See, for example, Rashi to Bavli, *Hulin* 116a and *Pesahim* 24a, Maimonides, Introduction to *Perush ha-Mishnayot* (but compare to his Introduction to *Mishne Torah*), and Nahmanides, *Hasagot ha-Ramban le-Sefer ha-Mitzvot, Shoresh* II. For a survey of the main medieval rabbinic works that address the issue, see *Sifra on Leviticus*, ed. L. Finkelstein, vol. 1, pp.178–86. As Finkelstein observes, one finds no evidence that the talmudic or even the Gaonic sages ever held to such a view (p. 185).

142. See, for example, R. Yair Bakhrakh, *Havot Yair* (Frankfurt, 1770), §192. Bakhrakh does not contest the view that the *midot* might have been originally revealed to Moshe at Sinai. For all we know, he submits, this might well have been the case, and, in any case, we have no evidence to the contrary. What he seriously objects to, is the idea that the rabbis can be said to have operated under such an assumption. Judging by the amount of disagreement recorded in the talmudic texts concerning the precise number of the *midot*, their meaning, scope of application, and so forth, argues Bakhrakh, one cannot but firmly, if hesitantly, conclude that if they were initially received, by the time the rabbis were writing, they had been forgotten.

143. See Yerushalmi, *Pesahim* vi, 33a: R. Yossai b. R. Boon in the name of R. Ba bar Mamal says: a person (may) reason by (means of) *gezera shava* in sup-

port of what he was taught (*lekayem talmudu*), but may not do so to refute what he was taught (*levatel talmudu*). Interestingly, the possibility of employing *gezera shava* in order to further develop halakha, in ways that neither coincide with nor refute former teachings, is not even considered.

144. See, for example, the fifteenth-century R. Yeshua b. Yosef Halevi of Talmiseon, *Halikhot Olam*, appended commentaries by R. Yosef Karo (*Kelalei ha-Gemara*) and R. Shlomo Algazzi (*Yavin Shemua*), and the seventeenth-century R. Aharon Even-Haim, *Midot Aharon* (appended to his *Korban Aharon* on *Sifra* to Leviticus)

145. Thus, for example, the Talmud distinguishes between three types of *gezerot shavot*. Those based on congruent biblical phrases which are exegetically "free" or "available"—that is to say, phrases not employed for other exegetical purposes, those in which one of them is "available" and the other is not, and those in which neither of the two phrases involved is "available." All agree that a *gezera shava* which is "free (*mufne*) from both sides" is immune to counterargument (by *kal va-homer* for instance)—they "are taught but not challenged" (*lemeidin ve-ein meshivin*). With respect to *gezerot shavot* that are "free from only one side" there exists a tannaitic debate: R. Yishma'el maintains that they too are immune to counterargument, while the sages hold that they are admissible but can be refuted (*lemeidin u-meshivin*). Finally, there exists an amoraic disagreement with regard to *gezerot shavot* that are free from neither side: R. Yehuda in the name of Shmuel maintains that they are prima facie inadmissible, while R. Aḥa in the name of R. Elazar is of the opinion that they are admissible but liable to be refuted. See Bavli, *Shabbat* 64a and esp. *Nida* 22b–23a. Had there existed a precondition requiring all *gezerot shavot* to be revealed at Sinai, there would have been no point at all in these distinctions. In the case of an authentically received *gezera shava*, the question of the "availability" of the biblical phrases involved would have been totally superfluous.

146. Not all problems are resolved by such a move. Thus, for example, the entire issue of "availability," referred to in the previous note, still remains problematic. If the words on which a *gezera shava* is grounded are divinely given, the question of their "availability" is wholly inconsequential. Conversely, if the halakha to be inferred is divinely given, the *gezera shava* should be irrefutable regardless of the "availability" of the proposed biblical phrases on which to ground it!

147. The reader is referred to Avi Sagi, *Elu va-Elu: A Study on the Meaning of Halakhic Discourse* (in Hebrew) (Tel Aviv: Hakibbutz Hameuchad, 1996). I am grateful to Professor Sagi for sharing this illuminating text with me prior to publication.

2. The Changing of the Guard:
Amoraic Texts and Tannaitic Legacies

1. The tannaitic midrashic compilations are, of course, phrase-by-phrase interpretations of Scripture, but none of the major tannaitic texts are explicitly fashioned as responses to or commentaries on prior *rabbinic* writings. In other words (with the possible, insignificant exception of *Abot de-Rabbi Nathan*), no tannaitic text sets itself up as partaking in an ongoing *tradition* of human learning, in a continuous, transgenerational effort to fathom God's word and wish. As for *Abot de-Rabbi Nathan*, even if the text did originate as a systematic commentary to

Mishna *Abot*, it lacks the keen interrogatory quality that characterizes all amoraic treatments of tannaitic materials. *Abot de-Rabbi Nathan* muses, recalls and narrates *apropos* the various sayings and aphorisms contained in *Abot*, but nowhere does it question the earlier Mishna nor ponder its meaning. Some even regard it as Tosefta, *Abot*. See for example D. T. Hoffmann, *Die erste Mischna und die Controversen der Tannaïm*, trans. S. Greenberg (Berlin: Buchdruckerei Nord-Osi, 1913), p. 30.

2. In its final and accepted form, the Mishna comprises six *sedarim* or orders, each of which is largely devoted to a general area of halakha and consists of a number of thematically ordered tractates. *Seder Zeraim* (Seeds or Planting) consists of ten tractates dealing with the various laws related to agriculture, especially in the Land of Israel, and one (*Berakhot*) devoted to the daily prayers and other prayers and blessings; *Seder Moed* (or *Zemanim*) consists of eleven tractates that deal with the laws of the Sabbath, the festivals and fast days, and one (*Shekalim*) mostly devoted to certain of the financial aspects of the Temple cult; *Seder Nashim* (Women) comprises five tractates that deal with the various laws related to marriage and divorce, one (*Nedarim*) deals with regulations concerning vows, and one (*Nazir*) devoted specifically to the Nazirite vows; *Seder Nezikin* (Injuries) consists of seven tractates that deal with various aspects of civil law, judicial procedure, criminal law, and the treatment of halakhic testimonies (tractate *Eduyot*), one (Avoda Zara) on idolatry, and one (*Abot*, the Fathers) comprising various sayings attributed to the Sages; *Seder Kodashin* (Holy Things) consists of ten tractates devoted to the laws of the sacrifices and other Temple rituals and one (*Hulin*) devoted mostly to the laws of non-ritual slaughtering; *Seder Tehorot* (Purifications) consists of twelve tractates that deal with the various laws of ritual cleanness and uncleanness. The Yerushalmi and Bavli are both compiled in the form of systematic commentaries to large portions of the Mishna. The Yerushalmi covers all of *Zeraim, Moed,* and *Nashim; Nezikin* with the exception of *Eduyot* and *Abot*; none of *Kodashin*; and one tractate of *Tehorot* (*Nida*). Of the sixty-three tractates comprising the Mishna, the Yerushalmi discusses thirty-nine. In terms of the number of tractates, the Bavli covers a little less of the Mishna than the Yerushalmi (in all thirty-seven), these, however, are distributed differently, and in most cases include much additional material. Of *Seder Zeraim*, the Bavli only comments on tractate *Berakhot; Moed* is fully covered with the exception of *Shekalim; Nashim* is fully covered; *Nezikin*, like the Yerushalmi, is covered apart from *Eduyot* and *Abot*; nine of the tractates of *Kodashin* are covered (omitting *Midot* and *Kinim*); and of *Teharot* only *Nida* is treated.

3. This does not include the unnamed redactors and narrators whose anonymous voices lend the talmudic *sugya* its final placing, form, and dialogical structure. In the case of the Bavli, this final and decisive stratum of the talmudic text is normally attributed to the anonymous *savoraim* of the sixth and seventh centuries A.D. who were responsible for the final form of the text.

4. In what follows I shall often use the term "Gemara" as shorthand for "the anonymous latter-day framer and narrator of the passage under consideration," who shall also be referred to interchangeably as "the *s'tam*" (the unnamed redactor) or "the framer of the *sugya*".

5. In chapters to come it will be come apparent that the Bavli's extraordinarily uniform vocabulary of talmudic 'give and take' is at face value heavily biased toward traditionalism. A closer look at key *sugyot*, however, reveals a fascinating discrepancy between the form and content of many such narratives. So

that even if one chooses to focus, as does the present study, on the final redaction of the text (see following note) it would still be quite erroneous to draw hasty conclusions even when the units of comparison appear to share identical vocabularies.

6. By this I mean the self-understanding and project of the Bavli as a whole, that is to say of those responsible for its final and ultimate version: those behind the voices of its anonymous narrators and moderators. Just how much of the Bavli's philosophy of talmudic discourse should be attributed to its final redactors, and how much to the rabbinic cultures to which they were heir, is an open and debated question. In recent years Jacob Neusner is perhaps the most outspoken adherent to the former view, arguing repeatedly that the extraordinary uniformity and rhetorical cogency of the "Talmud's one voice" strongly favor the view "that the whole—the unit of discourse as we know it—was put together at the end" (Jacob Neusner, *The Talmud: A Close Encounter* [Minneapolis: Fortress Press, 1991], pp. 59 ff., quotation at p. 65). Neusner admits that "we can never definitively settle the issue of whether a unit of discourse came into being through a long process of accumulation and agglutination or was shaped at one point" (p. 65), but the way the Bavli does manage to speak in a single voice, he argues, renders the latter view clearly the more likely in his opinion. Boyarin is, justifiably, far more skeptical and wholly rejects Neusner's conclusion. "We can assume in confidence," he writes, "neither that a given passage quoted from a particular authority represents an expression of that authority's time and place, nor that it doesn't and that it only belongs to the culture in which the text was put together." He too, therefore, is forced to treat the Bavli as a whole: "By default," he concludes, "I am generally constrained to write of rabbinic culture as a whole, even knowing that such discussion represents only a gap in our knowledge." Daniel Boyarin, *Carnal Israel*, p. 25. However, their holism regarding the Bavli notwithstanding, their subsequent approaches to the text remain worlds apart. While Neusner concludes from the Bavli's uniformity of language, that it speaks in one voice, Boyarin, concludes that despite a shared surface rhetoric the Bavli represents a polyphony even at its most basic redactory level. Needless to say, the present author sides quite decisively with the latter. All this, however, is by no means to say that the task of discerning earlier textual strata within the Bavli is an impossible one. It is a highly complex undertaking, but fascinating beginnings have been made. For one important such attempt see Richard Kalmin, *Sages, Stories, Authors, and Editors in Rabbinic Babylonia*, Brown Judaic Studies, vol. 300 (Atlanta: Scholars Press, 1994).

7. "If they had a tradition thereon," rules the traditionalist Tosefta, *Sanhedrin* 7:1, of the Great *Bet din* in the Hall of Hewn Stones, and if not—and only if not—"they would stand to be counted." See also Tosefta, *Hagiga* 2:4; Bavli, *Sanhedrin* 88b; and Yerushalmi, *Sanhedrin* i, 19a.

8. Neusner's *Close Encounter* is but one of his many publications devoted to this theme.

9. Nuesner, *Close Encounter*, p. 114.

10. Ibid., p. 80.

11. Ibid., p. 56. I do not, however, wholly endorse Neusner's criticism of the many historians and philologists who study the talmudic texts precisely for the sake of reconstructing the complex history of amoraic learning prior to their final framing. To that end, it seems to me methodologically sound to treat the text "archaeologically," seeking to undo its latter-day seams and expose its substantial

layers and linguistic strata. But given the fact that the only textual evidence of earlier amoraic learning resides in the *talmudim* themselves, it follows that in order to backtrack from the finished product to its possible precursors, one needs first to have a reasonably good idea of the contribution of its final framers. I agree, therefore, with Neusner that any study of the talmud—be it for the sake of looking at it, back from it, or forward to the ways it was later received—must first study it as a crafted whole. In this respect too Richard Kalmin's *Sages and Stories* is an exceptional contribution.

12. Nuesner, *Close Encounter*, p. 79.

13. The idea that an amora will never contradict a tanna—except when relying on a different tannaitic source—was first proposed as a general rule of talmudic scholarship in the tenth-century *Iggeret Rav Sherira Gaon*, ed. B. M. Levin (Haifa, 1921), p. 30. The Bavli and Yerushalmi themselves, however, contain no general statement to this effect. What the Bavli does offer (though not the Yerushalmi) are a number of specific comments that appear to premise the idea that an amora cannot formally and directly dispute an undisputed tannaitic ruling. On five occasions in which an opinion attributed to Rav is perceived to contradict views stated explicitly by the Mishna, the Gemara explains that since Rav is considered a tanna, he may challenge a mishna (*Rav tanna u-palig*). And on one occasion the same is said of R. Ḥiya. (For Rav, see: Bavli, *Eruvin* 50b; *Ketubot* 8a; *Gitin* 38b; *Bava Batra* 42a; *Sanhedrin* 83b. For R. Ḥiya: *Bava Metzia* 5a.) Elsewhere, however, when R. Yoḥanan, who is nowhere granted tannaitic status by the talmud, is found to have been in clear disagreement with an explicit *beraita*, Abaye suggests that he was probably not aware of its existence, and that had he been, he would have surely changed his mind. To this the anonymous narrator adds: "but then he might still have heard it and decided that the halakha should be different"! (Bavli, *Shabbat* 61a. See also Bavli, *Yoma* 43b.) The idea that according to the Bavli an amora cannot formally dispute the words of a tanna is also entailed by the way the Bavli occasionally explains the disputed words of a Sage by asserting that he "is a tanna and can therefore disagree" (cf. Bavli, *Shabbat* 64b; *Ta'anit* 14b.

14. See, for example, Bavli, *Eruvin* 46b.

15. Strictly speaking, the "great divide" established by the Bavli between the tannaitic and amoraic periods is somewhat artificial from the point of view of the avid traditionalist, who is equally committed to the reliably transmitted teachings of all preceding generations. To a significant degree the Bavli treats all generations of tannaim and all generations of amoraim as two extended peer groups, members of which are "permitted" as it were to criticize one another freely. There is seldom a problem for a fifth-generation amora, for instance, to criticize and contradict a second-generation amora. There is nothing particularly antitraditionalist, however, about this curious fact. The great generational changing of the guard is experienced in and by the Bavli across the one dividing line between the two great texts it comprises: the Mishna and Gemara. The Bavli is about the reception of the Mishna by later generations. And it is in relation to this one crucial "moment" that its philosophy of reception, interpretation, and amelioration of prior texts is presented. One might say that more than the Bavli is a study of the relationship between consecutive generations of Torah-*learners*, it is a study of the relationship between consecutive *compositions* by Torah-learners.

16. I first became aware of the potential significance of this remarkable *sugya* for understanding the program of the Bavli during a study session devoted to

human dignity in the course of the Twelfth Annual Philosophy Conference of the Shalom Hartman Institute, Jerusalem, July 1995. I am grateful to the members of our study group, Zvi Zohar, David Heyd, Manuel de Oliviera, and Mayer G. Freed, as well as to Moshe Halbertal who led the discussion thereafter.

17. If the amoraic statement is challenged with more than one tannaitic document, only the first of them is normally introduced by *meitivi* or *eitivei*, while successive incongruous *beraitot* are presented by *"ta shema"* (come and hear) or *t'nan* (we have learned). The main difference between objections introduced by *meitivi* and *eitivei* is that the latter almost always describes a move in a narrated dialogue between named parties. We are told by the anonymous narrator that amora *A* responded to the ruling attributed to amora *B* by citing the following tannaitic source. In such cases, the understandings and policies regarding transgenerational incongruities may not necessarily reflect those of the narrator, who may very well be faithfully citing a dispute he himself had nothing to do with. By contrast, objections introduced by *"meitivi"* seldom aspire to describe moves performed in face-to-face dialogue at all. Here it is far more obvious that we are being addressed exclusively by the framer of the exchange, who is not merely reporting that someone or other considered the following to be an objection, but that he himself regards it as such. It is natural therefore, that a study such as the present one, in which the epistemological presuppositions of the framers of the Talmud are its main concern, will concentrate primarily on transgenerational confrontations of the *meitivi*, rather than the *eitivei* variety.

18. This number does not include follow-up challenges to the same statements introduced by *ta shema* or *t'nan*. For convenience as well as for reasons outlined in the previous note, I shall refer to all such transgenerational objections as *meitivi*-type challenges, confrontations, or objections.

19. In talmudic Aramaic the word *s'tam* frequently denotes an anonymous (tannaitic) halakhic ruling. It is now used in the secondary literature as shorthand for the anonymous framer/narrator of an amoraic *sugya*. In what follows I shall refer to "the Bavli," the Gemara," "the framer of the *sugya*," and "the *s'tam*" interchangeably.

20. Adin Steinzaltz, *Guide to the Talmud: Concepts and Definitions* (in Hebrew) (Jerusalem: Keter, 1988), p. 115. The Soncino invariably translates: "an objection was raised."

21. The assumption that the framers of the Bavli might have at all considered the idea of constructing an archetypal *sugya* is in keeping with my suggestion at the outset to view the Bavli as a primarily didactic text. The *sugya* in question does not explicitly declare itself paradigmatic, however; still, as we shall see, the circumstantial evidence for this being the case is decisive.

22. That is to say, has not yet been buried.

23. To comfort the mourners.

24. If they stand two or more deep.

25. The main source for this principle, attributed to R. Yossi the Galilean, is a *beraita* cited in Bavli, *Suka* 26a, in which the principle is named as the reason for exempting the scribes and their assistants engaged in writing Torah scrolls, *tefillin*, and *mezuzot* "from the recital of the *shema* and the *tefilla* and from *tefillin* and from all the precepts laid down in the Torah." The formulation of that *beraita* is strikingly similar to the Mishna under consideration here. Rashi, for one, appeals to this principle, with regard to the exemption granted to the mourner himself (Bavli, *Berakhot* 18a, V "exempt from the recital of the *shema*"). The Tosafot,

however, disagree (*loc. cit.* 17b, V "And he does not recite the blessing"). According to the Tosafot the principle at play is respect for the dead. The point of contention between them is the following: if the reason for the exemption is respect for the deceased, then one is not merely exempt from the other duties, but is actually forbidden to perform them even if he manages to find time to do so. If, however, the reason for the exemption is the mourner's engagement in other duties, he should be allowed to perform both if he is able to. The parallel passages in both Yerushalmi, *Berakhot* iii 5d, and Minor Tractate, *Semahot* 10 (see next note), appear to confirm the Tosafot's position.

26. Minor Tractate, *Semahot* 10, opens similarly to the Mishna here: "For as long as his dead (relative) lies before him, a mourner is exempt from the recital of the *shema* and from the tefilla and from all the precepts mentioned in the Torah," but adds that if the mourner wishes to ignore the exemption and recite the *shema* anyway, he should refrain from doing so "out of respect for the dead." Indeed, the passage continues, "When the time for the recital of the *shema* arrives, the entire congregation recites (it) and he remains silent." In other words, this source makes it quite clear that a person whose dead lies before him will be held in violation of *kevod ha-met* if he continues to attend to his other religious duties, even though he is not currently engaged in performing any *specific* duty related to the deceased. *Kevod ha-met*, in other words, takes clear precedence, according to *Semahot*, over all other obligations.

27. Bavli, *Berakhot* 19b, citing Tosefta, *Berakhot* 2:10, with alterations.

28. Rashi explains that these are persons who come to the funeral not to comfort the mourner, but "for the occasion." In the original Tosefta, as later in Minor Tractate *Semahot*, the distinction made by R. Yehuda is that between those who come for the sake of the mourner and those who come "for the sake of (earning) respect (for themselves)"—*le-shum kavod*. For a discussion of the possible significance of the Bavli's rewording of the *beraita*, see below n. 31.

29. That is to say, a prohibited mixture of linen and wool. See Lev. 19:19.

30. The Bavli's rewording of the Tosefta thus subtly serves a double purpose. By avoiding use of the word "respect" in the citation from the Tosefta, the evident inconsistency between R. Yehuda's and Rav's rulings is concealed from the eyes of the less knowledgeable and attentive reader. Conversely, for those who are acquainted with the original Tosefta, by avoiding the word respect, the inconsistency is made all the more glaring. This is because one might then mistakenly draw an analogy between the two rulings, taking Rav to be claiming, similarly to R. Yehuda, merely that considerations of *self*-respect cannot take precedence over other religious duties.

At first blush it is not quite clear what exactly is the claim attributed to Rav. Should the question "What is the reason?" be considered as part of Rav's speech, or as an intervention on behalf of the *s'tam*? If the latter is the case, then the answer to it, along with the general principle concerning all profanations of God's name, might be claimed not to be an amoraic ruling at all—in which case the argument of this entire section would lose much of its force. The question is, of course, not what Rav actually said, but what the framer of this particular *sugya* understands Rav's statement to have been. It is clear from both the form and content of the *meitivi*-type interrogation of Rav's ruling that follows, that the entire speech is attributed to Rav. First, the term *meitivi* is exclusively reserved by the Bavli to transgenerational objections. If before the first objection is raised, the *s'tam* can be said to have grounded Rav's assertion by means of a tannaitic principle, the term *meitivi* would have been quite inappropriate. Second, all five tannaitic sources cited in objection to Rav seem directed specifically against the latter, rather than the first part of his speech. I take it, therefore, that the entire passage is attributed to Rav and subjected, in its entirety, to the *meitivi*-type confrontations that follow.

31. In which case, he would, of course, still remain in conflict with R. Yehuda, who, according to the *beraita*, explains the exemption granted to the inner rows in terms of the respect to be paid to the living mourning relative. But as a result of the subtle rewording of the Tosefta, and especially the *s'tam's* failure to even hint that such is the case, the matter is passed over wholly unnoticed. Steinzaltz, for instance, introduces the citation of Rav's ruling as follows: "Those comforting the [mourners] are exempt from the recital of *shema* out of respect for *the dead*. From which we arrive at a discussion of the general question: to what extent does the respect for people (*kevod ha-beriyot*) defer precepts of the Torah." Speaking of respect for the dead, as it were, what of respect for the living?!

32. All five of them not only register halakhic edicts that contradict Rav by allegedly proving that in order to show respect certain rules of the Torah are superseded, but all but one concern superseding rules of the Torah related directly to graves, graveyards, and funerals. One is again left with the distinct impression, created earlier by the insertion of the Tosefta passage ahead of Rav's ruling but not easy to prove, that behind these particular tannaitic rulings lurks a general tannaitic viewpoint founded on some form of analogy between respect for the living and the dead.

33. Because there is a grave in it.

34. See also Minor Tractate, *Semahot* 4:14. The reason given there for joining "the people" even if they chose the unclean way, is *"mipnei kevod ha-Shem"* (out respect for *God!*). Most commentators, however, amend the text to read: *"mipnei kevod ha-Am"* (out of respect for the people). See also Yerushalmi, *Nazir* vii 56a, and below n. 58 and text.

35. A field or place considered unclean, only by scribal injunction, because the grave or graves it is known to have once contained can no longer be located. The particular type of *bet ha-p'ras* referred to here is a grave which has been plowed over, so that bones may be scattered about.

36. A tanna of the second generation who was also a priest, and therefore normally prohibited from entering a graveyard.

37. Anything that overshadows.

38. In which case the uncleanness it overshadows does not remain contained within it.

39. For other citations of this tannaitic saying see: Bavli, *Shabbat* 81b, 94b; *Eruvin* 41b; *Megila* 3b; *Menahot* 37b; and Yerushalmi, *Berakhot* iii 6b. For a somewhat different rendition, see: Yerushalmi, *Nazir* vii 56a, *Kilayim* ix 32a. As for the saying itself. The Hebrew reads: "Great is human dignity *she-dohe et lo ta'ase she-ba-Torah."* The Soncino translates: "since it overrides *a* negative precept of the Torah," implying that the tanna indicates that human dignity overrides only one negative precept, neglecting, however, to say which one. But the Hebrew wording of the saying and some of its usages elsewhere, especially in the Yerushalmi (as we shall see shortly), suggest that the saying praises human dignity for overriding not *a* negative precept but *any* negative precept of the Torah.

40. And thus rendering it applicable in practice only to scribal decrees.

41. In that he stood to lose more from interrupting his own work than the other for the loss of his animal.

42. A nazirite who is also a priest.

43. Because those things must be done at specific times and cannot be postponed.

44. A *met mitzva* is an unclaimed body to which no one else is available to attend. The term literally means "(the burial of) a dead which is a religious obligation."

45. As in the previous case, the *s'tam* goes on to ask why, then, do we not make the case of the nazirite priest the basis for a general rule, contrary to that of Rav, applicable to all analogous cases? To which the enigmatic answer is given that one cannot draw an analogy between active and passive transgressions. It is unclear how this argument is meant to be taken. If it is intended to be a distinction between *prohibitions*—that is to say, between transgressing the Torah by performing a prohibited action and doing so by neglecting to perform a required one—then the answer is problematic because the prohibitions against a person mixing kinds and a nazirite making himself ritually unclean by contact with the dead are equally active. Rashi and others attempt to interpret the distinction as one between types of *transgressions* of prohibitions involved in each case. But, again, as the Tosafot point out, the two cases remain analogous: defiling oneself by attending to a body or knowingly continuing to wear prohibited mixed kinds in order not to disgrace someone equally require positive action. For this reason the Tosafot propose a different reason entirely for rejecting the proposal: namely, that the prohibition transgressed in this case is ungeneralizable not because it is transgressed passively, but because it is one not addressed to the entire people but only to nazirites and priests. It remains undisputed, however, that the possibility of rejecting Rav's principle in favor of one induced from the nazirite's obligation to attend to a *met mitzva* is flatly rejected. See also below n. 53.

46. The saying is attributed to "Rahva says in the name of R. Yehuda." (Rahva is a third-generation amora, disciple of R. Yehuda.) Interestingly, an early, gaonic citation of this saying attributes it to "R. Yehuda says in the name of Rav." See *Sheiltot de-rav Ahai Gaon* (Jerusalem: Mossad Harav Kook, 1986), *Bereishit*, xiv.

47. This is achieved by juxtaposing Prov. 19:17 and 14:31 while at the same time substituting *melave* (to accompany, to partake in a funeral procession) for *malve* (to make a loan) in the former.

48. I say almost intact because it turns out in the end at least to exclude scribal prohibitions and the two Torah-based exceptions. More on this below.

49. This impression is further corroborated by the fact that it is virtually impossible to square the Mishna with Rav's ruling by the tactics applied to the five other conflicting texts. The exemption granted by the Mishna cannot be limited to Scribal edicts because it explicitly mentions *tefillin*, and for the same reason cannot be written off as stating an exception expressly proscribed by Scripture.

50. However, the fact that in the case of Hillel the traditionalist is allowed the last word, as it were, does perhaps indicate a preference.

51. Of the 460 *eitivei* confrontations recorded by the Bavli, only three explicitly address the question to the tannaitic source under consideration and do so in exactly the same words as does the *s'tam* in *Berakhot* 19b. See Bavli, *Eruvin* 50a; *Hulin* 123b and 124a. To recall, as opposed to confrontations of the *meitivi* variety, *eitivei* transgenerational challenges are polemic moves attributed to specific,

named parties rather than to the anonymous s'tam (see n. 17 above). As explained above, these sugyot are less telling of the attitudes of their framers than those in which it is they who perform the questioning.

52. Two such readings come to mind immediately. One is to understand the s'tam's queries as queries regarding Rav's interpretation of the five tannaitic texts. Their premise is that an amora could not have contradicted explicit tannaitic rulings, and that whenever he seems to do so, he must have read them differently. Thus construed, the s'tam is asking how would Rav account for the fact that his principle was apparently not applied in these cases? One good reason for doubting such an interpretation is that two of the three nonmidrashic tannaitic sources are reinterpreted on the authority of amoraim who functioned much later than Rav. It seems quite clear that the s'tam's questions are directed at the tannaitic sources themselves rather than at Rav's possible understanding of them. A different, and far more plausible traditionalist reading of the s'tam's repeated question, was urged upon me privately by Daniel Boyarin. On such a reading the question "Why is this so?" is understood as a form of modus tollens argumentation: If Rav's principle is true, why is it ignored by the tannaitic rulings—it follows, therefore, that it cannot be true. This is a perfectly reasonable construal of the question when the five meitivi confrontations are treated in isolation. As we shall see, such a reading loses much of its force when placed in a slightly wider interpretative context.

53. Even this move is made surreptitiously. By presenting as an obvious solution to the problem, requiring no further comment, the relegation of the first three tannaitic rulings to transgressions of only Scribal decrees, the s'tam has to presuppose that it is self-evident that Rav's principle was never meant to apply to such transgressions in the first place. But Rav's original formulation does not obviously distinguish between types of religious obligation. There is nothing in the term "profanation of God's name" (hilul ha-shem) prima facie to exclude Scribal decrees. One has to assume that, like the tannaitic texts, our understanding of Rav's statement is also modified in the process of question and answer, if, of course, to a far lesser degree than its three tannaitic counterparts.

With respect to the two other tannaitic texts the story is far more complicated. To recall, after arguing that the two midrashei halakha register exceptions to rather than disconfirming instances of Rav's principle, the s'tam goes on to ask why neither of them is generalized and made the basis of a different principle opposed to that of Rav? In both cases the suggestion is turned down on grounds of incompleteness. It is impossible, argues the s'tam, to induce a rule as comprehensive as Rav's from either case, because, on the one hand, "we do not derive a ritual ruling from a ruling relating to property," nor, on the other hand, an active ruling from a passive one. The interesting question is: what, according to the Bavli, is the fate of these partial generalizations? Does the fact that the s'tam suggests in passing that they could have been made mean that they were actually made, and are, therefore, apt to be retained on the halakhic record? In other words, can the s'tam be taken to assert that rules opposed to Rav's indeed obtain with regard to passive transgressions of any law (Scribal or Scriptural) and even active transgressions of property law? If this is the case, then Rav's ruling has been extensively modified in the course of the two last confrontations and in the end applies only to active transgressions of ritual obligations laid down in the Torah (of which that of mixed kinds is one)—as if he might have meant the term hilul ha-shem, profanations of God's name, to apply only to transgressions of this kind. If so, it is also done surreptitiously. But it is not the quietness with which the move is made that is of interest, but whether it was meant to be made at all. Interestingly, the majority of post-talmudic halakhists and commentators, assert that, to varying degrees, the move is intended to be made. Rashi appears to endorse both partial generalizations, while the Tosafot, as we have seen (n. 45 above) reject the second. Like the Tosafot, Maimonides endorses the first (excluding from Rav's principle all Scriptural rituals related to property), but chooses to ignore the second. He summarizes the halakhic situation thus:

> If one sees his companion wearing a garment of diverse kinds forbidden by the Torah, even when the latter is walking in the street, he should immediately accost him and tear it off him. He should do so even if it is his Master who taught him wisdom, because respect for human dignity does not set aside a negative commandment explicitly stated in the Torah. Why then does [the Torah] so set aside in the case of the return of a lost article? *Because in the latter case the negative commandment involves property (money).* And why does it so set aside in the case of defilement by the dead? Because Scripture specifies "or for his sister," from which the Sages learned by tradition that this means: for his sister he may not be defiled, but for a *met mitzva*he may defile himself.
>
> Anything forbidden by Scribal law, however, may be set aside everywhere when human dignity is involved. Even though Scripture says: "thou shalt not deviate from the sentence," this negative commandment is set aside when human dignity is involved. Therefore, if one's companion is wearing a garment of diverse kinds forbidden only by Scribal law, he may not tear it and take it off him in the street, but should wait until he reaches his home. But if it is diverse kinds forbidden by the Torah, he must remove the garment immediately. (*Mishne Torah, hilkhot kilayim*, x, 29–30, my italics)

Maimonides does not explain his decision not to endorse the generalization, from "or for his sister," to all passive transgressions of Scriptural prohibitions. (For some of the difficulties he might have encountered see n. 45 above.) Others, however, endorse it unhesitantly (cf. the seventeenth-century halakhist and scholar R. Yair Bakhrakh, *Havot Yair*, §8, 9, and 10). But I know of no latter-day writer who suggests that neither of the two lower-level generalizations are actually made by the Bavli. And yet (from my admittedly biased perspective) I find it hard to believe that the *s'tam* could have meant them to be made. Generalizing from specific case(s) stated by the Torah to all analogous unstated cases (the *midot* of *hekesh* or of constructing a *binyan av*) is a standard mode of halakhic exegetical reasoning. Equally standard in talmudic literature is the way of shooting down such arguments, which is most frequently done by challenging the criteria of sameness on which the proposed analogy is founded. Now, had the *s'tam* only asked "why can't the case of the mixed kinds (referred to by Rav) be inferred from that of returning the lost animal?" and had answered: "because one cannot infer a matter of ritual from a one of property," one may still have been justified in concluding that a generalization to all matters of property was plausibly forthcoming. But the situation here is different. The case of the mixed kinds is presented by Rav as but one instance of a quite general principle ("wherever a profanation of God's name is involved no respect is paid (even) to a teacher"), which is itself read off a verse of Scripture ("There is no wisdom nor understanding nor counsel against the Lord"). Why, then, asks the *s'tam*, don't we instead learn (a general principle) from the phrase "hide thyself from them"? The answer is that such a principle would not be as general. In other words, the answer is that Rav's midrashic generalization is better than the one proposed because it covers more ground. Read thus, there is absolutely no halakhic reason for retaining a different ruling for transgressions of property law. By signaling an exception to Rav's principle, the verse in Deuteronomy is perfectly accounted for, so that there is also no exegetical need for a separate principle for property law. The same applies to "or for his sister." It seems to me that one needs to be biased toward traditionalism to a greater extent than the *sugya* itself will allow in order to maintain such a reading. Had the *s'tam* intended to preserve such generous interpretations of the last two tannaitic sources at the expense of Rav's principle, why did he not do so earlier when dealing with the former three? Why are the first three texts reinterpreted so drastically to allow Rav's principle the widest domain of application possible, while the latter two are amplified far beyond the modest exegetical claims they actually make in order to restrict it so drastically? Biased as it is in the opposite direction, I believe that the text fares better with the antitraditionalist reading proposed here.

54. The principle is cited and applied five times in the Bavli—*Shabbat* 81b, 94b; *Eruvin* 41b; *Megila* 3b; *Mena<u>h</u>ot* 37b—in order to justify infringements upon a variety of decrees in the name of respect for others. In all five cases the decrees involved are undeniably Scribal.

55. I use the term Yerushalmi as shorthand for "the narrator or redactor of the passage"—the Yerushalmi counterpart of the Bavli's *s'tam.*

56. The term here is *kevod ha-rabim*—the dignity of the many, or the majority—rather than *kevod ha-beriyot*—human dignity (lit. the dignity of people). Despite the verbal difference the two phrases are synonymous. See below n. 58.

57. Several commentators on the Yerushalmi, clearly uncomfortable with the idea that a priest may be allowed to defile himself merely in order not to offend the people he is walking with, have tended to amend the *beraita* by supplying ulterior religious motives for the excursion. See, for example, comments by both the *Pnei Moshe* and *Mar'eh ha-Panim* (both attributed to the eighteenth-century R. Moshe Margalit [d. 1881]) and *Perush me-Ba'al Sefer <u>H</u>aredim* (attributed to the sixteenth-century R. Eliezer Azkari [d. 1601]). To recall, unlike the Yerushalmi's two citations of the *beraita*, the version cited by the Bavli reads: "*If they have buried the body and are returning*, and there are two ways open to them," etc.

58. R. Zeira is a third-generation amora who emigrated from Babylonia and settled in Palestine in his prime. The term "human dignity" (*kevod ha-briyot*) employed in the Bavli's version and "dignity of the many" (*kevod ha-rabim*) used here are interchangeable (see above, n. 56) and possibly owe their origin to a scribal error (written in an abbreviated format, common in manuscript editions of talmudic texts, 'כבוד הרבי and 'כבוד הברי are easily confused). In any event, the version recapitulated almost verbatim in Yerushalmi, *Nazir* vii 56a, uses *kevod ha-briyot*. There is no question in my mind that the metahalakhic principle attributed by the Yerushalmi on all three occasions to R. Zeira is the same as the one confronted with that of Rav by Bavli, *Berakhot* 19b.

59. As we shall see shortly, the third mention of R. Zeira's principle in the Yerushalmi occurs in connection with the case most intimately associated with Rav's ruling, namely, that of discovering mixed kinds in public. As one would expect, commentators committed to harmonizing the two *talmudim* have done their utmost to interpret that occurrence of R. Zeira's principle in keeping with that of the Bavli. See below n. 61 and text.

60. As is the case throughout this part of the book, I use terms such as *existence* and *matter of fact* in the literary and not necessarily in the historical sense of the terms. Within the world of talmudic discourse, its *persona dramatis* and its history, *as they are constructed by the Yerushalmi*, such a faction identifies itself by name and by offering a direct answer to an explicit question. Whether such a faction holding these views ever existed is another matter entirely.

61. Cf. R. Moshe Margalit's *Mar'eh ha-Panim*.

62. In all, there are seven such rulings in the Yerushalmi. Two in favor of a position attributed to Rav—Yerushalmi, *Kilayim* ix, 32a, *Eruvin* viii, 25c; three in favor of the views of R. Shim'on b. Lakish—*Hagiga* ii, 78c, and twice in *Gitin* vii, 48c; one in favor of Shmuel—*Horayot* i, 45d; and one in favor of R. Yossi b. Hanina—*Nida* iii, 50c. Significantly, all four are *amoraim* of the first and second generation. More significantly, the latter three are never granted the special transitional tannaitic status granted to Rav on occasion by the Bavli (see below n. 77).

63. See also Yerushalmi, *Eruvin* viii, 25b. The same ruling, likewise attributed to Rav, is cited by the Bavli and also confronted with some of the same tannaitic material in Bavli, *Shabbat* 64b (twice) and 146b, *Beitza* 9a, and *Avoda Zara* 12a-b. As we shall see, the Bavli's treatment of these materials is very different.

64. The source is Tosefta, *Kilayim* 5:24.

65. In his commentary to Tosefta, *Kilayim* 5:24, Saul Lieberman explains that because wool was frequently dyed black, a linen garment with a visible black hem would look like mixed kinds and is therefore prohibited for appearances' sake.

66. Presumably, because they are only used in the privacy of one's home.

67. The source is Tosefta, *Avoda Zara* 7:3. But compare Bavli, *Avoda Zara* 12a.

68. See previous note.

69. The source is Mishna, *Hulin* 2:9. The Bavli, however, passes it over without noting its discrepancy with Rav's ruling. The Tosafot (Bavli, *Hulin* 41a), who tend normally to harmonize as far as possible all seemingly conflicting talmudic texts, are forced to admit that the two *talmudim* probably disagree about the precise meaning of this mishna.

70. Apparently collecting blood in this manner was practiced by heretics at the time.

71. But compare Bavli, *Shabbat* 64b-65a.

72. The first, unproblematic part of the *beraita* is Mishna, *Eruvin* 8:10; the additional last clause is a *beraita* cited again in Yerushalmi, *Eruvin* viii, 25b.

73. Just as it is prohibited to transport goods from one domain to another on the Sabbath, so one is not allowed to pour water from one's yard into the street. The halakha, however, calculates the amount of water normally disposed of per day as the amount contained in a four-cubit deep drain. With such a drain, water pouring in at one end would in fact not reach the other (but would 'push out' other water that was there prior to the Sabbath). Nonetheless, it is prohibited for appearance' sake.

74. The source texts of each of the last two *beraitot* record debates of which one side conforms with Rav's position. See above n. 71 and 72.

75. Here, the precise formulation of the Yerushalmi is: "All these (*mishnayot*) contradict Rav, *and* they have no existence" (*ve-let lehon kiyyum*). The "and" is ambiguous and could be taken to denote a conjunction or conditional with equal viability. The formulation of the same proposition in *Eruvin* viii, 25b is somewhat less ambiguous: "And Bar Kapra taught that if no one is present it is permitted." This contradicts Rav and has no validity (existence), *for* Rav says: "All that is pro-

hibited for appearance' sake, etc." The implication is that Bar Kapra's ruling is null and void *because* Rav thought differently.

76. The precise wording of this recurring Yerushalmi formulation is *ve-let lahon kiyyum*: and they have no existence, basis, ground. In modern parlance we would say: they are declared null and void. There is, however, a tendency among some philologists to soften the blow as it were. Thus, for example, M. Sokoloff, *Dictionary of Jewish Palestinian Aramaic* (Ramat Gan: Bar Ilan, 1990) renders the phrase: "and they have no answer." I find his translation wanting because it takes the absence of *kiyyum* to be a feature not of the six tannaitic sources but of the contradiction between them and the ruling attributed to Rav—in which case the sentence should have been in the singular: "All of these (*beraitot*) are in disagreement with Rav, and *it has* no answer".

77. The idea that *Rav tanna u-palig*, Rav enjoys transitional tannaitic status and may therefore dispute a mishna, is applied throughout the Bavli on five occasions (see above, n. 13). The Yerushalmi, on the other hand, makes no mention of such a principle as far as I can tell. However, even in the Bavli Rav's special status is referred to only as a last resort. Thus, for example, on four separate occasions positions attributed to Rav are declared problematic by the Bavli as a result of *meitivi* challenges to them from tannaitic sources. (The talmudic term is *teyuvta de-Rav*.) In none of these cases is the problem resolved by resorting to Rav's alleged license to dispute tannaitic authority, and the difficulty is dealt with by other means. See Bavli, *Shabbat* 40a; *Bava Metzia* 107a; *Menahot* 5a; *Bekhorot* 55a.

78. For an alternative reading of the phrase *ve-let lahon kiyyum*, see above n. 76.

79. As noted previously, in all of these cases the latter-day positions under discussion are those of first- or second-generation amaoraic authorities. In three of the cases the amora involved is Resh (R. Simeon b.) Lakish (Yerushalmi, *Hagiga* ii, 78c; *Gitin* vii, 48c [twice]), in one Rav (*Kilayim* ix, 32a and *Eruvin* viii, 25b), in one it is Rav's contemporary and frequent disputant Shmuel (*Horayot* i, 45d), and in another it is R. Yossi b. Hanina (*Nida* iii, 50c).

80. By this I mean that the *s'tam*, the anonymous narrator of the *sugya*, will never resolve a transgenerational incongruity by discarding the tannaitic source under consideration. As we shall see, he might suggest ways to amend its wording, at times even radically so, but he will never flatly reject it. This is not always the case, however, with regard to named authorities. The first generation Palestinian amora R. Yohanan is reported on seven different occasions to have declared tannaitic sources cited against him by a student unfit for consideration in his *bet midrash*. The colorful phrase he uses, which is attributed by the Bavli exclusively to him, is *pok t'nei le-vara*—"go and learn (this beraita) outside!" The *s'tam*, by contrast, will only cite *beraitot* that already possess a minimal level of authenticity as it were. For R. Yohanan, see Bavli, *Shabbat* 106a and *Bava Kama* 34b; *Eruvin* 9a; *Yoma* 43b; *Beitza* 12b; *Yevamot* 77b (twice) and *Sanhedrin* 62a-b.

81. Of the forty-three cases of *hisurei m'hasra* recorded in the Bavli, only twelve are motivated by transgenerational incongruities. The other thirty-one amendments of tannaitic texts are motivated by alleged inconsistencies perceived at the tannaitic level.

82. The phrase used by the Bavli in these cases is *matnitin yehida'a*. This tactic is employed on eight different occasions and, again, usually in order to solve intergenerational problems at the tannaitic level. See for example Bavli, *Shabbat* 140a, *Mo'ed Katan* 12b, *Yevamot* 104a.

83. Unless the text itself is indisputable, a genuine traditionalist, I suppose, would normally prefer this way of dealing with transgenerational inconsistencies. If the meaning of an incumbent earlier source is not prima facie ambiguous, it seems to make better sense to assume that it was misquoted rather than misunderstood. This is especially true for communities in which former texts are for the most part transmitted orally. There is something inherently forced in preserving the letter of a text while trying to claim that it wasn't meant to mean what it seems plainly to say. And yet an overwhelming majority of the transgenerational confrontations recorded by the Bavli are resolved in this way rather than that. Again, one is left with the thought that, in so frequently opting for the more strained resolution of such confrontations, the framers of the Bavli might have done so intentionally, thus cultivating an artificial tension between traditionalist framework and antitraditionalist content, in order to render the former less credible in the eyes of their better trained readers.

84. Bavli, *Shabbat* 64b (twice) and 146b, *Beitza* 9a and *Avoda Zara* 12a-b.

85. Because the person would then not necessarily appear to be going to, or to have come from, the place of forbidden worship.

86. The Hebrew phrase is: "*im eino nir'e*," which is ambiguous, as the Bavli will be quick to point out, translating equally well as "if he is not seen" and "if he does not *seem* (to be bowing)."

87. In doing so the Yerushalmi reformulates the conditional clause so as to avoid any misunderstanding, in place of the Tosefta's *u-be-makom she-eino nir'e* (and where he is not seen), the Yerushalmi submits *im haya be-makom tzanua* (if he was in a secluded place). Thus construed, Rav's view flatly contradicts these earlier teachings which, as we have seen, are subsequently dismissed.

88. The Bavli achieves this by a subtle change of phrase substituting for the Tosefta's *u-be-makom she-eino nir'e* (and *in a place where* he is not seen) the ambiguous phrase: *ve-im eino nir'e* (if *he* is not seen).

89. As mentioned above the Bavli contains 674 transgenerational queries introduced by the *s'tam* by means of the term *meitivi*. Like the one in *Berakhot* 19b, very many of them comprise follow-up challenges issuing from more than one tannaitic source.

90. As noted previously, the Bavli also contains transgenerational challenges leveled against amoraic rulings that are not introduced by the *s'tam* directly but by a named amoraic authority who is reported by the *s'tam* to have debated the ruling of his colleague. In these cases, the objection is introduced by the term "*eitivei.*" As I have argued in n. 51 above these *sugyot* obviously bear less significantly on the *s'tam*'s own position than those in which he himself appears to perform the polemic move. Still, even here the number of cases in which contender takes time out to explain the nature of the problem is exceedingly meager—three out of 460.

91. A prudent traditionalist could only use this format rhetorically, sarcastically—mockingly challenging the amora to explain why *R* never occurred to the tanna responsible for the *beraita*, but really meaning that the fact that the tanna refrained from arguing *R is proof enough* for the inappropriateness of *R*. But although the text admits in principle to such a reading, it is so totally at variance with the direct, matter-of-fact, respectful manner and tone of the Bavli's discursive style as to render it quite unrealistic.

92. The two fixed formulae are *teyuvta de-R. . . . ,* and *teyuvta de-R. . . . teyuvta.*

In the context of transgenerational inconsistencies, a single *teyuvta* means that in view of the tannaitic source, the words of R. . . . have been rendered seriously problematic. Single *teyuvtot* are always answered and resolved. A double *teyuvta*, on the other hand, normally signifies the refutation and subsequent dismissal of the amoraic ruling. Still, there are cases even of double *teyuvtot* in which the amora in question, or someone speaking on his behalf, will successfully propose a rebuttal. See, for instance, Bavli, *Sanhedrin* 27a.

93. With the exception of such specific cases in which the special privilege to contest a mishna is granted to Rav or R. Ḥiya. (See above n. 13.)

94. Such is the outcome of only about 5 percent of *meitivi*-type incongruities dealt with by the Bavli.

95. A. Cohen, for example, in his initial translation of *Berakhot* entitles the passage "Teachers Must be Honored," and renders R. Yehoshua b. Levi's statement: "In twenty-four places [it is taught] that the *Bet Din* excommunicates a person for [lack of] respect to a teacher." A. Cohen *The Babylonian Talmud: Tractate Berakot— Translated into English for the First Time, with Introduction, Commentary, Glossary and Indices* (Cambridge, Eng.: Cambridge University Press, 1921), pp. 123–24.

96. Ḥoni ha-Ma'agel, Ḥoni "the circle drawer," was a renowned ascetic who, in time of drought, is told to have drawn a circle and declare he would not step outside it until God had sent rain. See Bavli, *Ta'anit* 23a.

97. There is no question that this is the simple and immediate meaning of the statement. As far as the language of the talmud is concerned, the fact that the reason for a punishment is stated positively (i.e., "for respect of a teacher") rather than negatively (i.e., "for *dis*respect to a teacher) would have been immaterial. The talmud very often refers to *A* when really meaning transgressions of *A*. See, for instance, the identical format in Bavli, *Pesaḥim* 52a: "We impose the ban for the two Festival days of the Diaspora"—meaning, of course, that the ban is imposed for *violations* of the two days.

98. The tannaitic origin of this saying is the *Sifra* to Lev. 10:1, where, as in all its later citations, it comes accompanied with a typical story about a student of R. Eliezer b. Hyrqanus who apparently issued halakhic rulings not far from his master's place of residence. Hearing of this, Eliezer predicted that he would not last the weekend, and indeed by the end of the Sabbath he was dead. See also Bavli, *Eruvin* 63a; Yerushalmi, *Shvi'it* vi 36c; and *Gitin* i 43c. Such arrogance does not merit capital punishment, of course, but is only punishable by death by the Hand of God.

99. Bavli, *Sanhedrin* 100a. For a useful anthology of talmudic and post-talmudic sources related to these issues, see H. Ben-Menahem, N. Hecht., and Sh. Wosner, eds., *Controversy and Dialogue in Halakhic Sources* (in Hebrew) (Tel-Aviv: Alfil, 1991), vol. 1, chap. 11.

100. Cf. Bavli, *Berakhot* 27b and Minor Tractate *Kalla Rabbati* 2:15.

101. For the clearest statement of this principle, see Bavli, *Shevuot* 31a. For an explicit employment of Rav's principle to justify it, see Bavli, *Eruvin* 63a.

102. Bavli, *Berakhot* 20a. The lesson is made explicit by a pun. After being fined, R. Ada is said to have asked the woman her name. "*Matun*," she replied. "*Matun, Matun*," he muttered, "is worth four hundred *Zuzim*." "*Matun*" is, on the one hand, Hebrew for deliberate, moderate, measured, and, on the other, resembles the word for two hundred in Aramaic. Had I been more cautious, he mused, I would have saved having to pay. The lesson, then, is not to refrain from ripping

off a person's inappropriate attire, or contradicting one's master's rulings, but not to be a zealot about it—to think twice, consider alternatives, and to always take a second look.

103. *Eruvin* 63a tells of Ravina, disciple of R. Ashi, who, in the presence of his teacher, noticed a person publicly violate a rabbinic injunction (by tying his ass to the trunk of a palm on the Sabbath). Rather than wait for his master to respond, Ravina is said to have shouted to the man, and getting no reaction from him declared him excommunicated. "In doing so was I acting disrespectfully to you?" he asked R. Ashi. No, answered his teacher, in these cases we say: " 'There is no wisdom nor understanding nor counsel against the Lord'; wherever a profanation of God's name is involved no respect is paid (even) to a teacher." Interestingly, according to *Berakhot* 19b Rav's principle would *not* be expected to apply in the case of rabbinic prohibitions. See also Bavli, *Moed Katan* 17a.

104. On the citation of the incident concerning the Passover feast introduced by Todos of Rome, the *s'tam* remarks that R. Yehoshua referred specifically to "our Mishna," and this particular incident is related only in a *beraita*. "Are there no such cases in the Mishna itself," he asks, clearly implying the greater significance of the Mishna in comparison to other tannaitic sources. In view of the fact that a page later the Gemara will tacitly discriminate again between the Mishna and *beraitot*, allowing the former to be contradicted by Rav while fastidiously harmonizing Rav with the latter, I find this remark highly significant.

105. See Cohen's translation and commentary to Tractate *Berakhot* (n. 95 above), p. 125, n. 3. It is not uncommon for the rabbis to refer to, or to liken the Almighty to, a *rav*—Master. A relatively common turn of phrase used by the Bavli exclusively to indicate that in the cases of apparent discrepancy between the teachings of the Torah and those of a local authority one should abide by the former is: "When the words of the teacher and those of the pupil [are contradictory], whose words should be hearkened to; surely the teacher's!" This argument, states Bavli, *Sanhedrin* 29a, could have been marshaled by the serpent against the accusation of seducing Eve. See also Bavli, *Kidushin* 42b, *Bava Kama* 56a, *Temura* 25a. The Yerushalmi, on the other hand, uses the phrase but once and with reference to human teacher and pupil—Yerushalmi, *Nida* ii, 49d.

106. The case of R. Eliezer's banishment is seemingly clearer in this respect than that of Akavia. At one point in the story, Akavia is indeed accused of speaking disrespectfully of Shema'aya and Avtalyon. Not surprisingly perhaps this is the only part of the story reproduced here. The full story, related in Mishna, *Eduyot*, and dealt with in the previous chapter, does not give itself so easily to such a reading.

107. The case alluded to here only by catchword is the Yerushalmi's version of an incident mentioned briefly in Mishna, *Rosh Ha-Shana* 1:6. After forty sects of eyewitnesses had testified before the *Bet din* that they had sighted the new moon, R. Akiva is said to have held them back (because one sect suffices for the *Bet din* to declare the new month). Raban Gamliel sent him a message saying: "if you hold the community back (from carrying out a religious duty), you will cause them to fail in the future," meaning that those who are turned back are liable not to come forward at all in the future. To this the Yerushalmi adds (*Rosh Ha-Shana* i 57b): "and whoever holds the community back from carrying out a religious duty is to be excommunicated." Compare, however, Bavli, *Rosh Ha-Shana* 22a.

108. King Yerov'am is described throughout the talmudic literature as *the* ex-

ample of a person who knowingly led the nation in the ways of sin. See, for instance, Mishna, *Abot* 5:18; Tosefta, *Sanhedrin* 13:5; Yerushalmi, *Avoda Zara* i, 39a; Minor Tractate *Semahot* 8:13; Bavli, *Berakhot* 35b; *Rosh Ha-Shana* 17a; and esp. *Sanhedrin* 101a–102b.

109. See above n. 2.

110. Tractate *Uktzin* of Order *Tahorot* is notorious for its complexity, as are the disputations between Rav and Shmuel, R. Yehuda's teachers. For the former, see Bavli, *Horayot* 13b.

111. This refers not to *Uktzin* but to Mishna *Tahorot* 2:1, also of Order *Tahorot*.

112. Mishna, *Uktzin* 2:1.

113. Or by thirteen different methods. The implication is the same: our command of the world of learning is far superior to that of the generation of R. Yehuda.

114. On occasions of mourning and public and private fasting one always removes one's shoes. At times of drought R. Yehuda, then, despite his inferior learning, would only have to make the fist sign of self-humiliation for his plea to answered.

115. The entire exchange appears almost verbatim in Bavli, *Ta'anit* 24b and *Sanhedrin* 106b.

3. Understanding the Bavli

1. Except, of course, for a *sugya* like *Berakhot* 19b which seems to be geared deliberately to exploding its traditionalist facade and giving the game away. But even here, the object of the exercise strikes one as having been more subtly instructional than brusquely polemic.

2. I would like to register special thanks at this juncture to three friends and colleagues, philosopher Avi Sagi, historian and philosopher of science Gideon Fruedenthal, and philosopher of law Hanina Ben-Menahem, for forcefully presenting variants of this option in our long and hard discussions of several of the themes presented in this book.

3. Halbertal develops these distinctions in the beginning of part 3 of *People of the Book*, entitled "Canon and Curriculum."

4. In *Interpretive Revolutions*, Halbertal explores the difference between reading in and reading out of a formative text. A document may function as formative text without it necessarily having to serve as the *source* of the true or the good. As Halbertal shows, this is often the case in tannaitic *midrash halakha* where value judgments are used by the rabbis to adjudicate between rival readings of a verse.

5. "(The *halakhot* concerning) the dissolution of Vows hover in the air and have nought to rest on," states Mishna, *Hagiga* 1:8, colorfully, "and the *halakhot* concerning the Sabbath, Festal-Offerings, acts of trespass, are as mountains hanging by a hair—with scant scriptural basis but many *halakhot.*" See also Tosefta, *Eruvin* 8:23, *Hagiga* 1:9, and Bavli, *Nazir* 62a and *Hagiga* 10a–11a.

6. This is a matter rather different from the type of interpretative activity that accompanies every act of adjudication. Problems pertaining to the interpretation of the law arise each time the law is officially *applied*. The law at any time is of course a formative text of, and for, a variety of adjucative activities inside and outside the courts. Here, however, I am interested in the type of text that is formative of the law itself (as opposed to its application). Texts, the interpretation

of which ground and govern the world of legislative discourse. (Where legislative is taken in the broadest possible sense of the term to include both institutional and judicial legislation. To recall, in Jewish law, the Sanhedrin, the Great *Bet din*, functions as the High Judicial Court *and* the authoritative legislative institution.) I am aware that the distinction I am making is rudimentary, and that it is impossibe to differentiate fully between the two undertakings. For a major criticism along these lines see, for example, Ronald Dworkin, *Law's Empire* (Cambridge, Mass.: Harvard University Press, Belknap Press, 1986), chap. 1.

7. Proof of which some of them find in the tendency of late to codify all Common Law systems. Hans Kelsen is perhaps the most prominent legal theorist to insist that every system of law must assume some kind of *grundnorm*, or Basic Norm as he calls it. The question, however, is not the presence of such a fundamental grounding of the system as its openness in principle to revision. For Kelsen's views, see Hans Kelsen, *Introduction to the Problems of Legal Theory* (Oxford: Clarendon Press, 1992), esp. chap. 5.

8. The French Code Civil, later known as the Code Napoléon, was enacted in 1804. It was followed by a Code de Procédure Civile (1806), a Code Commercial and Code d'Instruction Criminelle (1808), a Code Pénal (1810), and a Code Forestier (1827). The German Civil Code (the Bürgerliches Gesetzbuch or BGB) was enacted in 1896 and came into force on January 1, 1900. For an insightful summary and critique of the general idea of codification see Ferdinand Fairfax Stone's centennial essay, "A Primer on Codification," *Tulane Law Review*, 29 (1955), pp. 303–10. For the French and German codes, see Andrew West, Yvon Dedevises, Alain Fenet, Dominique Gaurier, and Marie-Clet Heussaff, *The French Legal System: An Introduction* (London: Fourmat, 1992), and Ernst Joseph Cohn and Wolfgang Zdziebelo, eds., *Manual of German Law*, vol. 1 (London: British Institute of International and Comparative Law, 1968). Needless to say, the manner in which the great existing codes are treated in real life legislative situations is somewhat more complex than described here. For analyses of the function of continental codes see, for example, Chaim Perelman, *Logique juridique: Nouvelle rhetorique*, trans. Orah Gringard (in Hebrew) (Jerusalem: Magnes Press, 1984), part 1, chap. 3.

9. See above part 2, section 1, n. 1.

10. Over which one would normally say: "Who bringest forth bread from the earth."

11. Over which one would normally say: "Who creates the fruits of the vine."

12. This is a good example of a purely halakhic problem that pertains to legal definitions and criteria, suggesting that the system fails to cohere with respect to the halakhic status of this sort of fungi. It is a problem that has nothing to do with biblical exegesis.

13. The entire discussion is reproduced in a slightly different order but with the same conclusion in Bavli, *Nedarim* 55b.

14. The problem raised and solved by the *s'tam* does not at all pertain to a transgenerational tension. The legal incongruity involved obtains wholly within the limits of the tannaitic corpus.

15. See, for instance, Mishna, *Bava Metzia* 7:2; Bavli, *Bava Metzia* 89a, and *Hulin* 115b.

16. For other employments of precisely the same formula see, for example, Bavli, *Shabbat* 5a; *Nedarim* 9b; *Bava Batra* 144b; *Sanhedrin* 10b; *Bekhorot* 2b; *Nida* 5a.

17. See, for example, Bavli, *Pesaḥim* 12a. For a testimony to be considered refuted, it is not enough that other witnesses testify to the contrary. That would be a case of contradiction on which the court will be required to decide. A testimony is considered refuted by the halakha only when a second sect of witnesses testifies that "at the exact time that you claim to have been in place A and to have seen what you saw, you were really with us in place B." It is up to the court to ensure that testimonies are *refutable* by interrogating each member of a sect separately as to the precise date, time, and place of the occurrence. The very idea that evidence is considered viable only if it is refutable, and acquires credibility from the fact that it is not actually refuted, is itself refreshingly Popperian. On this peculiar aspect of the Jewish laws of evidence, see Y. Ben-Menahem and H. Ben-Menahem, "Popper's Criterion of Refutability in the Legal Context" in Aleksander Peczenik, Lars Lindhal and Bert van Roermund, eds. *Theory of Legal Science: Proceedings of the Conference on Legal Theory and the Philosophy of Science, Lund, Sweden December,* 1983, Synthese Library, vol. 176 (Dordrecht: D. Reidel, 1984), pp. 425–35.

18. Bavli, *Sanhedrin* 32a–b.

19. The *beraita* is Tosefta, *Makot* 1:2.

20. As the *s'tam* explains further down, the explicit mention of the sabbatical year is to indicate that such bonds are considered valid even if their dates are highly suspect. It would be very unlikely for a loan to be signed toward the end of the sabbatical year when a moratorium is declared on all standing debts that are not defended by a special contract (Hillel's well-known *Pruzbul*).

21. The Tosefta reads: "Their testimony (signature) is valid and the bond is valid, because we say perhaps they put off its date when writing (it)." The Bavli version of the *beraita* omits the explanatory clause and is stated in the form of a question. Here the answer—namely, that it is not unusual for the date on which a bond was actually witnessed and signed and that of its coming into force, not to coincide—is provided by the *s'tam* in Aramaic.

22. A bond legally comes into effect at the time that it states that it was signed. Apparently it was not uncommon for the officially stated date of the bond not to reflect the actual date of the signing and witnessing, if, for example, the persons involved would not be available for signing on the day the bond was meant to take effect. The problem is that such a testimony would never be able to withstand normal 'inquiry and examination'.

23. In fact, notes the *s'tam*, not only is the ruling of this Mishna in conflict with the Tosefta, it is also contradicted by another Mishna—*Shevi'it* 10:5—that rules that antedated bonds are in general valid while postdated ones are not. See also Bavli, *Rosh Ha-Shana* 2a and 8a, *Bava Metzia* 17a and 72a, *Bava Batra* 157b and 171b.

24. Bavli, *Pesaḥim* 12a; *Bava Kama* 75b; *Sanhedrin* 41a and 75a.

25. The idea being that if bonds were rendered too vulnerable to challenges of this sort, people would tend to avoid issuing loans for fear of losing their money.

26. There are close to seventy instances of rabbinic decrees that are presented in exactly this way. See, for example, Bavli, *Bekhorot* 51b; *Ḥulin* 122a; *Bava Batra* 48a; *Nedarim* 45a; *Ketubot* 57b and 102a; *Eruvin* 4b; *Shabbat* 145a and 146a and, of course, *Berakhot* 19b.

27. As for example the rabbinic restriction regarding coffins that we encountered in *Berakhot 19b*: "*Raba said: It is a rule of the Torah (devar Torah)* that a 'tent' which has a hollow space of a handbreadth (between its outside and what it con-

tains) forms a partition against uncleanness, but if it has not a hollow space of a handbreadth it forms no partition against uncleanness. Now most coffins do have a space of a handbreadth, but [the rabbis] decreed that those which had such a space [should form no partition] for fear that they should be confused with those which had no space."

28. For R. Ḥanina's move, see also Bavli, *Yevamot* 122b, and *Sanhedrin* 2b. For similar repudiations of tannaitic restrictions that use the same argument schema, see Bavli, *Gitin* 53a and 55a, *Bava Kama* 8a, and *Bava Batra* 175b.

29. R. Ḥanina's proposal is rejected, however. Limiting the *Bet din*'s ability to challenge bonds and to oversee the money market to that extent, argues the *s'tam*, would only have made matters worse. Under such circumstances, potential loaners, wary of fraud in the absence of court control, would be even less willing to lend to the needy.

30. Maimonides, in *Mishna Torah, hilkhot Sanhedrin* xii, 3 and *hilkhot edut* iii, 1, clearly ignores the Bavli's refutation of R. Ḥanina's suggestion and adopts his view. Although 'inquiry and examination' were originally meant to apply to all manner of legal testimony, he writes, the rabbis exempted property law from this restriction in order "not to shut the door in the face of" those in need of a loan. However, he adds (*hilkhot edut*, iii 2), the exemption applies only to contracts and bonds. Maimonides, on this point, clearly follows R. Ḥanina's view of the Mishna as a system liable to be rendered problematic and in need of revision, although he interprets the changes that were in fact introduced along the narrower lines proposed by Rava. The two other great codifications, the *Tur, Ḥoshen Mishpat* 30, and *Shulḥan Arukh* 30:1, both ignore R. Ḥanina's view altogether and state simply that in civil law inquiry and examination are not required unless foul play is envisaged—a combination of the opinions of Rava and R. Papa.

31. Stone, "A Primer on Codification," pp. 305, 308.

32. That is to say, commandments the observance of which depends upon a certain point of time—lit. "caused by time."

33. For the amoraic discussion see Bavli, *Kidushin* 33b-34a. The *beraita* in question is an interestingly modified version of Tosefta, *Kidushin* 1:10.

34. One intriguing difference between the *beraita* and Tosefta is that the latter counts *Tzitzit*, fringes, as an affirmative precept that is *not* limited to a specific time and, therefore, as one incumbent upon women.

35. Mishna, *Eruvin* 3:1 and 7:10.

36. It is forbidden on a holy day to walk in any direction beyond a certain distance or to move objects beyond four cubits in any domain other than a private one. If it is desired to walk a greater distance, or if objects are to be moved in a courtyard belonging to more than one tenant, or in an alley into which more than one courtyard opens, it is required to 'make' an *Eruv* (lit. a mixture or fusion). To put it simply, an *Eruv* is a symbolic meshing of domains. It consists of a specified quantity of foodstuffs deposited in a specified place and contributed by, or on behalf of, the people concerned. Different abodes are by this means regarded amalgamated or fused into one.

37. Bavli, *Eruvin* 27a. R. Yoḥanan's principle is cited and applied once more in Bavli, *Eruvin* 29a.

38. In *Eruvin* 27a the amoraic authorities named in connection with the application of R. Yoḥanan's principle of principles are the fourth-generation Abaye, the third- to fourth-generation R. Yirmiya, the sixth-generation Ravina, and the second- to third-generation R. Naḥman.

39. See Yerushalmi, *Terumot* i 40c; *Hagiga* i 74a; *Yevamot* ix 10a.

40. Yerushalmi, *Peah* ii 17a. Compare Bavli, *Bava Batra* 130b. I understand the word "learn" here to refer specifically to the process of making halakhic rulings on the basis of former rulings, and not to the learning process itself. Also, I understand the term "talmud" to denote the activity of *talmud-Torah* and not the name of a compilation.

41. Nancy Cartwright's keen analysis of Miles V. Klein's textbook on optics is a good example of the inconsistencies a good textbook will inevitably be found to harbor if one ignores its didactic function. Cartwright reads Klein's text as a failed attempt to approximate the truth on every page, rather than as a typically good attempt to introduce the reader to modern optical theory step by step. See Miles V. Klein, *Optics* (New York: John Wiley and Sons, 1970), and Nancy Cartwright, *How the Laws of Physics Lie* (Oxford: Clarendon Press, 1983), pp. 46–50.

42. That particular mishnaic passage—Mishna, *Kidushin* 1:7—consists of several general principles of this kind. Thus, for example, it states that all the duties of parents toward their children fall on the father and the mother is exempt, while male and female children are equally obligated towards their parents. For a fascinating discussion of how the Mishna's desire for economy "affects the vocabulary of halakha to the point of imprecision" and several illuminating examples of this tendency (including evidence of the Gemara being aware of it), see Jose Faur, *Golden Doves with Silver Dots*, pp. 89–96.

43. See for example Chanoch Albeck, *Introduction to the Mishna* (in Hebrew) (Jerusalem: Mossad Bialik, 1979), pp. 1–2, and Faur, *Golden Doves and Silver Dots*, pp. 52–53.

44. Of the great value laid by talmudic culture on perfect memorization of the halakhic system and evidence of the memorization drills that characterized the early stages of a novice's discipleship at the time, see Faur, *Golden Doves and Silver Dots*, pp. 89–90, and the bibliography cited therein.

45. I refer the reader again to Avi Sagi's important study and survey of post-talmudic Jewish attitudes to halakhic disagreement, pluralism and truth: *Elu va-Elu: A Study on the Meaning of Halakhic Discourse*.

46. See above section 1, n. 115.

47. In his commentary to Exod. 34:1, Berlin claims that the second set of tablets received by Moshe was different from the ones that were previously broken, in that they established a very different process of Torah-study than did the ones they replaced. The first reception of the Torah was meant to be complete, requiring no further development of the halakhic system (he says nothing of the other aspects of discerning the Word of God) except for simple analogy. The second, by contrast, comprised a set of foundational rulings and rules of inference (jointly called *hukim*—the laws) from which new rulings (*mishpatim*—propositions, theorems?) are inferred in the course of lively *pilpul* and debate. The difference between the first and second sets of tablets, he adds, is to an extent preserved in the different attitudes of the Yerushalmi and the Bavli.

48. By which I mean, again, the portions of the texts that are presented as tannaitic. As stated at the outset, this study negotiates the amoraic/tannaitic divide *as the framers of texts construct it for us*. We are interested here in the ways in which the amoraic redactors sought to define themselves in relation to their predecessors and not in discerning the true historical facts of the matter.

49. Before changing his ways and joining R. Yohanan's academy, Resh Lakish

is reputed to have kept company with criminals. In recent years this story has been read, to quote Boyarin, for "the anxiety about gender and the boundaries of gendered performance" that inhabits it, as the paradigmatic story "of Jewish male subjectivity . . . that is figured as at . . . the margins of the Roman cultural Empire." Boyarin's "Homotopia: The Feminized Jewish Man and the Lives of Women in Late Antiquity," pp. 41–81, presents a most powerful and compelling reading along these lines. What follows is not meant to encroach upon or to contest this aspect of the story, but to add to it an epistemological dimension that is equally · evident in the text and to an even greater degree intentional.

50. This strange metaphor probably bespeaks either the total commitment required of those who dedicate themselves to Torah or the fact that once such a commitment is made, there is no return.

51. As we noted in connection with the oven of *Akhnai*, only fully-fledged utensils, *kelim*, are liable to "receive" impurity—*tum'at kelim*. It is, therefore, crucial to determine at precisely what moment something ceases to be raw materials and becomes a utensil.

52. Mishna, *Kelim* 14:5, rules that swords are susceptible to ritual impurity only after they are filed, and knives only after they are sharpened. Many of the post-talmudic commentators have tried to harmonize the two sources by claiming that the Mishna (probably echoing Tosefta, *Kelim* [*Bava Metzia*] 3:10) deals only with swords and knives that have rusted over. These, they claim, can no longer be considered fully fledged utensils until their rust has been filed away. Maimonides, however, rules that the criterion for utensilhood in general follows the Mishna in both case, which in his opinion, seems not to have anything to do with rust. See Tosafot to Bavli, *Bava Metzia* 84a, Maimonides, *Mishna Torah*, *Hilkhot Kelim* 8:2.

53. The exact wording is: "in matters of robbery, a robber understands." Resh Lakish's shady past appears to be common knowledge in the Bavli, but there seems to be some disagreement regarding its precise nature. Here the term *listim*—robber—is used. In Bavli, *Gitin* 47a, he is reputed to have hired himself out as a gladiator.

54. The exact wording is: "Let R. Elazar b. Pedat go, whose *halakhot* are sharp."

55. Here the original phrase is: *u-mimeyla ravha shema'ata*, that is to say: "and so the halakha profits or prospers."

56. As I shall explain shortly, it is my contention that the story was originally written as a midrashic rendition of Exod. 33:13–23, in which case it is about Moshe's second ascent to Sinai to receive the Torah with which he was already well acquainted.

57. The ritual scroll of the Torah consists of a text lacking punctuation and cantillation marks as well as vowel dots. The letters, however, are affixed with coronets. To the best of my knowledge, the coronets themselves are decorative and nowhere enter midrashic or halakhic considerations. They serve in the story to create two of its central images: that of Moshe arriving 'on the High' and finding the Written Torah not yet ready to be given, and that of midrashically inferring a multitude of halakhic rulings from the minute nuances of the written text—for which R. Akiva was renowned.

58. The precise phrase is "turn thee (or return) to behind of thee."

59. The phrase is *tashash koho*—"he lost his strength"—the term is used by the Bavli to indicate severe loss of control.

60. An explanation for Moshe's inability to comprehend Akiva's teachings,

commonplace among traditional commentators, is to point out that at the time, according to the story, he had not yet received the Torah and therefore was in no position to know what Akiva was talking about. (See, for example, Rashi to Menaḥot 29b.) The problem is that if that was all R. Yehuda in the name of Rav was trying to tell their readers, the story would have been utterly pointless. On the other hand, the story specifically states that the events took place prior to Moshe's the reception of the Torah. My suggestion below to associate the story with Exod. 33:13–23 will conveniently situate it after the reception and eventual breaking of the first tablets and immediately before his receiving of the second. On such a showing Moshe would have already known the Torah he was about to receive for the second time.

61. The Hebrew wording is *kakh alta be-maḥashava le-fanai* "thus it arose in (my) thought before (or, in front of) me." I stress this because of the way the author of the legend plays with the terms "in front" and "behind." Moshe is asked to "turn to his behind," failed to understand what he perceived there, and is informed by the Almighty, quite abruptly, to be quiet and to accept the thoughts or decisions related to what was "before" Him.

62. According to talmudic lore R. Akiva was cruelly executed by the Romans for violating the prohibition issued against Torah-study. See Yerushalmi, *Berakhot* ix 14b, *Sotta* v 20c, but esp. Bavli, *Berakhot* 61b.

63. The idea that the story originated as a midrashic reading of Exod. 33:13–23 was first suggested to me in private conversation by Dr. Y. Licht.

64. For an intriguing reading of the burning bush encounter along these lines see Gabriel Motzkin, " 'Eheye' and the Future: 'God' and Heideger's Concept of 'Becoming' Compared," in Aharon R. E. Agus and Jan Assmann, eds., *Ocular Desire: Sehnsucht des Auges* (Berlin: Akademie Verlag, 1994), pp. 173–82.

65. Notice his clear allusion to the burning bush encounter and to the fact that his request for privileged foresight was then turned down.

66. The English translations miss much of this, which is more apparent in the Hebrew. Harold Fisch translates these verses thus: "And he said, I will make all my goodness (morality) pass before thee, and I will proclaim the name of the Lord before thee; and will be gracious to whom I will be gracious, and will show mercy on whom I will show mercy. . . . Thou canst not see my face; for no man shall see me, and live. And the Lord said, Behold, there is a place by me, and thou shalt stand upon a rock: and . . . while my glory passes by, that I will put thee in a cleft of the rock, and will cover thee with my hand while I pass by: and I will take away my hand and thou shalt see my back: but my face shall not be seen." The entire passage plays on the difference between seeing the Almighty's face and seeing His back, seeing forwards and seeing backwards. But the verb "to see," *ra'o* in Biblical Hebrew, invariably means to perceive, to comprehend, to understand. By substituting knowing or comprehending for seeing and setting the Almighty's reply in the context of Moshe's other requests of this kind, we arrive at the reading I have proposed.

GENERAL INDEX

Abaye 159–160, 173, 230 n.13, 246 n.38
R. Abba 124
R. Abraham b. David (RABAD) 216 n.61
R. Ada 134, 152, 160
Agassi, Joseph xi, 202 n.31, 203 n.36
R. Aḥa 227 n.145
Akavia b. Mehalal'el 76–78, 150, 154–
    155, 157, 216 n.72, 242 n.106
R. Akiva 65, 83, 104–106, 157, 192–196,
    218–219 n.81, 226 n.138, 242
    n.107, 249 n.62
Alon, Gedaliahu 206–207 n.11
antitraditionalism, antitraditionalists, anti-
    traditionalist voice xix–xx, 43–47,
    57–63, 104, 184, 189–196
  and constructive skepticism xx, 43–45,
    59, 87–88, 209 n.12
  and pluralism 189–192
  and the extent of judicial discretion 67–
    68, 77–78, 94–95
  and the role of counterargument 218–
    219 n.81
  and the Houses of Hillel and Shammai
    66–67, 188, 211–212 n.18, 212
    n.19, 213–214 n.43, 218–219 n.81
  and the status of minority opinion 73–
    77, 84, 216 n.61
  and the supremacy of reason 100–102,
    104–106
  construal of gezera shava 107–109, 225
    n.132, 227 nn.145,146
  critically assessed 191–192
  manifesto 88–95
  of Mishna, Eduyot 71–78, 84, 95, 188,
    213–214 n.43, 217–218 n.80
  of the Jabne reforms 70–71, 72–78
  toned down by the Bavli xx, 45–46,
    51, 121–162, 164, 185–188
Antonius Caracallus (Marcus Aurelius)
    48–49
R. Ashi 45, 242 n.103

Bacon, Francis
  "two books" metaphor xv
  philosophy of science 11, 199 n.8
Baitner, Azaria 214 n.47, 219 n.88
Bakhrakh, R. Yair 226 n.142, 235–237
    n.53
Bavli, Babylonian Talmud
  as didactic text 52–53, 167, 182–186,
    188

  as model textbook and model logbook
    combined 182–186
  its different readerships, its double-talk
    147–149, 155–159, 161–162, 166–
    168, 180–189, 198 n.9, 240 n.83
  its (redactors') possible objectives 51–
    53, 165–171, 180–186, 229 n.6
  its uniform vocabulary 228–229 n.5
  the traditionalist idiom of its narration
    117–118, 120, 129–134, 155–159,
    163, 240 n.83
    certain to mislead beginners 163–
      166, 185–188
    certain to mislead die-hard tradition-
      alists 164
    inaudibly exploded in Berakhot 19b
      125–134, 142–148, 157–162
    not a polemical device 164–166
    possible impact on seasoned readers
      166, 185–188
Ben-Menahem, Hanina xi, 243 n.2
Ben-Shalom, Israel 214 n.45
Berlin, R. Naftali, Tzvi, Yehuda (NaTZYB)
    186–188, 247 n.47
binyan av (generalization to generic spe-
    cies) 235–237 n.53
Boyarin, Daniel xi, 44, 84–88, 198 n.9,
    205 n.2, 209 n.12, 214 n.45, 217
    n.79, 220 n.94, 229 n.6, 235 n.52,
    247–248 n.49

Cartwright, Nancy 247 n.41
Cavell, Stanley 16
codes, legal 172, 177–180, 182–183
Cohen, L. Jonathan 200 n.21, 203 n.36,
    205 n.47
Collingwood, R. G. xviii, 201–202 n.28,
    204 n.46
Common Law systems 244 n.8
Comte, Auguste 199 n.5
concubine in Gibeah 209 n.12
constructive skepticism 17–39, 180–181,
    191–192, 209 n.12
  versus antitraditionalism 43–45
  and antitraditionalism xx, 43–45, 59,
    87–88, 189–192, 209 n.12
conventionalism, conventionalist 55–56,
    57–59, 85–86, 88, 209 n.12

Daston, Lorraine xi
Davidson, Donald 220–221 n.102

decline of the generations 212 n.20
Descartes, René
  theory of vortices 6
  philosophy of science 10
Dewey, John 15, 199 n.5
discontinuity of the Temple cult 48–49,
    206–207 n.11, 207 nn.14,15
R. Dosa b. Hyrqanus 218–219 n.81
Duhem, Pierre 199 n.5

Efron, Noah xi
Einstein, Albert 6, 39
R. Elazar b. Azaria, 218–219 n.81
  and the Jabne reforms 65–66, 69–71,
    72, 77–78, 80, 95–96, 215 n.50,
    216–217 n.73
  backed by R. Yehoshua b. Hanania 93–
    94, 213 n.34
  reading of *Qohelet* 12:11 89–93, 222
    n.112
  successor of Raban Gamliel 65–66, 72,
    89, 213 n.32, 221 n.108
Elazar b. Hanokh 76, 150, 154–157
R. Elazar b. Hisma 89, 93–94
R. Elazar b. Pedat 150, 190–192, 227
    n.145
R. Elazar b. Zadok 124
R. Eliezer b. Hyrqanus
  on his deathbed 223 n.117
  archtraditionalist of the Jabne legends
    63–64, 82, 87, 89, 94, 98, 109,
    165, 210 n.16, 212 nn.21,25, 218–
    219 n.81
  conflict at Jabne 78–88, 150, 217–218
    n.80, 219 n.85
  excommunicated 83–84, 88, 150, 154–
    158, 242 n.106
  and respect for one's teacher 151
Elkana, Yehuda xvii
Elon, Menahem 208 n.2
R. Evyatar 209 n.12
Ezra the Scribe 49, 55, 156

Faur, José 209 n.11, 247 nn.42,44
Feyerabend, Paul 88
Finkelstein, Louis 223–224 n.121, 226
    n.141
Fisch, Harold x, xi, xviii, 207–208 n.19,
    220 n.99
Freed, Mayer G. xi, 230–231 n.16
Fresnel, Augustine 18
Friedman, Michael 8
Fruedenthal, Gideon 243 n.2
Funkenstein, Amos xv–xvi

Raban Gamliel 154–155, 157, 213 n.30,
    242 n.107
  removed from office 64–66, 69, 89
  reinstated alongside Elazar b. Azaria 89,
    213 n.32, 221 n.108
gezera shava (inference from verbal congru-
    ity) 98–103, 224 n.124, 225
    n.131, 226–227 n.143, 227 n.145
  traditionalist versus antitraditionalist
    construals 107–109, 225 n.132,
    226 n.141, 227 n.146
Giere, Ronald N. 202 n.31
goal-directed systems 22–39, 201 n.26
  and realism 22–23
  and theory of problems 22–25
  as the site of rational action 30–39, 95–
    96, 185, 190–192, 201 n.26

Halbertal, Moshe xi, 209 n.12, 220 n.96,
    230–231 n.16
  canonization and religious authority
    207 n.16
  "central" versus "formative" canons
    168–169, 171–172, 176, 243 n.4
R. Hanina b. Papa 151, 175–177, 179
hekesh (reasoning from analogy) 235–237
    n.53
Heller, R. Yom Tov Lipmann 215–216 n.60
Herod 48
Heyd, David 230–231 n.16
Hillel, 72–74, 215 n.58
  and *b'nei Beteira* 96–104, 106, 107–
    108, 109, 113–114, 211 n.17, 213–
    214 n.43, 223–224 n.121, 224
    n.125
  as antitraditionalist 96–104, 107, 225
    n.132
  as converted to traditionalism 100–104
  as traditionalist 102–104
  House of xxi
    antielitism 65, 212 n.19
    revise their position in the face of
      criticism 75–76, 210–211 n.17,
      219 n.87
R. Hisda 151
R. Hiya 230 n.13, 241 n.93
Hobbes, Thomas 86, 220 n.99
Hollis, Martin 35–37
Honi ha-Ma'agel 150, 154–156
Houses of Hillel and Shammai xxi, 96,
    164, 188, 214 n.45
  antitraditionalist construal of 75–76
  the debates and their significance 210–

211 n.17, 218–219 n.81, 219 n.87,
223 n.120
traditionalist construal of 66–67, 104,
213–214 n.43
Hume, David
skeptical biperspectivalism 15–17, 20
hypothetico-deductivism 8, 199 n.8
models of induction refuted 8–9, 18

(ritual) impurity
of utensils (tum'at kelim) 79, 81, 189,
217 n.75, 248 n.51
contracted from corpses and graves
124–125, 133–134, 234 n.45, 245–
246 n.27
inductivism, inductivist 8–9, 12–13, 17, 19
instrumentalism 7, 12–13, 59–60, 199 n.5

Jabne xxi
Bavli and Yerushalmi versions compared 223 n.118
dispute on Torah study 50, 80–88, 95–96
establishment of 49–50
stories 50, 51, 61, 63–96, 113–114,
164, 188, 210 n.16, 221 n.108,
223 n.117
Jarvie, Ian C. 202 n.31
Jewish responses to science 197 n.4

R. Kahana 125, 134
kal va-homer (inference from the minor to
the major) 98–103, 107, 224
n.125, 225 n.131
Kalmin, Richard 229 n.6, 229–230 n.11
Kant, Immanuel 12
Karo, R. Joseph 246 n.30
Shulhan Arukh and its reception 52,
172, 208 n.2
Kekes, John 28, 202–203 n.33
Kellner, Menachem 212 n.20
Kelsen, Hans 244 n.8
Kepler, Johannes
laws of planetary motion 8–9
Klein, Miles V. 247 n.41
Koestler, Arthur 200 n.16
Kolbrener, William xi
Kraemer, David xvii, 221 n.103
Kuhn, Thomas 10, 12, 20, 25, 200–
201 n.24

Lagrange, Joseph Louis 13, 199 n.5
Lakatos, Imre
foundationalist historiography of science 12–15

methodology of science 12–13, 20, 200–
201 n.24
philosophy of mathematics 52
theory of scientific rationality 13–14,
21, 26–27, 31, 203 n.34
Leiberman, Saul 238 n.65
Licht, Yehuda 249 n.63
logical positivism 7–9, 199 n.5
Popper's response to 8, 9

Mach, Ernst 199 n.5
Maimonides
Mishne Torah and its reception 52, 172,
208 n.2
on gezera shava 226 n.141
on legal testimony 246 n.30
on the discovery of mixed kinds 235–
237 n.53
on the relative superiority of Batei din
215–216 n.60
reading of Mishna, Sanhedrin 11:2
214 n.44
superiority of Torah study 205–206 n.6
Mishna
ahistorical nature of 77, 216 nn.71,72
as formative or central text 169–180
as legal code 52–53, 172–180
as textbook 180–186
its code-like form 52, 180–183
its structure and coverage by the Bavli
and Yerushalmi 228 n.2
redacted by R. Yehuda "The Patriarch"
ix, 45, 48
mixed kinds
if discovered in public 123, 134–137,
151, 234 n.45, 237 n.59
the appearance of 138, 238 n.65

Nagel, Thomas 15
R. Nahman 246 n.38
Nahmanides 225 n.132, 226 n.141
National Aeronautics and Space Administration (NASA) 5
R. Nathan 82
R. Nehemia 54, 208 n.5
Neusner, Jacob 216 n.71
construal of the Bavli's project 114–
116, 229 n.6, 229–230 n.11
essentialist approach xvi
failure of talmudic Judaism to render
science x, xiii–xvii
similarities and dissimilarities between
talmudic and western philosophy
197–198 n.6

Newtonian physics
  optics 6
  overthrow of 5–6, 8, 9–10, 38–39
    philosophical responses to 5–10, 38–39
  success of 9
  theory of gravitation, astrophysics 6, 9,
    24–25
R. Nisim Gaon 219 n.85

Oppenheimer, Aharon xi, 207 n.12

R. Papa 159–160, 176
Pasteur, Louis 4
Peirce, Charles Sanders 15
Perry, Theodore Anthony 222 n.112
Polanyi, Michael 10, 29–30
Popper, Karl, R. xviii–xix, 14, 202 n.30,
    204 n.40, 209 n.12
  and logical positivism 8, 205 n.3
  epistemology 201–202 n.28
  model of scientific progress 19–22,
    26–27
  refutation of induction 8–9, 15, 17
  scientific realism 18–19, 21, 26, 200
    n.23, 201 n.25
  theory of corroboration 203 n.36
  theory of science 12, 17–22, 87–88,
    187, 245 n.17
  theory of verisimilitude 200 n.22
pragmatism, pragmatist
  theory of truth 5, 57, 86, 198–199 n.4
problems
  and theories 31–34, 202–203 n.33
  contextualized 26–28
  defined 22–23, 30
  realist construal of 22–23, 201–202
    n.28, 202–203 n.33
  scientific 18–20
  solution versus dissolution of 24–25
  versus questions 201–202 n.28
problem seeking and solving, troubleshoot-
    ing 18–39
  and teleology 22
  as rational action 28–39, 180–181,
    185, 204 n.39
  as measure of progress 18–22, 25–28
  in general 22–39
  logic of 22–27, 30, 32–34
progress
  as goal of rational endeavor 30–39
  as retrospective category of evalu-
    ation 31
  defined with reference to goal-directed
    systems 25–28, 30
  erotetic model of 18–22, 25–28, 30. See

    also problem seeking and solving,
    troubleshooting
  in science 1–22, 26
  the problem of 5, 7–9
  as treated by relativists 11–12
  Lakatos's account of 13–14
  Popper's account of 18–22
  value neutrality of 27–28, 34–35
  with respect to multiple goals 25–26
Putnam, Hilary xi, 16, 199 n.9
Putnam, Ruth Anna 200 n.17

Qohelet (Ecclesiastes)
  as irony x
  significance to the Jabne reformers 89–
    93, 222 n.112
quantum mechanics 5, 7

Rashi (R. Shlomo b. Yitzhak) 92, 150,
    186, 194, 218–219 n.81, 222–223
    n.115, 225 n.132, 226 n.141, 231–
    232 n.25, 232 n.28, 234 n.45, 235–
    237 n.53
Rav b. Shaba 125, 133, 134, 161
Rav 159, 192–193, 195–196, 207 n.17,
    224 n.128, 243 n.110
  enjoying near tannaitic status 138–139,
    173, 230 n.13, 239 n.77, 241 n.93
  ruling regarding prohibitions for appear-
    ance' sake 137–142
  ruling regarding respect 122–137, 142–
    162, 232–233 n.30, 237 n.59
Rava 175–176
Ravina 242 n.103, 246 n.38
rationality, rational action, rational dis-
    course
  and belief 35–37
  and error 31–37
  and ethics 34–35
  and goal-directed criticism 19, 28–39,
    43, 180–181, 190–192, 201 n.26,
    202–203 n.33, 204 nn.39,40,41
  and teleology 29
  and Torah study 59–60, 205 n.1
  as exclusively a category of voluntary
    and deliberate action 14, 28–29,
    34, 36–37, 202 nn.30,31, 203 n.35
  as wholly prospective category of ac-
    tion 14, 31, 34
  criteria of adequacy for 28–35
  defined 30, 34
  in science 1–39, 180–181
    problem of 4–5, 9,
    Lakatos's account of 13–14, 203 n.34

Popperian account of 19–20, 28–39, 202 n.30, 209 n.12
instrumental construal of 202 n.31
nourished by two traditions 37–39, 181–183
recursive theory of 34
realism, realists
and goal-directed action 22–24
and Torah study 57–63, 85–86, 209 n.12
scientific 18–19, 21–22, 198 n.1, 199 n.5, 200 n.23, 201 n.25
rebellious elder 67–68, 77, 214 n.44
relativism, relativists 10–12
and wholesale skepticism 16–17
philosophical challenge of 11–12, 14–15
relativity, theory of 5, 7–8, 9, 24–25
Resh (R. Simeon b.) Lakish
and R. Yoḥanan 189–192
respect, honor, reverence, dignity
and the suspension of other religious duties 121–128, 131, 133–135, 142–147, 231–232 n.25
for kings 124–125, 131
for the dead 121–128, 131, 231–232 n.25, 233 n.31
for the living, the many 121–128, 131, 133–135, 157–158, 233 n.31, 237 n.57
for one's teacher 123–124, 142–143, 147, 150–159
Rorty, Richard 199 n.4, 220–221 n.102, 221 n.104
Ruderman, David xi, 197 n.4
Russell, Bertrand 23

Sadducees, Boethusians 50, 164–165, 207–208 n.19
Sagi, Avi xi, 209 n.12, 210 n.15, 227 n.147, 243 n.2, 247 n.45
Safrai, Shmuel 206–207 n.11
science, scientific, scientific knowledge
and talmudic Judaism xiii–xviii, 87–88, 205 n.1, 209–210 n.14
Neusner's account xiii–xvii, 197–198 n.6
Steinsaltz and Funkenstein's account xv–xvi
as conjecture 19–22, 200 n.23
as model of rational and progressive endeavor xvii–xix, 2–39, 180–181
enlightenment views of 5–7
foundationalist theories of 11–15, 17–22
Lakatos's approach 12–14, 200–201 n.24, 203 n.34
Popper's approach 17–22, 87–88,

200 nn.22,23, 201 n.25, 203 n.36, 220–221 n.102
history versus philosophy of 12–15
realism 18–19, 21–22, 198 n.1, 199 n.5
relativist accounts of 10–12, 14, 16–17
teaching, curriculum 38–39, 181–182, 184–185
library: textbooks and logbooks 181–182
tradition 37–39
troubleshooting 18–22, 180–181, 220–221 n.102
Shammai 72–74, 213–214 n.43, 215 n.58
Shapira, Haim 223 n.118
Shapira, Ron xi
Shema'aya and Avtalyon 73–74, 98–99, 100, 215 n.58, 242 n.106
R. Sherira Gaon 230 n.13
Shmuel 79, 124, 159, 179–180, 207 n.17, 223 n.120, 227 n.145, 243 n.110
Silman, Yoḥanan 209 n.12
skepticism 200 n.18
wholesale versus retail discussions of 15–17, 20
constructive versus destructive 17, 20–21
Skinner, Quentin 35–37, 203–204 n.37, 204 n.46
Soloveitchik, Haim 208 n.3
Snow, C. P. xvi
Steinsaltz, R. Adin xv–xvi, 120, 233 n.31
Stern, David 89–90, 221 n.105
Stone, Ferdinand Fairfax 177

R. Tarfon 104–106, 226 n.139
the "two books" metaphor xv, 1, 56
Todos of Rome 150, 154, 157
Torah study (see also antitraditionalism; traditionalism)
comprising exegetical and halakhic efforts xxii, 1, 170–171, 210 n.15, 243 n.5, 244 n.12
epistemological presuppositions for 57–63, 209 n.12
in historical perspective 192–196
ontological presuppositions for 57, 59–60, 209 n.12
rules of inference, the midot of 58, 98, 101–103, 107–109, 127, 187, 224 n.125, 225 nn.131,132, 226 nn.141,142, 226–227 n.143, 234 n.45, 235–237 n.53, 247 n.47. See also binyan av; gezera shava; hekesh; kal va-ḥomer
"superior to all" 47–49, 205–206 n.6

Written versus Oral Torah 55–63, 168–
    170, 208–209 n.11
Torricelli, E. 32
traditionalism, traditionalist 44–45, 57–63
    and the finality of halakhic rulings 74–
        76, 118–119
    and the Houses of Hillel and Shammai
        66–67, 188, 211–212 n.18, 212
        n.19, 213–214 n.43, 217–218
        n.80, 218–219 n.81
    and the limiting of judicial discretion to
        halakhic lacunae 66–69, 75, 77–
        78, 114, 217–218 n.80
    and the role of counterargument 218–
        219 n.81
    and the status of minority opinion 74–
        77, 216 n.61
    and the supremacy of firsthand testi-
        mony 100–102, 104–106
    construals of *gezera shava* 107–109, 225
        n.132, 227 nn.145,146
    limited to halakha 53, 186–187
    of *b'nei Beteira* 97–102
    of Tosefta, *Eduyot* 71–77, 83–84, 95, 96,
        188, 213–214 n.43, 217–218 n.80
    variants of 58–63, 67, 106
transgenerational confrontations, *meitivi-*
        type *sugyot*
    antitraditionalist construal of 118–119,
        130, 145–149
    in the Bavli 119–162, 185–186, 239
        n.80
    in the Yerushalmi 137–142, 148, 156,
        185
    resolution of
        by declaring the tannaitic source a
            minority opinion 139
        by declaring the tannaitic source gar-
            bled 139–140, 148, 173, 240 n.83
        by reinterpreting the tannaitic source
            124–127, 129–130, 132–133, 140–
            142, 148, 173, 240 n.83
        by rejecting the amoraic position 129–
            130, 148, 240–241 n.92
        by rejecting or modifying the tannai-
            tic source 137–142, 148, 173, 239
            nn.79,80
        by treating the tannaitic ruling as an
            exception 126–127, 148
    traditionalist construal of 118–119,
        129–130, 135–137, 145–149, 230
        n.13, 240 nn.83,91
    traditionalist narration versus antitradi-
        tionalist content of

explained 163–186
exposed 120, 125–126, 128–133,
    136–142, 143–149, 155–159
evaluated 186–188
truth
    analytical 199 n.5
    as achievable objective of Torah study
        57–63, 209 n.12. *See also* tradition-
        alism
    as not an objective of Torah study 55–
        56, 57–59, 86–88, 209 n.12. *See
        also* conventionalism
    as regulative objective of science 19, 21
    as regulative objective of Torah study
        57–63, 86–88, 209 n.12. *See also*
        antitraditionalism
    knowing versus believing 205 n.47
    pragmatist theory of 5, 57, 86, 198–
        199 n.4

van Fraassen, Bas 198 n.1

wave versus emission theories of light 18
Whewell, William 15
Williams, Michael 15–16, 200 n.18
witnesses, testimonies 174–177, 245 n.17

R. Yehoshua b. Hanania 64, 218–219
    n.81, 219 n.85
    antitraditionalist extraordinaire of the
        Jabne legends 81–88, 89–95
    humiliated by Raban Gamliel 64–65,
        213 n.30
    support by of Elazar b. Azaria 93–94,
        213 n.34
R. Yehoshua b. Levi 150–158, 160
R. Yehuda (b. Ilai) 54, 73–75, 76, 122,
    216 n.61, 232–233 n.30
R. Yehuda (b. Yehezqel) 79, 123, 124,
    159–160, 192–193, 195–196, 207
    n.17, 224 n.128, 227 n.145
R. Yehuda the Patriarch ix, 45, 48–49,
    207 nn.12,14,17
R. Yermia 81–82, 85–87
Yerov'am son of Nevat 157
R. Yirmiya 246 n.38
R. Yishma'el 215 n.50, 227 n.145
R. Yohanan 230 n.13, 239 n.80
    and Resh Lakish 189–192
    and the status of the Mishna's general
        rulings 178–180, 183, 246 n.38
R. Yohanan b. Baroka 89, 93–94
Raban Yohanan b. Zakai 49, 63, 64, 206–
    207 n.11, 214 n.45

R. Yona 179–180
R. Yonatan 209 n.12
R. Yossi b. Dormaskit 63–64, 94, 212
    n.25
R. Yossi the Galilean 231–232 n.25

Young, Thomas 18
R. Y(eh)uda b. Pazi 223 n.120

R. Zeira 133–135, 142–143, 161, 179–180
Zohar, Zvi xi, 230–231 n.16

INDEX OF TALMUDIC REFERENCES

Mishna

Abot
1:1   61
2:8   63, 64
5:18   242–243 n.108
6:5   48

Bava Metzia
7:2   244 n.15

Beitza
2:7   154

Bekhorot
4:4   197 n.1

Berakhot
3:1   121–126, 127–128, 130–132,
         143–145, 149, 153, 160
6:3   172

Eduyot
1:1–2   72
1:3   72–73
1:4–6   73–74
1:12–14 75–76, 210–211 n.17
2:4   215 n.50
2:5   72
2:8   72
3:1   154
3:12   213 n.37
5:6–7   76–78, 154
7:7   79, 154

Eruvin
3:1   178
7:10   178
8:10   238 n.72

Hagiga
1:8   243 n.5

Hulin
2:9   238 n.69

Kelim
5:10   79, 154
9:2   210–211 n.17
14:5   189

Kidushin
1:7   177–179, 183, 247 n.42

Peah
1:1   205 n.6

Pesahim
6:1   97

Rosh Ha-Shana
1:6   242 n.107

Sanhedrin
7:1   174–176
11:2   67–69, 71, 77–78, 94, 188

Shvi'it
10:5   245 n.23

Ta'anit
3:8   154

Tahorot
2:1   159

Uktzin
2:1   159

Yadayim
3:5   222 n.112
4:3   212 n.25

Yevamot
15:2   210–211 n.17

Tosefta

Aholot
2:7–8   218–219 n.81
3:3   218–219 n.81
4:2   197 n.1
15:12   226 n.137

Avoda Zara
7:3   138, 141–142

Beitza
2:15   154

Berakhot
1:1–6   197 n.1

2:10   122–126, 127–128, 131–132,
                  143–145, 149, 160

Eduyot
1:1   71–72, 75
1:3   74
1:4   74–76, 82–83, 91, 104, 118
1:6   75–76
1:7–8   72
2:1   217 n.78, 217–218 n.80
2:2   219 n.87
2:10   218–219 n.81

Eruvin
8:23   243 n.5

Hagiga
1:9   243 n.5
2:4   211–212 n.18, 229 n.7
2:9   66–67
3:33   226 n.137

Kelim
3:10   248 n.52

Kidushin
1:10   246 n.33

Kilayim
5:24   138, 238 n.65

Makot
1:2   174–176

Makhshirin
2:14   226 n.137

Mikva'ot
1:9   226 n.137

Nida
1:3   197 n.1

Pesahim
4:11   96–97, 102–104

Sanhedrin
7:1   62, 66–71, 75, 77–78, 82–83,
                  91, 94, 104, 188, 229 n.7
7:6–7   218–219 n.81
9:11   224 n.125
13:5   242–243 n.108

Shabbat
13:5   226 n.137

Sota
7:9–12   221 n.106, 222 n.114

Zebahim
1:8   226 n.136

Mekhilta to Exodus
(references are to chapter and verse)

13:2   221 n.106
17:14   206 n.8

Sifra to Leviticus
(references are to chapter and verse)

Beraita de-Rabbi Yishma'el   108, 224
                  n.125
1:5   104–106
10:1   241 n.98

Sifri to Numbers
(references are to chapter and verse)

10:8   104–106
27:12   206 n.8

Sifri to Deutronomy
(references are to chapter and verse)

1:1   206 n.9
1:6   206 n.8
1:7   47–48
3:25   206 n.8
34:1   206 n.8

Abot de-Rabbi Nathan

A iii   212 n.19, 213 n.35
A xxv   212 nn.23,26

Yerushalmi

Avoda Zara
i, 39a   242–243 n.108

Beitza
ii, 61c   225–226 n.135

Berakhot
iii, 5d   231–232 n.25
iii, 6b   133, 233 n.39
iv, 7d   213 nn.31,32, 215 n.50, 223
                  n.118
ix, 14b   249 n.62

Eruvin
viii, 25b   238 nn.62,63,72, 238–239
               n.75, 239 n.79

Gitin
i, 43c   241 n.98
iii, 45b   210–211 n.17
vii, 48c   238 n.62, 239 n.79

Ḥagiga
i, 74a   179
i, 75d   221 n.106
ii, 77d   211–212 n.18
ii, 78a   225–226 n.135
ii, 78c   238 n.62, 239 n.79

Horayot
i, 45d   238 n.62, 239 n.79
iii, 47d   226 n.136

Ketubot
v, 30b   206 n.8

Kilayim
ix, 32a   134–140, 233 n.39, 239 n.79

Megila
i, 70c   207 n.14
i, 72b   226 n.136
ii, 74b   207 n.13

Moed Katan
iii, 81c–d   156–158, 217 n.74, 223 n.118

Nazir
vii, 56a   133, 233 nn.34,39, 237 n.58

Nida
ii, 49d   242 n.105
iii, 50c   238 n.62, 239 n.79

Peah
ii, 17a

Pesaḥim
vi, 33a   96–97, 100–104, 107, 211
               n.17

Rosh Ha-Shana
i, 57b   242 n.107

Sanhedrin
i, 19a   66–67, 68, 211–212 n.18,
               229 n.7

Shabbat
xvi, 15c   226 n.137

Shvi'it
vi, 36c   241 n.98

Suka
ii, 53b   210–211 n.17

Ta'anit
iv, 69c   207 n.14

Terumot
i, 40c   179

Yevamot
i, 3b   210–211 n.17
vi, 7d   207 n.14
ix, 10a   179

Yoma
i, 38d   226 n.136
iv, 41d   206 n.8
vii, 44d   218–219 n.81

Bavli

Avoda Zara
2b   55, 56
10b   207 n.13
12a–b   140–142, 146, 238 n.63

Bava Kama
5a   230 n.13
8a   246 n.28
34b   239 n.80
56a   242 n.105
74b   64
75b   245 n.24
94b   214–215 n.48

Bava Metzia
2a–5b   117
17a   245 n.23
33b   214–215 n.48
59a–b   78–88, 94–95
72a   245 n.23
84a   189–192
86a   208 n.10
89a   244 n.15
107a   239 n.77

Bava Batra
11a   225–226 n.135
42a   230 n.13

48a   245 n.26
130b   247 n.40
144b   244 n.16
157b   245 n.23
171b   245 n.23
175b   246 n.28

Beitza
9a   140, 238 n.63
12b   239 n.80
20a   225–226 n.135

Bekhorot
2b   244 n.16
30b   214–215 n.48
36a   212 n.28
51b   245 n.26
55a   239 n.77

Berakhot
17b   231–232 n.25
18a   128, 231–232 n.25
19a   79, 150–159, 216 n.70, 217
           n.74
19b–20a   121–162, 186, 241–242
           n.102, 243 n.1, 245 n.26,
           245–246 n.27
27b   241 n.100
27b–28a   64–66, 69–71, 72, 77–78,
           80, 83, 89, 93, 152, 213
           n.32
28b   223 n.117
32a   208–209 n.11
35b   242–243 n.108
40b   172–174
57a   208–209 n.11
61b   249 n.62

Eruvin
4b   245 n.26
9a   239 n.80
13b   209 n.12, 210–211 n.17, 219
           n.87
27a   179, 246 n.38
29a   246 n.37
41b   133, 233 n.39
46b   230 n.14
50a   234–235 n.51
50b   230 n.13
63a   152, 241 nn.98,101

Gitin
6b   208 n.10, 209 n.12
38b   230 n.13
47a   248 n.53

53a   246 n.28
55a   246 n.28
56b   206 nn.8,11

Hagiga
3a–b   63–64, 88–95, 213 n.34
10a–11a   243 n.5
24a   214–215 n.48

Horayot
13b   243 n.110

Hulin
6b   225–226 n.135
41a   238 n.69
85a   108
115b   244 n.15
116a   226 n.141
122a   245 n.26
123b–124a 234–235 n.51

Ketubot
8a   230 n.13
38b   108
57b   222–223 n.115, 245 n.26
87b   218–219 n.81
102a   245 n.26

Kidushin
33b–34a   178–180
40b   205–206 n.6
42b   242 n.105
66b   226 n.137

Megila
3b   133, 233 n.39
5a–b   207 n.14
25b   218–219 n.81

Menahot
5a   239 n.77
29b   192–193, 195–196
37b   133, 233 n.39
65b–66b   220 n.97

Mo'ed Katan
12b   239 n.82
17a   242 n.103

Nazir
22b   197 n.1
62a   243 n.5

Nedarim
9b   244 n.16

45a   245 n.26
55b   244 n.13

Nida
5a   244 n.16
22b–23a   227 n.145
25b   214–215 n.48

Pesaḥim
12a   174, 245 n.17
19a   214–215 n.48
24a   226 n.141
49b   208–209 n.11
52a   241 n.97
66a   96–104, 113, 129, 211 n.17,
           225 n.132
66b   224 n.128
71b   218–219 n.81
74b   216 n.61
94b   197 n.1

Rosh Ha-Shana
2a   245 n.23
8a   245 n.23
17a   242–243 n.108
20a   64
22a   242 n.107
25a   64–65

Sanhedrin
2b   246 n.28
10b   244 n.16
27a   240–241 n.92
29a   242 n.105
32a–b   174–176
33a   197 n.1
41a   245 n.24
62a–b   239 n.80
68a   63, 64, 223 n.117
75a   245 n.24
83b   108, 230 n.13
88b   66–67, 211–212 n.18, 229 n.7
100a   151
101a   223 n.117
101a–b   242–243 n.108
106b   243 n.115
108b   54

Shabbat
5a   244 n.16
17a   226 n.137
40a   239 n.77
47a   218–219 n.81
61a   230 n.13
64a   227 n.145

64b   140, 230 n.13, 238 n.63
64b–65a   238 n.71
77a   218–219 n.81
81b   133, 233 n.39
83b   214–215 n.48
94b   133, 233 n.39
104a   239 n.82
106a   239 n.80
116a   226 n.137
127a–b   205 n.6
123   214–215 n.48
145a   245 n.26
146a   245 n.26
146b   140
153b   218–219 n.81

Shevuot
7a   108
31a   152

Sota
47b   211–212 n.18

Suka
2a–9b   117
26a   231–232 n.25
27b–28a   63

Ta'anit
14b   230 n.13
23a   241 n.96
24b   243 n.115

Temura
16a   225 n.132
25a   242 n.105

Yevamot
16a   211 n.17, 218–219 n.81
77b   239 n.80
84a   225–226 n.135
104a   239 n.82
122b   246 n.28

Yoma
12b   218–219 n.81
23b   218–219 n.81
39b   206 n.8
43b   230 n.13, 239 n.80
66b   212 n.24
76a   226 n.139

Zebaḥim
13a   226 n.136

Genesis Rabbah
(references are to chapter and verse)

1:1	55
1:4	55
2:8	206 n.8
2:11	206 n.8
6:18	54
8:16	54
31:12	208 n.4
34:7	208 n.4
35:1	208 n.5

Bamidbar (Numbers) Rabbah
(references are to chapter and verse)

7:48	221 n.106

Qohelet Rabbah
(references are to chapter and verse)

12:1	221–222 n.111

Minor Tractates

Kala Rabbati

2:15	241 n.100

Semaḥot

4:14	233 n.34
8:13	242–243 n.108
10:1	121, 232 n.26

MENACHEM FISCH is Senior Lecturer in the Cohn Institute for the History and Philosophy of Science and Ideas, Tel Aviv University, and Fellow of the Shalom Hartman Institute for Advanced Judaic Studies, Jerusalem. He has published widely on philosophy of science, the history of early Victorian science, and talmudic thought. He is author of *William Whewell Philosopher of Science* and coeditor (with Simon Schaffer) of *William Whewell: A Composite Portrait.*

Lightning Source UK Ltd.
Milton Keynes UK
UKOW06n1135310516

275327UK00008B/97/P